D0852891

ENVIRONMENTAL ENGINEERING

Second Edition

ENVIRONMENTAL ENGINEERING
Second Edition

P. Aarne Vesilind

Department of Civil and
    Environmental Engineering
Duke University
Durham, North Carolina

J. Jeffrey Peirce

Department of Civil and
    Environmental Engineering
Duke University
Durham, North Carolina

Ruth F. Weiner

Huxley College of
    Environmental Engineering
Western Washington University
Bellingham, Washington

Butterworths

Boston   London   Singapore   Sydney   Toronto   Wellington

Copyright © 1988 by Butterworth Publishers, a division of Reed Publishing (U.S.A.) Inc.
All rights reserved.

No part of this publication may be reproduced, stored in a retrieval system, or transmitted, in any form or by any means, electronic, mechanical, photocopying, recording, or otherwise, without the prior written permission of the publisher.

Library of Congress Cataloging-in-Publication Data
Vesilind, P. Aarne.
  Environmental engineering.
  Includes bibliographical references and index.
  1. Environmental engineering. 2. Sanitary engineering. I. Peirce, J. Jeffrey. II. Weiner, Ruth F. III. Title
  TD146.V47   1988   628   87-30913
  ISBN 0-409-90050-8

British Library Cataloguing in Publication Data
Vesilind, P. Aarne
  Environmental engineering.—2nd ed.
  1. Environmental engineering
  I. Title II. Peirce, J. Jeffrey II. Weiner, Ruth F.
  620.8   TD145
  ISBN 0-409-90050-8

Butterworth Publishers
80 Montvale Avenue
Stoneham, MA 02180

10 9 8 7 6 5 4 3 2 1

Printed in the United States of America

Contents

Preface

In environmental engineering *everything matters*. The natural sciences, the social sciences, and the humanities can be just as important to the discipline of environmental engineering as classical engineering skills such as mathematics and fluid mechanics. For many environmental engineers, this attribute of the profession provides the challenge and reward in their careers. For universities, however, the fact that everything matters creates havoc in the design of courses and course content. The problem is most critical in deciding what material to include in an introductory book on environmental engineering.

We respond to this challenge by organizing this text into the five areas most important to environmental engineers: water resources, air quality, solid and hazardous wastes (including radioactive waste), noise, and social and ethical considerations. We begin with a short introduction on the roots of environmental engineering and present the concept of risk and safety in Chapter 2. These principles surface in almost all of the subsequent chapters. The last chapter deals with environmental ethics, an important and emerging component of engineering in general. The ethics material can be used for the last few class meetings, or it can be introduced at any point in the course.

This book is intended for junior or senior level engineering students who have already been introduced to fluid mechanics and water resources engineering. Basic concepts of hydraulics are covered with the assumption that the students have a fluids background. The amount of material in this book can readily be covered in a one-semester course.

P. Aarne Vesilind
J. Jeffrey Peirce
Ruth F. Weiner

Chapter 1

Environmental Engineering

Environmental engineering is a new profession with a long and honorable history.

This is not a contradiction. The descriptive title of "environmental engineer" was not used until the 1960s, when academic programs in engineering and public health schools broadened their scope and required a more accurate title to describe their curricula and their graduates. The roots of this profession, however, go back as far as recorded history.

These roots reach into four major disciplines: civil engineering, public health, ecology, and ethics. From each, the environmental engineering profession draws knowledge, skill, professionalism, and concern for the greater good.

CIVIL ENGINEERING

Skills in agriculture spawned the development of a cooperative social fabric and the growth of communities. As farming efficiency increased, a division of labor was possible, and communities began to build permanent public and private structures. Defense of these structures and the land became paramount, and other structures were built purely for defensive purposes. In other societies, the conquest of neighbors also required the construction of machines of war. Builders of these facilities became known as military engineers, and the term engineer continued to imply military involvement well into the eighteenth century.

In 1782, John Smeaton, builder of roads, structures, and canals in England, recognized that since his profession was in the construction of public facilities, he should correctly be designated a *civil* engineer. This title was widely adopted by engineers engaged in public works.[1]

1

The first engineering curriculum in the United States was established at The United States Military Academy at West Point in 1802. The first engineering course outside the Academy was offered in 1821 at the American Literary, Scientific, and Military Academy, which later became Norwich University. The Rensselaer Polytechnic Institute conferred the first truly *civil* engineering degree in 1835. In 1852, the American Society of Civil Engineers was founded.[2]

Among the public facilities for which the civil engineer became responsible were water supply and wastewater drainage. The availability of water has remained a critical component of civilizations.* Ancient Rome, for example, had water supplied by nine different aqueducts, with cross sections from 2 to 15 m (7 to 50 feet), up to 80 km (50 miles) long.

As cities grew, the demand for water became increasingly acute. In the eighteenth and nineteenth centuries, common people in cities lived under abominable conditions, with water supplies grossly polluted, expensive, or nonexistent.

In London, the water supply was controlled by nine different private companies, and water was sold to the public. Often the poorer people could not afford to pay for the water and begged or stole it. During epidemics, the privation was so great that people went into plowed fields and drank water out of depressions in the ground. Droughts caused water supplies to be curtailed, causing great crowds to form at the public pumps, wanting their "turn" at the pump.[3]

In the New World, the first public water supply system consisted of wooden pipes, bored and charred, with metal rings shrunk on the ends to prevent splitting. The first such pipes were installed in 1652, and the first citywide system was constructed in Winston-Salem, North Carolina, in 1776. The first American water works was built in the Moravian settlement of Bethlehem, Pennsylvania. A wooden waterwheel, driven by the flow of Monocacy Creek, powered wooden pumps that lifted spring water to a wooden reservoir on top of a hill where it was distributed by gravity.[4] One of the first major water supply undertakings was the Croton Aqueduct, started in 1835 and completed six years later. This engineering marvel brought clear water to Manhattan Island, which had an inadequate supply of groundwater.[5]

Although the quantities of water provided by municipal systems might have been adequate, the quality of water was often suspect. As one observer described it:[6]

> The appearance and quality of the public water supply were such that the poor used it for soup, the middle class dyed their clothes in it, and the very rich used it for top-dressing their lawns. Those who drank it filtered it through a ladder, disinfected it with chloride of lime, then lifted out the dangerous germs which survived and killed them with a club in the back yard.

---

*A fascinating account of the importance of water supply to a community through the ages may be found in James Michener's *The Source*.

The earliest known acknowledgment of the effect of impure water is found in *Susruta Samhitta*, a collection of fables and observations on health, dating back to 2000 B.C., which recommended that water be boiled before drinking. The filtration of water became commonplace only toward the middle of the nineteenth century. The first successful filter for water supply was in Parsley, Scotland, in 1804, and many less successful attempts to clarify water by filtration followed.[7] A notable failure was the system for New Orleans, which was to filter water from the Mississippi River. The water proved to be so muddy that the filters clogged up too fast for the system to be workable. This problem was not alleviated until aluminum sulfate (alum) began to be used as a pretreatment. Although the use of alum to clarify water as a pretreatment to filtration was first proposed in 1757, it wasn't until 1885 that this was convincingly demonstrated. Disinfection of water with chlorine was started in 1902 in Belgium and followed in 1908 in Jersey City, New Jersey. The years 1900 to 1920 saw dramatic drops in deaths from infectious diseases, owing in part to the effect of cleaner water supplies.

Human waste disposal in early cities also presented both a nuisance and a serious health problem. Often the method of disposal consisted of nothing more than flinging the contents of chamberpots out the window (Figure 1–1).

Stormwater was considered the main "drainage" problem, and it was in fact illegal in many cities to discharge wastes into the ditches and storm sewers. Eventually, as water supplies developed,* the storm sewers became used for both sanitary waste and stormwater. Such "combined sewers" exist in some of our major cities even today.

The first system for urban drainage in America was constructed in Boston around 1700. There was a surprising amount of resistance to the construction of sewers for waste disposal. Most American cities even at the end of the nineteenth century had cesspools or vaults. The most economical means of waste disposal was to pump these out at regular intervals and cart the waste to a disposal site outside the town. The engineers argued that although sanitary sewers were capital intensive, they provided the best means of wastewater disposal in the long run. Their argument won, and there was a remarkable period of sewer construction between 1890 and 1900.

Actually, the first separate sewerage systems in America were built in the 1880s in Memphis, Tennessee, and Pullman, Illinois. The Memphis system was a complete failure. It used small pipes that were to be flushed periodically. No manholes were constructed, and thus cleanout became a major problem. The system was later removed and larger pipes, with manholes, were installed.[4]

Initially, all sewers emptied into the nearest watercourse, without any treatment. As a result, many of these lakes and rivers became grossly polluted, and as an 1885 Boston Board of Health report put it, "large territories are at

---

*To hold down the quantity of wastewater discharge, the city of Boston in 1844 passed an ordinance prohibiting the taking of baths without doctor's orders.

Figure 1–1. Human excreta disposal, from an old woodcut. [Source: Reyburn, W. *Flushed with Pride* (London: McDonald, 1969).]

once, and frequently, enveloped in an atmosphere of stench so strong as to arouse the sleeping, terrify the weak and nauseate and exasperate everybody."

Wastewater treatment first consisted only of screening for the removal of the large floatables to protect sewage pumps. Screens had to be cleaned manually, and the wastes were buried or incinerated. The first mechanical screens were installed in Sacramento, California, in 1915, and the first mechanical comminutor for grinding up the screenings was installed in Durham, North Carolina. The first complete treatment systems were operational by the turn of the century, with land spraying of the effluent being a popular method of wastewater disposal.

The engineering for these facilities was the responsibility of the civil engineer, since these were constructed public works. There was, however, little appreciation of the broader perspectives of environmental pollution control and management. These considerations historically have come from the public health professions, from the science of ecology, and perhaps most subtly, from the philosophy of ethics.

PUBLIC HEALTH

Life in cities during the middle ages and through the industrial revolution was difficult, sad, and short. In 1842, the *Report from the Poor Law Commissioners on an Inquiry into the Sanitary Conditions of the Labouring Population of Great Britain* described the sanitary conditions in this manner:

> Many dwellings of the poor are arranged round narrow courts having no other opening to the main street than a narrow covered passage. In these courts there are several occupants, each of whom accumulated a heap. In some cases, each of these heaps is piled up separately in the court, with a general receptacle in the middle for drainage. In others a pit is dug in the middle of the court for the general use of all the occupants. In some the whole courts up to the very doors of the houses were covered with filth.

The great rivers in urbanized areas were in effect open sewers. The River Cam, like the Thames, was for many years grossly polluted. There is a tale of Queen Victoria being shown over Trinity by the Master, Dr. Whewell, and saying, as she looked down over the bridge: "What are all those pieces of paper floating down over the river?" To which, with great presence of mind, he replied: "Those, ma'am, are notices that bathing is forbidden."[8]

During the middle of the nineteenth century, medical knowledge was still primitive, and public health measures were inadequate and often counter-productive. The germ theory was not as yet appreciated, and great epidemics swept over the great cities of the world. Some intuitive measures taken by the public health agencies did, however, have a positive effect. Removal of corpses during epidemics and appeals for cleanliness undoubtedly helped the public health.

The 1850s witnessed what is now called the "Great Sanitary Awakening." Led by tireless public health advocates like Sir Edwin Chadwick, proper and effective measures began to evolve. Possibly the single most important investigation of a public health problem was John Snow's classic epidemiological study of the 1849 cholera epidemic in London. By using a map of the area and identifying the residences of the people who contracted the disease, he was able to pinpoint the cause of the epidemic as the water from a public pump on Broad Street. Removal of the handle from the Broad Street pump eliminated the source of the cholera organism, and the epidemic subsided.* Ever since, waterborne diseases have become one of the major concerns of the public health. The reduction of such diseases by providing safe and pleasing water to the public has been one of the dramatic successes of the public health profession.

---

*Interestingly, it wasn't until 1884 that Robert Koch proved that *Vibrio comma* was the micro-organism responsible for cholera.

Today, the concerns of public health encompass not only water, but all aspects of civilized life, including food, air, toxic materials, noise, and other environmental insults.

ECOLOGY

The recognition that all life forms are inextricably dependent on one another and on the physical environment is a fairly recent phenomenon. The word "ecology" was not even invented until the mid-1800s, and the study of ecology as a branch of natural sciences was not widespread until a few decades ago.

The science of ecology defines "ecosystems" as groups of organisms that interact with each other and the physical environment and that affect the population of the various species in the environment. For example, the simplest type of ecosystem consists of two animals such as the hare and the lynx. If the hare population in a specific locality is high, the lynx have an abundant food supply, procreate, and increase in population until they outstrip the availability of hares. As the lynx population decreases because of the unavailability of food, the hares increase since the number of predators is fewer, and the cycle repeats (Figure 1–2). Such a system is dynamic in that the numbers of each population are continually changing, but over a long time span it is at a steady-state condition, known as "homeostasis."

A slightly more complex example includes three species: the sea otter, the sea urchin, and kelp. The kelp forests along the Pacific coast consist of 60-m (200-foot) streamers fastened to the ocean floor. Kelp is an economically

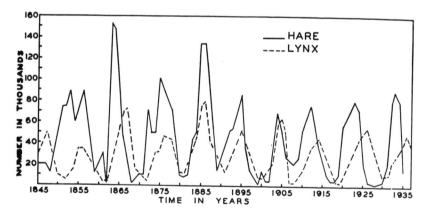

Figure 1–2. The hare and lynx homeostasis. [Source: MacLurich, D.A. "Fluctuations in the Numbers of the Varying Hare," University of Toronto Studies, Biological Sciences No. 43. Reproduced in Odum, G. *Fundamentals of Ecology*, 3rd ed. (Philadelphia: W.B. Saunders, 1971).]

valuable plant since it is the source of algin (used in foods, paints, cosmetics, etc.), and its harvesting is regulated to preserve the forests. A few years ago, kelp started to disappear mysteriously, leaving barren ocean floors. The mystery was solved when it was recognized that sea urchins feed on the kelp, weaken the stems, and cause them to detach and float away. The sea urchin population had increased because the population of their predators, the sea otters, had been reduced drastically. The solution to the problem was to protect the sea otter and allow its population to increase, thus reducing the number of sea urchins and maintaining the kelp forests.

Some ecosystems are fragile, whereas others are resistent and able to withstand even serious perturbations. One characteristic of a resilient ecosystem is that more than one species fills a "niche" within that system. That is, many species perform the same function. For example, in the hare/lynx system, if the hare were the only source of food for the lynx (a niche), the destruction of the hares would result in the eventual eradication of the lynx as well. If, however, the lynx used other small animals for food, the system might survive the loss of the hares.

Engineers must appreciate the fundamental principles of ecology and design in consonance with the environment to reduce the adverse impact on fragile ecosystems. For example, one of the most fragile of all ecosystems is the deep oceans, yet ocean disposal of hazardous waste is seriously advocated by some engineers. The inclusion of ecological principles in engineering decisions is a major component of the environmental engineering profession.

ETHICS

Historically, engineering did not concern itself with ethical decisions. Ethics as a framework for making decisions appeared to be irrelevant to engineering, since the engineer did precisely what employer or client required. Indeed, even today, some engineers revel in the supposed freedom from concern for ethical questions since engineering is applied science, and science is value-free.[9]

Unfortunately (or perhaps fortunately), engineers are daily confronted with questions that demand responses for which purely technical tools are inadequate. This is especially true for engineers engaged in environmental and pollution control activities, since their work interfaces a newly emerging field of "environmental ethics."[10]

Among the interesting questions asked in the search for environmental ethics is the origin of our attitude toward the environment. This problem, and much of the following discussion, is covered in greater detail in the last chapter. Suffice for now to recognize that Western traditions encourage private ownership of land and other resources and the exploitation of the environment. This tradition is contrary to the view of the American Indians and other cultures that consider land and natural resources as being a public trust. European settlers, arriving in the New World from countries where all land was owned by wealthy aristocrats, considered it their right to similarly own and exploit land.* And indeed, the natural resources were so plentiful that there grew a "myth of super-abundance," where the likelihood of running out of any resource (including oil!) was considered remote.[11]

Some dissenting voices were raised during the nineteenth century rape of the environment. Henry David Thoreau, Ralph Waldo Emerson, and later John Muir, Gifford Pinchot, and President Theodore Roosevelt all contributed to the growth of awareness in environmental concern. The first explicit statement of a need for an environmental ethic was penned by Aldo Leopold in the 1930s. Since then, many people have contributed thoughtful and well-reasoned arguments toward the development of a comprehensive and useful ethic for judging questions of conscience and value.

Since 1970, environmental and ecological awarenesses have been incorporated into public attitudes and are thus an integral part of engineering planning. No public works project, however small, is undertaken without a thorough assessment of its environmental impact and an exploration of alternatives. The environmental component of an engineer's work increases the difficulty of the questions engineers are called on to answer.

ENVIRONMENTAL ENGINEERING AS A PROFESSION

The purpose of colleges and universities is to allow students to mature intellectually and socially and to prepare themselves for careers that will earn them a livelihood. One of the main considerations used by students, especially in engineering, in selecting a profession is that the skills learned be marketable. There are, however, two other factors that should go into the equation by which career decisions are calculated.

First, the vocation should ideally be an avocation as well. It should be a job that is "fun," a job that is approached with enthusiasm even after many years in the profession. The people who are doing exactly what they would like to do, people who made a fortunate career decision, are surprisingly rare. Most engineers in fact get bored with their work, search around for other jobs, and often leave the profession altogether. In the selection of a career, therefore, the enjoyment factor is immensely important.

Second, it is necessary to be proud of what one has accomplished. Building a bridge to allow access to a previously isolated community may be a positive and personally satisfying undertaking. On the other hand, building a dam on a

---

*An exception to this general rule is found in states within the boundaries of the Northwest Purchase, notably Wisconsin. Included in the Purchase agreement between France and America is the condition that state constitutions must ensure that the water and air resources must be held in trust by the state for the people "as long as the wind blows and the water flows." Witness the incalculable number of public accesses to lakes and rivers in Wisconsin.

river that wipes out a wild whitewater stream and that creates a useless, highly eutrophied body of water cannot be a source of pride, regardless of what the boss says, or what political favors and deals are involved. In other cases, the social benefits from one's job are even more clearly lacking. Consider, for example, the people involved in the production of cigarette commercials. After completing a professional career on Madison Avenue, is it possible to note with pride that one wrote the world's best cigarette advertisement?

Environmental engineers are employed in all industries and by virtually all agencies of federal, state, and local government that deal with public works projects, roads, water and air quality, and management of natural resources. Much of the research for government policy is done under contract by consulting firms, who also hire many engineers. Environmental assessment also requires engineering input.

Environmental engineering has a proud history, and a bright future. It is a career that may be profitable, challenging, enjoyable, and satisfying, and perhaps most importantly, environmental engineers are committed to high standards of interpersonal and environmental ethics. Environmental engineers try to be part of the solution, while recognizing that *all* people are part of the problem.

REFERENCES

1. Kirby R.S., S. Withington, A.B. Darling, and F.G. Kilgrour. *Engineering in History* (New York: McGraw-Hill, 1956).
2. Wisely, W.H. *The American Civil Engineer, 1852–1974* (New York: The American Society of Civil Engineers, 1974).
3. Ridgeway, J. *The Politics of Ecology* (New York: E.P. Dutton & Co., 1970).
4. American Public Works Association. *History of Public Works in the United States, 1776–1976* (Chicago: American Public Works Association, 1976).
5. Lankton, L.D. "1842: The Old Croton Aqueduct Brings Water, Rescues Manhattan from Fire, Disease," *Civil Engineering* 47:10(1977).
6. Smith, G. *Plague on Us* (Oxford: Oxford University Press, 1941).
7. Baker, M.N. *The Quest for Pure Water* (New York: American Water Works Association, 1949).
8. Raverat, G. *Period Piece*, as quoted in Reyburn, W., *Flushed with Pride* (London: McDonald, 1969).
9. Florman, S. *The Existential Pleasures of Engineering* (New York: St. Martins Press, 1976).
10. *Environmental Ethics*, a professional journal published quarterly by the University of Georgia, Athens, GA.
11. Udall, S. *The Quiet Crisis* (Reading, MA: Addison–Wesley, 1968).

Chapter 2
Risk Analysis

In a nutshell, the job of the environmental engineer is to reduce the risks of environmental hazards, both long- and short-term. To accomplish this objective in an orderly fashion, the risks associated with various hazards must be quantified and evaluated.

In this chapter, risk analysis is introduced as a necessary tool of environmental engineering, crossing all media boundaries.

RISK

A number of laws regulate release of pollutants into the environment. However, major issues must be addressed. For instance, how is a pollutant identified as such, so that steps may be taken to control it? A substance is considered a pollutant because it is perceived to have an adverse effect on the environment and, in particular, an adverse effect on human health. However, it is sometimes difficult to determine whether there is an effect, or whether the effect has been deleterious or detrimental to the organism that is affected.

For example, we are now certain that cigarette smoke is unhealthy. Specifically, we have identified inhaled cigarette smoke as contributing significantly to lung cancer, chronic obstructive pulmonary disease, and heart disease. Notice that we do not say that cigarette smoking *causes* these health problems, because we have not identified the causes—the etiology—of any of them, at least in the sense that we have identified the poliomyelitis virus as the cause of polio. How, then, has cigarette smoking been identified as a contributing factor, if it cannot be identified as the cause? Cigarette smoking serves as a good example of how the health effects of pollutants are determined, although cigarette smoke is not regulated the way other pollutants are regulated.

In the latter half of the twentieth century, with the discovery of antibiotics and sulfa drugs, infectious diseases ceased to be a primary cause of death. The life span of people in the developed countries of the world lengthened considerably, and heart disease and cancer became leading causes of death. It was thus observed, in the early 1960s, that lifelong heavy cigarette smokers often died from lung cancer.

EXPRESSION OF RISK

Let us investigate the details of such an observation: how do you identify the cause of death, and how is it related to a habit like heavy smoking? The cause of death is listed on death certificates, which are the source of much epidemiological information. To relate death from lung cancer to smoking, one must show that more lung cancer deaths occur in smokers than in nonsmokers. Such a showing can be made by determining the *standard mortality ratio* (SMR), which is defined as

$$SMR = observed\ deaths/expected\ deaths$$

The number of "expected deaths" in the above equation is the number of deaths from the particular disease, lung cancer, in this case, that happen without any identifiable cause. In the general population of smokers and nonsmokers, there are a certain number of lung cancer deaths. Even in nonsmokers, there are a certain, albeit small, number of lung cancer deaths. In this instance, then, the SMR may be defined as

$$SMR = D_s/D_{ns}$$

where D_s = lung cancer deaths in a population of smokers
         D_{ns} = lung cancer deaths in a nonsmoking population of the same size

In this instance, the SMR is approximately 11/1. Since the SMR is greater than 1.0, then we can say that cigarette smoke contributes to lung cancer or, as it is usually phrased, smoking cigarettes increases the risk of death from lung cancer. To be precise, the risk of death from lung cancer is 11 times as high for a heavy smoker as for a nonsmoker. Determination of the SMR tells something about the *epidemiology* of smoking, but not about its *etiology*.

There are three characteristics of epidemiological reasoning in this example that are important:

● Everyone who smokes heavily will not die of lung cancer.
● Some nonsmokers die of lung cancer.
● Therefore, one cannot unequivocally relate any given individual lung cancer death to cigarette smoking.

Risk may be expressed in other ways as well. For example, in the United States there are 350,000 deaths each year from lung cancer and heart disease

Table 2–1.   Adult Deaths/10^5 for 1985

Cause of Death	Deaths/10^5 Population
Cardiovascular disease	408
Cancer	193
Chronic obstructive pulmonary disease	31
Motor vehicle accidents	18.6
Alcoholism	11.3
All causes	870

that are attributable to smoking. The United States has a population of 226 million. The risk of death associated with the effects of cigarette smoking may thus be expressed as deaths per 100,000 population, or

$$350,000/226 \times 10^6 = 155/100,000$$

In other words, a heavy smoker in the United States has an annual probability (or risk) of 155 in 100,000, or 1.55 in 1,000, of dying of lung cancer or heart disease. Table 2–1 presents some typical statistics for the United States.[1]

A third way to present risk is as deaths per 1,000 deaths. For example, in 1985, there were 2.084×10^6 deaths in the United States. Of these, 350,000 or 168 deaths per 1,000 deaths, were related to heavy smoking.

These figures show that meaningful risk analyses can be conducted only with very large populations. Moreover, the health risk posed by most pollutants is observed to be considerably lower than the risks cited in Table 2–1 and may simply not be observed in a small population. Chapter 18 cites several examples of statistically valid risks from air pollutants.

The Environmental Protection Agency (EPA) has adopted the concept of unit risk in discussions of potential risk. *Unit risk* is defined as the risk to an individual from exposure to a concentration of 1 $\mu g/m^3$ of an airborne pollutant, or 10^{-9} g/L of a waterborne pollutant. *Unit lifetime risk* is the risk to an individual from exposure to the above concentration for 70 years, or a lifetime, while *unit occupational risk* implies exposure for 40 hours per week for 50 years, or a working lifetime.

Example 2.1

The EPA has calculated that the unit lifetime risk from exposure to ethylene dibromide (EDB) in drinking water is $0.85/10^5$ population. What is the risk to an individual who, for 5 years, drinks water with an average EDB concentration of 5 pg/L?

$$\text{Risk} = (\text{concentration})(\text{unit risk})(\text{exposure time})/70 \text{ yr}$$

$$\text{Risk} = (5 \times 10^{-12} \text{ g/L})(0.85 \times 10^{-5} \text{ L}/10^{-9} \text{ g})(5 \text{ yr}/70 \text{ yr})$$
$$= 3.0 \times 10^{-8}$$

or approximately one chance of death in 33 million.

ASSESSMENT OF RISK

Risk of death (mortality risk) is easier to determine for populations in the developed countries than risk of illness (morbidity), because all deaths and their apparent causes are reported. Death certificate data may still be misleading: an individual who suffers from high blood pressure but is killed in an automobile accident becomes an accident statistic rather than a cardiovascular disease statistic. In the United States until very recently, statistics for deaths caused by occupational exposure could be determined only for males, because the majority of women did not work outside the home for much of their adult lives.

These particular uncertainties may be overcome in assessing risk from a particular cause by isolating the influence of that particular cause. This requires studying two populations whose environment is virtually identical, except that the risk to be studied is present in one population and not in the other. Such a study is called a *cohort study* and may be used to determine morbidity as well as mortality risk. One cohort study[1], for example, showed that residents of copper smelter communities, who were exposed to airborne arsenic, had a higher incidence of a certain type of lung cancer than residents of similar industrial communities where there was no airborne arsenic.

Retrospective cohort studies are almost impossible to perform because of uncertainties in data on habits, other exposures, etc. Cohorts must be well matched in cohort size, age distribution, lifestyle, and other environmental exposures and must be large enough for an effect to be distinguishable from the deaths or illnesses that occur anyway.

EXPOSURE AND LATENCY

Most malignant neoplasms (cancers) grow very slowly and are found (expressed) many years after the exposure to the responsible carcinogen. The length of time between exposure to a pollutant and expression of the adverse effect is called the *latency period*. Malignant neoplasms occurring in adults have apparent latency periods ranging in length from about 10 years to about 40 years. Relating a cancer to a particular exposure is fraught with inherent inaccuracy: it is exceedingly difficult to isolate the effect of a single carcinogen when examining 30 or 40 years of a person's life. Thus, many carcinogenic effects are simply not identifiable over the lifetime of an individual.

There are a few instances in which a particular neoplasm is found only on exposure to a given agent (e.g., a certain type of hemangioma is found only on exposure to vinyl chloride monomer), but for most cases, the connection between exposure and effect is far from clear. Many carcinogens are identified through animal studies, but one cannot always extrapolate from animal results to human results. Finally, chronic low-level exposure may (or may not) have different effects than acute high-level exposure, even when the total dose is the same. The cumulative uncertainty surrounding the epidemiology of pollutants has resulted in a conservative posture towards regulation and control. That is, if there is any evidence, even inconclusive evidence, that exposure to a substance results in adverse health effects, release of that substance into the environment is regulated and controlled.

DOSE-RESPONSE EVALUATION

The effect of a pollutant (an organism's response to the pollutant) always depends in some way on the amount or dose of the pollutant to the organism. The magnitude of the dose, in turn, depends on the pathway into the organism. Pollutants have different effects depending on whether they are inhaled, ingested, or absorbed through the skin or whether exposure is external. Ingestion or inhalation determines the biochemical pathway of the pollutant in the organism. In general the human body detoxifies an ingested pollutant more efficiently than an inhaled pollutant.

The relationship between the dose of a pollutant and the organism's response can be expressed in a dose-response curve, as shown in Figure 2–1. Some characteristic features of the dose-response relationship are:

1. *Threshold.* The existence of a threshold in health effects of pollutants has been debated for many years. With reference to Figure 2–1, there are four basic dose-response curves possible for a dose of a specific pollutant (e.g., carbon monoxide) and the response (e.g., reduction in the blood's oxygen-

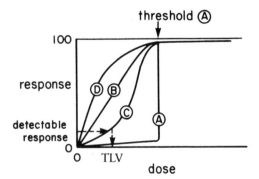

Figure 2–1. Possible dose-response curves.

carrying capacity). Curve A shows that if this dose-response relationship holds, there is no effect on human metabolism until a critical concentration is reached. This critical concentration is called the *threshold* and is indicated in the figure. Curve A is a linear dose-response curve.

Curve B suggests that there is a detectable response for *any* finite concentration of the pollutant; this curve is also linear but shows no threshold.

Curve C is a sigmoidal dose-response curve and is characteristic of many pollutant dose-response relationships. Although curve C has no clearly defined threshold, the point at which a response can be detected (shown in the figure) is called the *threshold limit value* (TLV). Occupational exposure guidelines are frequently set at the TLV. Curve C is sometimes called a "sublinear" dose-response relationship. Curve D represents a "supralinear" relationship, which is found when low doses appear to provoke a disproportionately large response. The sum of the doses given to the organism from all pathways is the *total body burden* and includes concentrations of pollutants remaining in the body from previous expo-sures. The rate at which pollutants are eliminated from the body is measured by the physiological half-life.

2. *Time Versus Dosage.* Most pollutants require time to react, and thus the time of contact is as important as the level. The best example of this is the effect of carbon monoxide. CO reduces the oxygen-carrying capacity of the blood by combining with the hemoglobin and forming carboxyhemoglobin. At about 60 percent carboxyhemoglobin concentration, death results from lack of oxygen. The effects of CO at sublethal concentrations are usually reversible. Because of the time-response problem, ambient air quality stan-dards are set at maximum allowable concentrations for a given time (see Chapter 20).

3. *Synergism.* Synergism is defined as an effect that is greater than the sum of the parts. For example, black lung disease in coal miners occurs only when the miner is also a cigarette smoker. Coal mining by itself or cigarette smoking by itself will not cause black lung, but the synergistic action of the two puts miners who smoke at high risk. The opposite of synergism is *antagonism*: a phenomenon that occurs when two pollutants counteract each other's effects.

4. LC_{50} *and* LD_{50}. Dose-response relationships for humans are generally determined from health data or epidemiological studies. Human volunteers obviously cannot be subjected to doses of pollutants that produce major or lasting health effects. Toxicity may, however, be determined by subjecting nonhuman organisms to increasing doses of pollutants until the organism dies. The term LD_{50} refers to the dose that is lethal for 50 percent of the experimental animals used (LC_{50} refers to lethal concentration instead of lethal dose). LD_{50} values are of most use in comparing toxicities, as for pesticides and agricultural chemicals; no direct extrapolation is possible,

either to humans or to any species other than the one used for the LD_{50} determination.

5. *Bioaccumulation and Bioconcentration.* The term *bioaccumulation* is used when a substance is concentrated in one organ or type of tissue. Iodine, for example, bioaccumulates in the thyroid gland. The dose to an organ may thus be considerably greater than the body burden or whole-body dose.

    Bioconcentration occurs with movement up the food chain. A study[2] of a Lake Michigan ecosystem found the following bioconcentration of DDT:

0.014 ppm (wet weight) in bottom sediments

0.41 ppm in bottom-feeding crustacea

3 to 6 ppm in fish

2,400 ppm in fish-eating birds

POPULATION RESPONSES

Individual responses to pollutants are not identical. Dose-response curves will differ from one person to another; in particular, thresholds will differ. In general, threshold values in a population follow a Gaussian distribution. Figure 2–2 shows the distribution of odor thresholds for hydrogen sulfide in a typical population.

Individual responses and thresholds also depend on age, sex, and general state of physical and emotional health. As might be anticipated, healthy young adults are less sensitive to pollutants than are elderly people, chronically or acutely ill people, and children. Allowable releases of pollutants are, in theory, restricted to amounts that assure protection of the health of the entire population, including its most sensitive members. In many cases, however, such restriction would mean zero release.

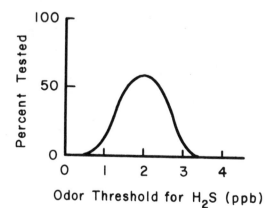

Odor Threshold for H$_2$S (ppb)

Figure 2–2. Distribution of odor thresholds in a population.

The levels actually chosen take technical and economic control feasibility into account and usually are set below threshold level for 95 percent or more of the entire population. For nonthreshold pollutants, however, no such determination can be made. In these instances, there is no release level for which protection can be assured for anyone, so that comparative risk analysis is necessary.

Many pollutants, including carcinogens, are considered to have no measurable threshold. The best available control for such pollutants still entails a residual risk. There is a continuing need in our industrial society, therefore, for accurate quantitative risk assessment. We must also remain aware that we have not precisely identified carcinogens or mutagens, nor do we understand their mechanism of action. We can only identify apparent associations between most pollutants and a given health effect.

In almost all cases, doses to the general public are so small that excess mortality and morbidity are not identifiable. In fact, almost all of our knowledge of adverse health effects of pollutants comes from occupational exposure, in which doses are orders of magnitude higher.

PROBLEMS

2.1 Using the data given in the chapter, calculate the SMR (from all diseases) for heavy smoking in the United States.

2.2 The EPA has determined the lifetime unit risk for cancer for low-energy ionizing radiation to be 2.8×10^{-4} per rem of radiation. The allowed level of ionizing radiation (EPA standard) above background is 25 mrem per year. Average nonanthropogenic background is about 100 mrem per year. How many excess cancers would result in the United States each year if the entire population were exposed at the level of the EPA standard? How many cancers may be attributed to background? If 10 percent of these are fatal each year, what percentage of the annual cancer deaths in the United States may be attributed to exposure to background radioactivity?

2.3 The allowed occupational dose for ionizing radiation is 5 rem per year. By what factor does a worker exposed to this dose over a working lifetime increase his or her risk of cancer?

2.4 By what factor was DDT bioconcentrated at each level of the food chain in Lake Michigan in the example given in the text?

2.5 The EPA has set standards for the emission of arsenic from non-ferrous smelters that project a maximum unit cancer incidence risk of 150 in 10,000 (per $\mu g/m^3$ of airborne arsenic) to a population of 190,000 persons living within 20 km of a smelter. Assume that the population exposed is exposed throughout life and that death from this type of cancer occurs 25 years after incidence. What would you consider an acceptable level of ambient airborne arsenic from such a smelter, in $\mu g/m^3$, based on the risk? Suppose a resident lived within this radius of the smelter for 10 years, on the average. Would your answer change? By how much?

2.6 Using the data of the preceding problem, estimate an acceptable workplace standard for ambient arsenic.

2.7 It is widely assumed that the Love Canal and Times Beach incidents, in which toxic chemicals were found to have leaked into the environment over a period of years, resulted in adverse health effects to the residents of these areas. Design an epidemiological study that would determine whether there actually were adverse health effects resulting from these incidents. Assume that the leaks occurred over 5 years and that there were leaks into the air, soil, and drinking water.

2.8 How does synergism make it difficult to establish cause/effect relationship between air pollutants and disease? Give at least one example.

REFERENCES

1. National Center for Health Statistics, U.S. Department of Health and Human Services, 1985.
2. Hickey, J.J., et al. "Concentration of DDT in Lake Michigan," *Journal of Applied Ecology* 3:141(1966).

Chapter 3
Water Pollution

Although people intuitively relate filth with disease, the fact that pathogenic organisms can be transmitted by polluted water was not recognized until the middle of the nineteenth century. Probably the most dramatic demonstration that water can indeed transmit disease was the Broad Street pump-handle incident.

A public health physician named John Snow, assigned to attempt to control a cholera epidemic, realized that there seemed to be an extremely curious concentration of cholera cases in one part of London. Almost all of the people affected drew their drinking water from a community pump in the middle of Broad Street. Even more curious was the fact that the people who worked and lived in an adjacent brewery were not afflicted. Although this seemed to demonstrate the health benefits of beer, welcome news to most students, Snow recognized that the absence of cholera in the brewery might be because the brewery obtained its water from a private well and not the Broad Street pump.

Snow's evidence convinced the city council to ban the obviously polluted water supply, which was done by simply removing the pump handle, thus effectively preventing the people from using the water. The source of infection was stopped, the epidemic subsided, and a new era of public health awareness related to water supplies began.

The concern with water pollution was, until recently, a concern about health effects. In many countries it still is. In the United States and other developed countries, however, water treatment and distribution methods have for the most part eradicated the transmission of bacterial waterborne disease. We now think of water pollution not so much in terms of health as in terms of its effect on the aquatic ecosystem, and fish and shellfish in particular. We also consider conservation, aesthetics, and the preservation of natural beauty and resources. People have an inexplicable affinity for water, and the fouling of lakes, rivers, and oceans is intrinsically unacceptable to the concerned citizen.

SOURCES OF WATER POLLUTION

The United States has more than 40,000 factories that use water, and their industrial wastes are probably the greatest single water pollution problem.

Organic wastes from industrial plants, at present-day treatment levels, are equal in polluting potential to the *untreated* raw sewage of the entire population of the United States. In most cases the organic wastes, as potent as they might be, are at least treatable, in or out of the plant. Inorganic industrial wastes are much more difficult to control and potentially more hazardous. Chromium from metal-plating plants is an old source of trouble, but mercury discharges have only recently received their due attention.

As important as these and other well-known heavy metals might be, many scientists are much more concerned with the unknown chemicals. Industry is creating a fantastic array of new chemicals each year, all of which may eventually find their way into water. The decomposition of these chemicals in water and how the chemicals react with each other are only poorly understood, as are the respective acute and chronic toxicity.

Another industrial waste is heat. Heated discharges may drastically alter the ecology of a stream or lake. This alteration is sometimes called beneficial, perhaps because of better fishing or an ice-free docking area. The deleterious effects of heat, in addition to promoting modifications of ecological systems, include a lessening of dissolved oxygen (DO) solubility and increases in metabolic activity. Dissolved oxygen is vital to healthy aquatic communities, and the warmer the water, the more difficult it is to get oxygen into solution. As the level of DO decreases, metabolic activity of aerobic (oxygen-using) aquatic species increases, thus increasing oxygen demand. It is a small wonder, therefore, that the vast majority of fish kills owing to oxygen depletion occur in the summer.

Municipal waste is a source of pollution second in importance only to industrial wastes. Around the turn of the century, most discharges from municipalities received no treatment whatsoever. In the United States, sewage from 24 million people was flowing directly into our watercourses. Since that time, the population has increased, and so has the contribution from municipal discharges. It is estimated that presently the population equivalent* of municipal discharges to watercourses is about 100 million. Even with the billions of dollars spent on building wastewater treatment plants, the contribution from municipal pollution sources has not been significantly reduced. We seem to be holding our own, however, and at least are not falling further behind.

---

*Population equivalent is the number of people needed to contribute a certain amount of pollution. For example, if a town has 10,000 people, and the treatment plant is 50 percent effective, then their discharge has a population equivalent of 5,000 people. Similarly, if an industry discharges 1,000 lb of solids per day, and if each person contributes 0.2 lb/day into domestic wastewater, the industrial waste may be expressed as being the equivalent of $1,000/0.2 = 5,000$ people.

One problem, especially in the older cities on the east coast of the United States, is the sewerage systems. When the cities were first built, the engineers realized that sewers were necessary for both stormwater and sanitary wastes, and they saw no reason why both stormwater and sanitary wastes should not flow in the same pipelines. After all, they both ended up in the same river or lake. Such sewers are now known as *combined sewers*.

As years passed and populations increased, the need for the treatment of sanitary wastes became obvious, and two-sewer systems were built, one to carry stormwater and the other, sanitary waste. Such systems are known as *separate sewers*.

Almost all of the cities with combined sewers have built treatment plants that can treat the *dry weather flow*, or in other words, sanitary wastes. As long as it doesn't rain, they can provide sufficient treatment. When the rains come, however, the flows swell to many times the dry weather flow and most of it must be bypassed directly into a river or lake. This overflow contains sewage as well as stormwater and has a high polluting capacity. All attempts to capture this excess flow for subsequent treatment, such as storage in underground caverns and rubber balloons, are expensive. However, the cost of the alternative solution, separating the sewers, is estimated at a staggering $100 billion for the major cities in the United States.

In addition to industrial and municipal wastes, water pollution emanates from many other sources. Agricultural wastes, should they all flow into streams, would have a population equivalent of about 2 billion. The problems are intensifying with the increase in the size and number of feedlots: cattle pens constructed for the purpose of fattening up the cattle before slaughter. These are usually close to slaughterhouses (hence cities), and a large number of animals are packed into a small space. Drainage for these lots has an extremely high pollutional strength, as does drainage from intensive duck and chicken cultivation. There is an extremely high potential for water pollution whenever animal waste is concentrated, even in aquaculture.

Sediment from land erosion may also be classified as a pollutant. Sediment consists of mostly inorganic material washed into a stream as a result of farming, construction, or mining operations. The detrimental effects of sediment include interference with the spawning of fish by covering gravel beds; interference with light penetration, thus making food more difficult to find; and direct damage to gill structures. In the long run, sediment could well be one of our most harmful pollutants.

The concern with pollution from petroleum compounds is relatively new, starting to a large extent with the Torrey Canyon disaster in 1967. Despite maps showing submerged rocks, the huge tanker loaded with crude oil plowed into a reef in the English Channel. Almost immediately, oil began seeping out, and both the French and British became concerned. Rescue efforts failed, and the Royal Air Force attempted to set the oil on fire, with little success. Almost all of the oil eventually leaked out and splashed on the beaches of France and England. The French started the backbreaking chore of spreading straw on the beaches, allowing the straw to adsorb the oil, and then collecting and burning the oil-soaked straw. The English used detergents to disperse the oil and then flushed the emulsion off the beaches. Time has shown the French way to be best, since the English detergents have now been shown to be potentially more harmful to coastal ecology than the oil would have been.

Although the Torrey Canyon disaster was the first big oil spill, many have followed it. It is estimated that there are no fewer than 10,000 serious oil spills in the United States every year. In addition, the contribution from routine operations such as flushing oil tankers may well exceed all the oil spills.

The acute effect of oil on birds, fish, and microorganisms is reasonably well cataloged. What is not so well understood, and is potentially more harmful, is the subtle effect on other aquatic life. Anadromous fish, for example, have been known to find their home stream by the specific smell (or taste) of the water, caused in large part by the hydrocarbons present. If people continue (albeit unintentionally) to pour hydrocarbons into salmon runs, it is possible that the salmon will become so confused that they will refuse to enter their spawning stream.

Another form of industrial pollution, much of it willed to us by our ancestors, is acid mine drainage. The problem is caused by the leaching of sulfur-laden water from old abandoned mines (as well as some active mines). On contact with air, these compounds are soon oxidized to sulfuric acid, a deadly poison to all living matter.

It should be amply clear, therefore, that water can be polluted by many types of waste products.

Water pollution problems (and hence their solutions) can be best understood by first describing them in the context of an ecosystem and then studying one specific aspect of that ecosystem: the biodegradation of organics.

ELEMENTS OF AQUATIC ECOLOGY

Plants and animals in their physical environment make up an *ecosystem*. The study of such ecosystems is *ecology*. Although we often draw lines around a specific ecosystem to be able to study it more fully (e.g., a farm pond) and in so doing assume that the system is totally self-contained, this obviously is not true, and we must remember that one of the tenets of ecology is that "everything is connected with everything else."

Within an ecosystem there exist three broad categories of actors. The *producers* take energy from the sun and nutrients such as nitrogen and phosphorus from the soil and through the process of photosynthesis produce high-energy chemicals. The energy from the sun is thus stored in the chemical structure of their organic molecules. These organisms are often referred to as being in the first *trophic* level and are called *autotrophs*.

A second group of organisms are the *consumers* who use some of this energy by ingesting the high-energy molecules. These organisms are in the second level in that they directly use the energy of the producers. There may be several more trophic levels as the consumers use the level above as a source of energy. A simplified ecosystem showing various trophic levels is shown in Figure 3–1, which also illustrates the progressive use of energy through the trophic levels.

The third group of organisms, the *decomposers* or decay organisms, use the energy in animal wastes and dead plants and animals and in so doing convert the organic molecules to stable inorganic compounds. The residual inorganics then become the building blocks for new life, with the sun as the source of energy.

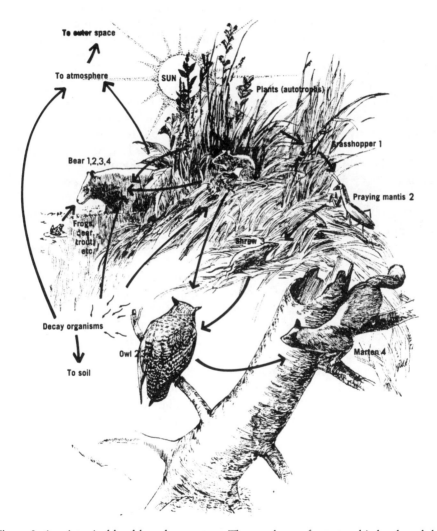

Figure 3–1. A typical land-based ecosystem. The numbers refer to trophic level, and the arrows show progressive loss of energy for life systems. [Source: Turk, A. et al. *Environmental Science* (Philadelphia: W.B. Saunders, 1974).]

Ecosystems exhibit a flow of both energy and nutrients. Energy flow is in only one direction: from the sun and through each trophic level. Nutrient flow, on the other hand, is cyclic. Nutrients are used by plants to make high-energy molecules, which are eventually decomposed to the original inorganic nutrients, ready to be used again.

The entire food web, or ecosystem, stays in dynamic balance, with adjustments being made as required. Such a balance is called *homeostasis*. For example, a drought one year may produce little grass, thus exposing field mice to predators such as owls. The mice, in turn, spend more time in burrows, thus not eating as much, and allowing the grass to reseed for the following year. External perturbations, however, may upset and even destroy an ecosystem. In the previous example, the use of a herbicide to kill the grass might also destroy the field mouse population, and in turn diminish the number of owls. It must be recognized that although most ecosystems can absorb a certain amount of insult, a sufficiently large perturbation can cause irreparable damage.

The amount of perturbation a system is able to absorb without being destroyed is tied to the concept of the *ecological niche*. The combination of function and habitat of an organism in an ecological system is its niche. A niche is not a property of a type or organism or species, but is its best accommodation with the environment. In the example above, the grass is a producer that acts as food for the field mouse, which in turn is food for the owl. If the grass were destroyed, both the mice and owls might eventually die out. But suppose there were *two* types of grass, each equally acceptable as mouse food. Now if one died out, the ecosystem would not be destroyed because the mice would still have food. This simple example demonstrates an important ecological principle: the stability of an ecosystem is proportional to the number of organisms capable of filling various niches. A jungle, for example, is a very stable ecosystem, whereas the tundra in Alaska is extremely fragile. Another fragile system is deep oceans—a fact that should be a consideration in the disposal of hazardous and toxic materials in deep ocean areas. Inland watercourses tend to be fairly stable ecosystems, but certainly not totally resistant to destruction by outside forces. Other than the direct effect of toxic materials such as heavy metals and refractory organics,* the most serious effect of water pollution for inland waters is the depletion of dissolved (free) oxygen. All higher forms of aquatic life exist only in the presence of oxygen, and most desirable microbiologic life also requires oxygen. Generally, all natural streams and lakes are aerobic (containing DO). If a watercourse becomes anaerobic (absence of oxygen), the entire ecology changes to make the water unpleasant or unsafe.

Problems associated with pollutants that affect the DO levels cannot be appreciated without a fuller understanding of the concept of decomposition or biodegradation, part of the total energy transfer system of life.

---

*Refractory organics are man-made organic materials, such as the pesticide DDT, which decompose very slowly in the environment.

BIODEGRADATION

Plant growth, or photosynthesis, may be represented by the equation

$$CO_2 + H_2O \xrightarrow[\text{\& nutrients}]{\text{sunlight}} HCOH + O_2$$

In this representation formaldehyde (HCOH) and oxygen are produced from carbon dioxide and water, with sunlight as the source of energy.* If the formaldehyde and oxygen are combined and ignited, an explosion results. The energy that is released during such an explosion is stored in the carbon-hydrogen-oxygen bonds of formaldehyde.

As discussed above, plants (producers) use inorganic chemicals as nutrients and, with sunlight as a source of energy, build high-energy molecules. The animals (consumers) eat these high-energy molecules and during their digestion process some of the energy is released and used by the animals. The release of this energy is quite rapid, and the end products of digestion (excrement) consist of partially stable compounds. These compounds become food for other organisms and are thus degraded further but at a slower rate. After several such steps, very low energy compounds are formed that can no longer be used by microorganisms for food. Plants then use these compounds to build more high-energy molecules, and the process starts all over. The process is symbolically shown in Figure 3–2.

It is important to realize that many of the organic materials responsible for water pollution enter watercourses at a high energy level. It is the biodegradation, or the gradual use of this energy, by a chain of organisms that causes many of the water pollution problems.

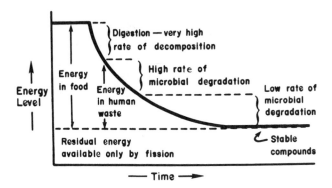

Figure 3–2.   Energy loss in biodegradation.

---

*Of course, formaldehyde is not the end product of photosynthesis, but it is an organic molecule and happens to provide a simple equation.

AEROBIC AND ANAEROBIC DECOMPOSITION

Decomposition, or biodegradation, may take place in one of two distinctly different ways: aerobic (using free oxygen) or anaerobic (in the absence of free oxygen).

The basic equation of aerobic decomposition of complex organics, expressed as $C_xH_yN_z$, is

$$\left(x + \frac{y+x}{2}\right)O_2 + C_xH_yN_z + O_2 \rightarrow xCO_2 + \frac{y}{2} H_2O + zNO$$

$$+ \text{ stable products}$$

Carbon dioxide and water are always two of the end products of aerobic decomposition. Both are stable, low in energy, and are used by plants in the process of photosynthesis. If sulfur compounds are involved in the reaction, the most stable end product is SO_4^{2-}, the sulfate ion. Similarly, phosphorus ends up as PO_4^{3-}, orthophosphate. Nitrogen goes through a series of increasingly stable compounds, finally ending up as nitrate. The progression is

$$\text{Organic Nitrogen} \rightarrow NH_3 (\text{ammonia}) \rightarrow NO_2^- (\text{nitrite}) \rightarrow NO_3^- (\text{nitrate})$$

Because of this distinctive progression, nitrogen has been in the past and to some extent is still used as an indicator of pollution.

A schematic representation of the aerobic cycle for carbon, sulfur, and nitrogen compounds is shown as Figure 3–3. This figure illustrates only the basic facts and is a gross simplification of the actual steps and mechanisms involved.

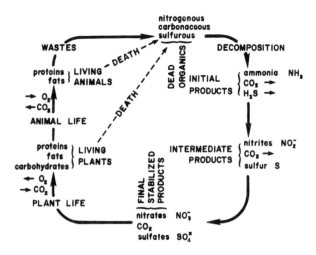

Figure 3–3. Aerobic nitrogen, carbon, and sulfur cycles.

A second type of biodegradation is anaerobic, performed by a completely different set of microorganisms, to which oxygen is in fact toxic. The basic equation of anaerobic decomposition is

$$C_xH_yN_z \rightarrow CO_2 + CH_4 + \text{other partially stable compounds}$$

Note that many of the end products shown are biologically unstable. CH_4, for example, is methane, a high-energy gas commonly called marsh gas, physically stable but still able to be decomposed biologically. Nitrogen compounds stabilize only to ammonia (NH_3), and sulfur ends up as evil-smelling hydrogen sulfide (H_2S) gas. Figure 3–4 is a schematic representation of anaerobic decomposition. Note that the left half of the cycle, the photosynthesis by plants, is identical to the aerobic cycle in Figure 3–3.

Biologists often speak about various compounds as "hydrogen acceptors." The hydrogen atoms torn from high-energy organic molecules must be attached to various compounds. In aerobic decomposition, oxygen serves this purpose and is thus known as the hydrogen acceptor. It accepts the hydrogen atoms to form water.

In anaerobic decomposition, free oxygen is not available, and the next preferred hydrogen acceptor is nitrogen, thus forming ammonia, NH_3. If free oxygen is not available, ammonia cannot be converted to nitrites or nitrates. If nitrogen is not available, the next preferred hydrogen acceptor is sulfur, thus forming hydrogen sulfide, H_2S, the chemical responsible for the notorious rotten egg smell.

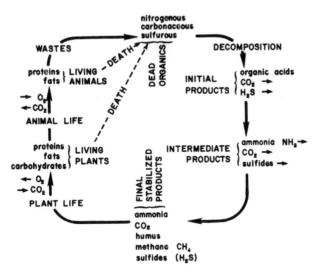

Figure 3–4.   Anaerobic nitrogen, carbon, and sulfur cycles.

EFFECT OF POLLUTION ON STREAMS

When a high-energy organic material such as raw sewage is discharged to a stream, a number of changes occur downstream from the point of discharge. As the organics are decomposed, oxygen is used at a greater rate than before the pollution occurred, and the DO level drops. The rate of reaeration, or solution of oxygen from the air, also increases, but this is often not great enough to prevent a total depletion of oxygen in the stream. When this happens, the steam is said to become anaerobic. Often, however, the DO does not drop to zero, and the stream recovers without experiencing a period of anaerobiosis. Both of these situations are depicted graphically in Figure 3–5. The dip in DO is referred to as a *dissolved oxygen sag curve*.

The effect of a certain waste on a stream's oxygen level may be estimated mathematically. The basic assumption is that there is an oxygen balance at any point in the stream and that this level is dictated by: (1) how much oxygen is being used by the microorganisms, and (2) how much oxygen is supplied to the water through reaeration (O_2 dissolved into the water from the atmosphere). The rate of oxygen use, or oxygen depletion, may be expressed as:

$$\text{Rate of deoxygenation} = -k_1'z$$

where $z =$ the amount of oxygen still required at any time t, or the biochemical demand for oxygen remaining in the water, mg/L*

$k_1' =$ deoxygenation constant, a function of the type of waste material decomposing, temperature, etc., days^{-1}.

The value of k_1' is measured in the laboratory, as discussed in the next chapter. Integrating the above equation yields

$$z = L_o e^{-k't}$$

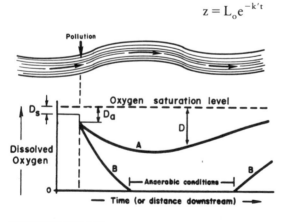

Figure 3–5. Dissolved oxygen downstream from a source of organic pollution. Curve A depicts an oxygen sag without anaerobic conditions; curve B shows DO levels when pollution is sufficiently strong to create anaerobic conditions. D_a is the oxygen deficit in the stream after the stream has mixed with the pollutant.

---

*mg/L (milligrams per liter) is a common way of expressing the concentration of chemicals in water. It is a weight/volume measurement, with the stated milligrams of a chemical per liter of water.

where L_o is the *ultimate oxygen demand*. Since the long-term need for oxygen is L_o and the amount of oxygen still needed at any time t is z, the amount of oxygen *used* at any time t must be

$$y = L_o - z$$

This relationship is shown in Figure 3–6. The term y is defined later in this text as the *biochemical oxygen demand* (BOD) and expressed as

$$y = L_o(1 - e^{-k't})$$

Although y is commonly termed oxygen *demand*, it is more correctly described as the DO *used*. In this text we use the terms *demanded* and *used* synonymously.

The reoxygenation of a watercourse may be expressed as:

$$\text{Rate of reoxygenation} = k_2'D$$

where   D = deficit in DO, or the difference between saturation (maximum DO the water can hold) and the actual DO, mg/L
   k_2' = reoxygenation constant, days^{-1}.

The value of k_2' is obtained by conducting a study on a stream by using a tracer. If this cannot be done, a generalized formula may be used:

$$k_2' = \frac{3.9v^{1/2}[(1.037)^{(T-20)}]^{1/2}}{H^{3/2}}$$

where   T = temperature of the water, °C
   H = average depth of flow, m
   v = mean stream velocity, m/sec

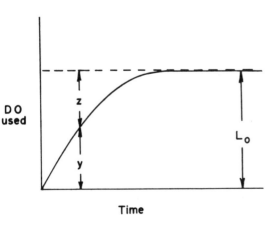

DO used

Time

Figure 3–6. The dissolved oxygen used at any time t (y) plus what is still needed (z) is equal to the ultimate oxygen demand (L_o).

Alternatively, k_2' values may be estimated by using generalized tables, such as Table 3–1.

If a stream is loaded with organic material, the simultaneous action of deoxygenation and reoxygenation forms the DO sag curve, first discussed by Streeter and Phelps in 1925.

The shape of the oxygen sag curve, as shown in Figure 3–5, is the result of adding the rate of oxygen use and the rate of supply. If the rate of use is great, as in the stretch of stream immediately after the pollution, the DO level drops because the supply rate cannot keep up with it. As the rate of oxygen use decreases (fewer high-energy readily decomposable organics) and the difference between oxygen saturation level and actual DO (the deficit) is great, the supply begins to keep up with the use, and the DO will once again reach saturation levels.

This may be expressed mathematically as:

$$\frac{dD}{dt} = k_1'z - k_2'D$$

where all the terms are as defined above. The rate of change in the deficit (D) depends on the concentration of decomposable organic matter, or the need by the microorganisms for oxygen (z) and the deficit at any time t. The need for oxygen at any time t was previously expressed as:

$$z = L_o e^{-k_1't}$$

where L_o = ultimate oxygen demand, or the maximum oxygen required in mg/L. Substituting this into the above expression and integrating,

$$D = \frac{k_1'L_o}{k_2' - k_1'} (e^{-k_1't} - e^{-k_2't}) + D_a e^{-k_2't}$$

where D_a = the initial oxygen deficit, at the point of pollution, after the stream flow has mixed with the source of pollution, mg/L
D = deficit at any time t, mg/L

Table 3–1.   Reaeration Constants

	k_2' at 20°C[a], (days^{-1})
Small ponds or backwaters	0.1–0.23
Sluggish streams	0.23–0.35
Large streams, low velocity	0.35–0.46
Large streams, normal velocity	0.46–0.69
Swift streams	0.69–1.15
Rapids	>1.15

[a]For temperatures other than 20°C, $k_{2_T}' = k_{2_{20}}' = 1.024^{T-20}$.

Note that

$$D_a = \frac{D_s Q_s + D_p Q_p}{Q_s + Q_p}$$

where D_s = oxygen deficit in the stream directly upstream from the point of pollution, mg/L
Q_s = stream flow, m^3/sec
D_p = oxygen deficit in the pollutant stream, mg/L
Q_p = flow rate of pollutant, m^3/sec

The deficit equation is often expressed in common logarithms, and since

$$e^{-k't} = 10^{-kt} \text{ when } k = 0.434 \, k'$$

then

$$D = \frac{k_1 L_o}{k_2 - k_1} (10^{-k_1 t} - 10^{-k_2 t}) + D_a (10^{-k_2 t})$$

The most serious concern of water quality is of course the point in the curve where the deficit is the greatest, or the DO concentration the least. By setting the $dD/dt = 0$, we can solve for the critical time as:

$$t_c = \frac{1}{k_2' - k_1'} \ln \left[\frac{k_2'}{k_1'} \left(1 - \frac{D_a (k_2' - k_1')}{k_1' L_o} \right) \right]$$

where t_c = time downstream when the DO is the lowest.

Example 3.1
    Assume that a large stream has a reoxygenation constant k_2' of 0.4/day and a flow velocity of 5 miles/hr and that at the point of the pollution discharge the stream is saturated with oxygen at 10 mg/L. The wastewater flow rate is small compared with the stream flow, so the mixture is assumed to have a DO at saturation and an oxygen demand of 20 mg/L. The deoxygenation constant is 0.2/day. What is the DO level 30 miles downstream?

    Velocity = 5 miles/hr, hence it takes 30/5 = 6 hours to travel 30 miles
        $v = 6/24 = 1/4$ days
        $D_a = 0$ since the stream is saturated

$$D = \frac{(0.2)(20)}{0.4 - 0.2} [e^{-0.2(0.25)} - e^{-0.4(0.25)}] = 20[e^{-0.05} - e^{-0.1}] = 1.0 \text{ mg/L}.$$

The DO is thus the saturation level minus the deficit, or $10 - 1.0 = 9.0$ mg/L.

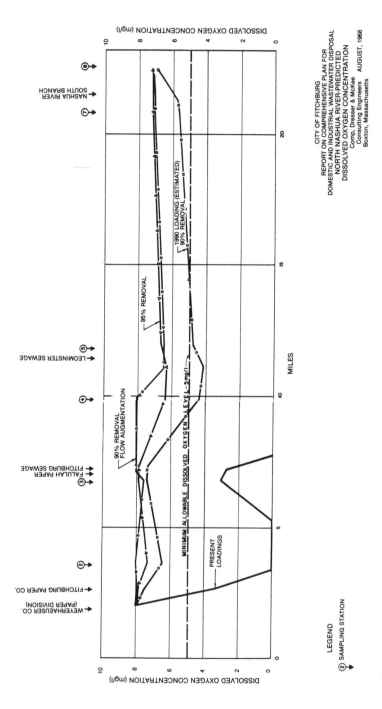

Figure 3–7. Actual DO sag (going to zero; anaerobic conditions) and the projected DO sags if various types of wastewater treatment are provided. [Courtesy of Camp Dresser & McKee, Boston, MA.]

An actual DO sag curve is shown in Figure 3–7. Note that the stream goes anaerobic at about mile 3.5, recovers, then drops back to zero after receiving effluents from a city and a paper mill. Also shown on the figure is the expected DO sag if 95 percent of the demand for oxygen is removed from all discharges.

Stream flow is of course variable, and the critical DO levels can be expected to occur when the flow is the lowest. Accordingly, most state regulatory agencies base their calculations on a statistical low flow, such as a 7-day, 10-year low flow, or the seven consecutive days of lowest flow that may be expected to occur once during a 10-year interval. This is calculated by first estimating the lowest 7-day discharge for each year, assigning these rankings, m as m = 1 for the most severe (least flow) and m = n (where n is the number of years) for the least severe case. The probability of a flow equal to or more than a low flow occurring is calculated as $m/(n + 1)$, and plotted versus the flow. Usually log-probability paper is used since this gives the best straight line fit. Then $m/(n + 1) = 0.1$ is read as the 10-year low flow.

Example 3.2
 Calculate the 10-year, 7-day low flow given the data below.

Year	Lowest flow 7 consecutive days (m^3/sec)	Ranking (m)	Lowest flow in order of severity (m^3/sec)	$\dfrac{m}{n+1}$
1965	1.2	1	0.4	0.071
1966	1.3	2	0.6	0.143
1967	0.8	3	0.6	0.214
1968	1.4	4	0.8	0.285
1969	0.6	5	0.8	0.357
1970	0.4	6	0.8	0.428
1971	0.8	7	0.9	0.500
1972	1.4	8	1.0	0.571
1973	1.2	9	1.2	0.642
1974	1.0	10	1.2	0.714
1975	0.6	11	1.3	0.785
1976	0.8	12	1.4	0.857
1977	0.9	13	1.4	0.928

These data are plotted on Figure 3–8, and the minimum 7-day, 10-year flow is read off as $0.5 \text{ m}^3/sec$.

When the rate of oxygen usage overwhelms the rate at which the oxygen can be supplied, the stream may become anaerobic. An anaerobic stream is easily identifiable. Since oxygen is no longer available to act as the hydrogen acceptor, ammonia and hydrogen sulfide are formed, among other gases. Some

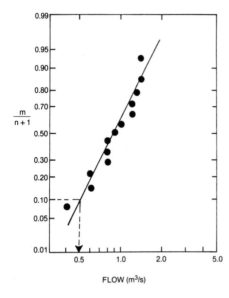

Figure 3–8. Plot of 10-year, 7-day low flows, for Example 3.2.

of these gases will dissolve readily, but others will attach themselves as bubbles to hunks of black solid material, known as benthic deposits (or simply sludge), and buoy this material to the surface. Anaerobic streams are thus recognized by floating sludge solids and the formation of gas that bubbles to the surface. In addition, the H_2S emitted will advertise the anaerobic condition for a considerable distance. Two other telltale signs are the color of the water, generally black, and the presence of long filamentous fungus growths, which cling to rocks and gracefully wave slimy streamers downstream.

It is logical to assume that such outward changes may also be evidenced by the effect on aquatic life. Indeed, the types and numbers of species change drastically downstream from the point of gross pollution. The increased turbidity, settled solid matter, and low DO all contribute to a decrease in fish life. Fewer and fewer species of fish are able to survive, but those types of fish that do survive find that food is plentiful, and they often multiply in numbers. Carp and catfish can survive in quite foul water and can even gulp air from the surface if necessary. Trout, on the other hand, need very pure, cold water to survive and are notoriously intolerant of pollution.

The numbers of other types of aquatic life are also reduced. Such characters as sludge worms, bloodworms, or rat-tailed maggots abound, and their numbers may be staggering—as many as 50,000 sludge worms per square foot. The variation of both the number of species and the total number of organisms downstream from a source of pollution is illustrated in Figure 3–9.

The diversity of species may be quantified by using an index, such as

$$\bar{d} = \sum_{i=1}^{s} \left(\frac{n_i}{n}\right) \log_2\left(\frac{n_i}{n}\right)$$

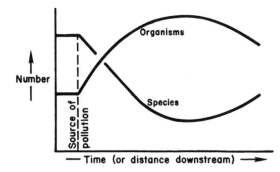

Figure 3–9. The number of species and the total number of organisms downstream from a point of organic pollution.

where $\bar{d}$ = diversity index
         n_i = number of individuals in the ith species
         n = total number of individuals in all S species

In one study, the diversity index was calculated above and below a sewage outfall, and the results were as shown in Table 3–2.

It was mentioned earlier that nitrogen compounds may be used as indicators of pollution. The changes in the various forms of nitrogen are shown in Figure 3–10. The first transformation, in both aerobic and anaerobic decomposition, is the formation of ammonia, and thus the concentration of ammonia increases as organic nitrogen decreases. Similarly, the concentration of nitrate nitrogen will finally increase to become by far the dominant form of nitrogen.

Table 3–2. Diversity of Aquatic Organisms

Location	Diversity Index $(\bar{d})$
Above the outfall	2.75
Immediately below the outfall	0.94
Downstream	2.43
Further downstream	3.80

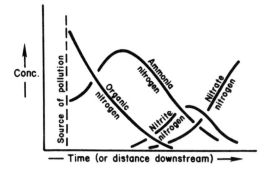

Figure 3–10. Typical variations in nitrogen compounds downstream from a point of organic pollution.

It is important to remember that the reactions of a stream to pollution outlined above occur when a rapidly decomposable organic material is the waste. The stream will react much differently to an inorganic waste, say from a metal-plating plant. If the waste is toxic to aquatic life, both the kind and total number of organisms will decrease below the outfall. The DO will not fall, and might even rise. There are many types of pollution, and a stream will react differently to each. Even more complicated is a situation in which two or more wastes are involved.

EFFECT OF POLLUTION ON LAKES

The effect of pollution on lakes differs in several respects from the effect on streams. For one thing, light and temperature have significant influences on a lake and must be included in any limnological* analysis. Light is the source of energy in the photosynthetic reaction, and the penetration of light into the lake water is important. This penetration is logarithmic. If, for example, at a 1-foot depth the light is 10,000 footcandles,** at 2 feet it might be 1,000, at 3 feet 100, and at 4 feet only 10 footcandles. Only the top two feet of a lake generally experience light penetration, and hence all photosynthetic reactions occur in that zone.

Temperature often has a profound effect on a lake. Water is at a maximum density at 4°C (water both colder and warmer is lighter; ice, therefore, floats). Water is also a poor conductor of heat and retains heat quite well.

Lakes usually undergo a seasonal variation in water temperature. These temperature-depth relationships are illustrated in Figure 3–11. During the winter, assuming the lake does not freeze, the temperature is often constant with depth. As warmer weather approaches the top layers begin to warm up. Since water is a poor conductor of heat and warmer water is lighter, a distinct temperature gradient is formed known as *thermal stratification*. These strata are often very stable and last through the summer months. The top layer is called the *epilimnion*, the middle the *metalimnion*, and the bottom the *hypolimnion*. The inflection point in the curve is called the *thermocline*. Circulation of water occurs only within a zone, and thus there is only limited transfer of biological or chemical material (including DO) across the boundaries.

As the colder weather approaches, the top layers begin to cool, become more dense, and sink. This creates circulation throughout the lake, a condition known as *fall turnover*. Often a spring turnover also occurs.

The biochemical reactions in a natural lake may be represented schematically as in Figure 3–12. A river feeding the lake would contribute carbon, phosphorus, and nitrogen, either as high-energy organics or as low-energy

---

*Limnology is the study of lakes.
**The intensity of lights is measured in footcandles.

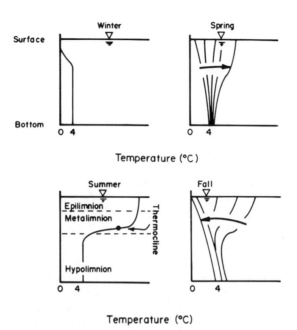

Figure 3–11.   Typical temperature-depth relationships in a lake.

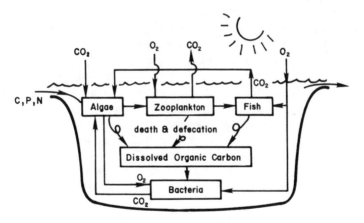

Figure 3–12.   Schematic representation of lake ecology.

compounds. The *phytoplankton* or *algae* (microbial free-floating plants) take C, P, and N and using sunlight as a source of energy make high-energy compounds. Algae are eaten by *zooplankton* (tiny aquatic animals), which are in turn eaten by larger aquatic life such as fish. All of these forms of life defecate, thus contributing to a pool of *dissolved organic carbon*. This pool is further fed by the death of aquatic life. Bacteria utilize dissolved organic carbon and produce

CO_2, in turn used by the algae. Additional CO_2 is provided from the respiration of the fish and zooplankton as well as the CO_2 dissolved directly from the air.

In an unpolluted lake the supply of incoming C, P, and N is sufficiently small to limit the production of algae, and productivity of the entire ecological system is limited. But what happens when excessive amounts of C, P, and N are introduced to the lake?

The plentiful supply of nutrients promotes the uncontrolled growth of algae. When the algae, along with the zooplankton and fish, die, they drop to the bottom and become another source of carbon for the bacteria. Aerobic bacteria will use all available DO in decomposing this material and may use enough oxygen to deplete all available DO, thus creating anaerobic conditions.

As more and more algae are produced in the epilimnion, the bacteria in the lower portions of the lake will utilize more and more oxygen, and the metalimnion might also become anaerobic. All of the aerobic biological activity would thus concentrate in the upper few feet of the lake, the epilimnion. All this activity causes turbidity, thus decreasing light penetration and in turn limiting algal activity to the surface layers. The amount of DO contributed by the algae is therefore decreased. Eventually, when the epilimnion is also anaerobic, all aerobic aquatic life disappears, and the algae, because of limited light penetration, concentrate on the surface of the lake, forming large green mats or *algal blooms*. These algae will also die and eventually fill up the lake, producing what we now know as a peat bog.

This entire process is called *eutrophication*. It is a continually occurring natural process. But what may have taken thousands of years to occur naturally may be accomplished in only a decade if enough nutrients are introduced into a lake as a result of human activities.

It is worth repeating that eutrophication is a natural process. The stages of a lake's natural aging are the *oligotrophic* stage, during which both the variety and number of species grow rapidly; the *mesotrophic* stage, during which there is a dynamic equilibrium of species in the lake; and the *eutrophic* stage, during which less complex organisms take over and the lake appears to become gradually choked with weeds. However, the addition of nutrients can speed up eutrophication considerably. Addition of phosphorus is particularly effective, since phosphorus is often the chemical that limits algal growth.*

Where do these nutrients originate? One source is excrement, since all human and animal wastes include C, P, and N. This source, however, is small compared with synthetic detergents and fertilizers. It is estimated that of the total P discharged to our lakes, one-half comes from agricultural runoff, one-fourth comes from detergents, and one-fourth comes from all other sources.

---

*Phosphorus "limits" algal growth in that it is a constraint against unlimited growth, much like the gas pedal on your car "limits" the speed of your car. The car is able to go faster—all the components are there and could function at a higher rate—but the gas pedal is a constraint against higher speed. Dumping excess phosphorus into a lake is like tromping down on the gas pedal.

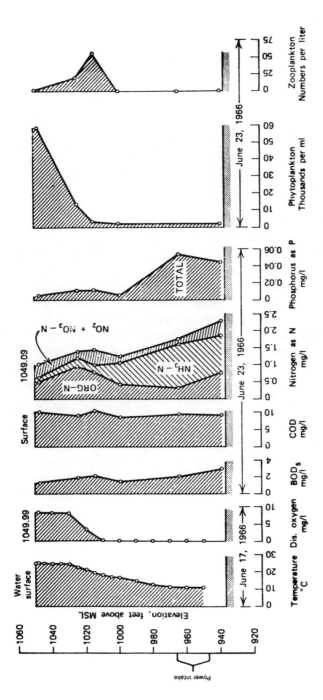

Figure 3–13. Actual water quality profiles for a water supply reservoir. [Source: Berthouex, P. and D. Rudd, *Strategy of Pollution Control* (New York: John Wiley & Sons, 1977).]

It seems unfortunate that the presence of phosphates in detergents has received so much unfavorable attention when runoff from fertilized land is a much more important source. The conversion to nonphosphate detergents would be of limited value if other sources are not controlled.

It is generally believed that a P concentration between 0.01 and 0.1 mg/L is sufficient to promote accelerated eutrophication. Effluents from sewage treatment plants often contain from 5 to 10 mg/L of P. A river flowing through farm country might carry from 1 to 4 mg/L of P. In moving streams this high P concentration is not a problem since streams are continually flushed out and the algae do not have time to accumulate. Eutrophication may thus occur only in lakes, ponds, estuaries, and sometimes in very slowly moving rivers.

Incidentally, it has been estimated that only about 15 percent of the people in the United States should be concerned about phosphate discharges from their municipalities. About 85 percent of the wastewater flows into moving streams and rivers where eutrophication is not a problem.

There is also some question as to the validity of blaming phosphorus for accelerated eutrophication. Generally a P:N:C ratio of 1:16:100 is required for growth. It takes 16 parts N and 100 parts C for every part P for algae to grow. If there are more than 16 parts N and 100 parts C for every P, then N and C are said to be in excess, and thus P must be limiting growth. Although this is the case with most lakes, there are those that have very high P levels and have few problems with algal blooms. Conversely, some lakes have experienced serious algal problems while carrying extremely low P levels. In addition, recent evidence suggests that nitrogen limits algal growth in brackish waters such as bays and estuaries. Suffice it to say, the blame for the accelerated eutrophication of many of our lakes cannot at this time be placed on any one chemical or product. An interaction among the many pollutants involved is the probable answer.

Actual profiles in a lake for a number of parameters are shown in Figure 3–13. Based on the above discussion, it should be clear why the temperature is warmer on the top, why the DO drops to zero, and why there is high phosphorus and nitrogen concentration at the deeper areas, while the phytoplankton (algae) are all on the surface.

HEAVY METALS AND TOXIC SUBSTANCES

In 1970, scientists alerted the nation to the growing problem of mercury contamination of lakes, streams, and marine waters. The manufacture of chlorine and lye from brine—the chlor-alkali process—was identified as a major source of mercury contamination. Elemental mercury is methylated by aquatic organisms, and methylated mercury finds its way into fish and shellfish and thus into the human food chain. Methyl mercury is a powerful neurological poison. Methyl mercury poisoning was first identified in Japan in the 1950s as "Minamata disease": mercury-containing effluent from the Minamata Chemical Company was found to be the source of mercury in food fish.

Arsenic, copper, lead, and cadmium are often deposited in lakes and streams from the air, in the vicinity of nonferrous smelters. These substances are also constituents of mine drainage and industrial effluent. The effluent from electroplating and metal refining contains a number of heavy metal constituents. Heavy metals, copper in particular, may be toxic to fish as well as harmful to human health.

In the 1960s, 70s, and 80s, there were a large number of incidents of contamination of the surface waters of the United States with toxic and carcinogenic organic compounds. The sources of these include effluent from petrochemical industries and agricultural runoff, which contains both pesticide and fertilizer residues. Trace quantities of chlorinated hydrocarbon compounds in drinking water may also be attributed to the chlorination of organic residues by the chlorine that is added as a disinfectant.

EFFECT OF POLLUTION ON OCEANS

Not many years ago, the oceans were considered infinite sinks; the immensity of the seas and oceans seemed impervious to assault. This is no longer the case. We now recognize seas and oceans as fragile environments and are able to measure the detrimental effect of our actions.

The water in the oceans is the most complicated chemical solution imaginable, and there is evidence that it has changed very little over millions of years. Because of this constancy, however, marine organisms have become highly specialized and intolerant to environmental change. Oceans are thus fragile ecosystems, quite susceptible to pollution.

A relief map of the ocean bottom reveals that there are two major areas of what we think of ocean: the continental shelf and the deep oceans. The continental shelf, and especially the areas near major estuaries, are the most productive in terms of food supply. These also receive the greatest pollution load. Many major estuaries have become so badly polluted that they have been closed to commercial fishing. Some large areas, such as the Baltic and Mediterranean, are also in danger of becoming permanently damaged.

Although ocean disposal of wastewater is severely restricted in the United States, many major cities all over the world still discharge all of their untreated sewage into the oceans. This is usually done by pipelines running considerable distances from the shore and discharging through diffusers to achieve maximum dilution. The controversy continues as to the wisdom of using the oceans for wastewater disposal, and what the long-term consequences might be.

CONCLUSION

The sources, causes, and effects of water pollution are legion. We have discussed only a limited number. The effects of pollution on oceans or groundwater or the effects of inorganic poisons such as mercury and cadmium on aquatic life also deserve attention but are ignored here because of the lack of space, and not because they are not important.

PROBLEMS

3.1 The aerobic cycle, shown as Figure 3–3, is only for nitrogen, carbon, and sulfur. Phosphorus should also have been included in the cycle since it exists as organic phosphorus in living and dead tissue and decomposes to poly-phosphates [such as $(P_2O_7)^{4-}$, $P_3O_{10})^{5-}$, etc.] and finally to the inorganic orthophosphate, PO_4^{3-}. Draw a phosphorus cycle similar to Figure 3–3.

3.2 Draw typical oxygen sag curves for the following wastes:

1. potato waste (high BOD)
2. electroplating waste (Cr, Co, etc.)
3. brick waste (clay)

Assume that the stream has a DO of 5 mg/L at the point of the pollution and a temperature of 20°C.

3.3 Some researchers have suggested that the empirical analysis of some algae has the following chemical composition (by weight):

$$C_{106}H_{181}O_{45}N_{16}P$$

Suppose an analysis of a lake water yields the following:

$$C = 62 \text{ mg/L}$$
$$N = 1.0 \text{ mg/L}$$
$$P = 0.01 \text{ mg/L}$$

Which elements would be limiting the growth of algae in this lake? (Show your calculations.)

3.4 A stream feeding a lake has an average flow of 1 ft³/sec and a phosphate concentration of 10 mg/L. The water leaving the lake has a phosphate concentration of 5 mg/L.

a. How much phosphate is deposited in the lake every year?
b. Where does this phosphorus go? (The outflow is less than the inflow, so it has to go somewhere.)
c. Would you expect the average phosphate concentration to be higher near the surface or the bottom?
d. Would you expect this lake to have accelerated eutrophication problems? Why?

3.5 Suppose you have a 55-gallon drum full of distilled water, and you add one frog weighing 0.2 lb. What is the concentration of frogs in terms of milligrams per liter? Show all your calculations.

3.6 What could be done to maximize the silting in a reservoir?

3.7 Given a make-believe high-energy molecule $H_{14}C_2SN_2O_3$, show how it decomposes anaerobically and how these end products in turn decompose aerobically to stabilized sulfur and nitrogen compounds.

3.8 The temperature soundings for a lake are as follows:

Depth (ft)	Water Temperature (°F)
surface	80
4	80
8	60
16	40
24	40
30	40

Draw a graph of depth versus temperature and label the hypolimnion, epilimnion, and thermocline.

3.9 If an industrial plant discharges solids at a rate of 5,000 lb/day, and if each person contributes 0.2 lb/day, what is the population equivalent of the waste?

3.10 Draw the DO sag curves you would expect in a stream from the following wastes. Assume the stream flow equals the flow of wastewater. (Do not calculate.)

Waste	Source	Demand for oxygen, BOD (mg/L)	Suspended solids, SS (mg/L)	Phosphorus (mg/L)
A	Dairy	2,000	100	40
B	Brick mfg.	5	100	10
C	Fertilizer mfg.	25	5	200
D	Plating plant	0	100	10

3.11 Starting with nitrogenous dead organic matter, follow N around the aerobic and anaerobic cycles by writing down all the various forms of nitrogen.

3.12 Why does eutrophication rarely occur in a stream?

3.13 Name six ways you can recognize an anaerobic stream (without instrumentation).

3.14 Given a make-believe molecule $(C_2Hg)_2S$, show how it would degrade under anaerobic conditions (use as many oxygens as needed).

3.15 Suppose a stream with a velocity of 1 ft/sec, a flow of 10 million gallons per day (mgd), and an ultimate BOD of 5 mg/L was hit with treated sewage at 5 mgd with an ultimate BOD (L_o) of 60 mg/L. The temperature of the stream water is 20°C, at the point of the sewage discharge the stream is 90 percent saturated with oxygen, and the wastewater is at 30°C and has no oxygen (see Table 3–1). Measurements show the deoxygenation constant $k_1' = 0.5$ and the reoxygenation $k_2' = 0.6$, both as days^{-1}. Calculate: (a) the oxygen deficit 1 mile downstream, (b) the minimum DO (the lowest part of the sag curve), and (c) the minimum DO (or maximum deficit) if the ultimate BOD of the treatment plant's effluent was 10 mg/L. Use a computer program if possible.

3.16 An industry discharges sufficient quantities of organic wastes to depress the oxygen sag curve to 2 mg/L 5 miles downstream, at which point the DO begins to increase. This DO is too low, and the state regulatory agency wants to fine them for noncompliance with standards. One solution is to build a wastewater treatment plant. This is expensive, and so the industry looks around for other means of resolving their problem.

A salesman tells the plant engineer about freeze-dried bacteria, which come to life in water and are especially adapted to the kind of organic waste the plant is discharging. The salesman suggests that the plant engineer dump the bacteria into the river at the point of the industrial waste discharge and that this will solve the problem.

Will it? Why or why not? Draw the DO sag curve before and after he adds the bacteria.

3.17 An industry discharges into a river with a velocity of 0.1 m/sec, a temperature of 26°C, and an average depth of 4 m. The ultimate oxygen demand (L_o) immediately below the effluent discharge point where the waste has mixed with the river water is 14.5 mg/L and the DO is 5.4 mg/L. Laboratory studies show that $k_1' = 0.11$ day^{-1} at 26°C. Find the point of critical DO and the minimum DO.

LIST OF SYMBOLS

D = deficit in DO, mg/L
D_a = initial DO deficit, mg/L
d = diversity index
H = depth of stream flow, m
k_1 = deoxygenation constant, ($\log_{10}$) sec^{-1}
k_1' = deoxygenation constant, ($\log_e$) sec^{-1}
k_2 = reoxygenation constant, ($\log_{10}$) sec^{-1}
k_2' = reoxygenation constant, ($\log_e$) sec^{-1}
L_o = ultimate biochemical oxygen demand, mg/L
m = rank assigned to low flows
n = number of individuals in all S species

n = number of years in low-flow records
n_i = number of individuals in species i
T = temperature, °C
t = time, sec
t_c = critical time, time when the minimum DO occurs, sec
v = velocity, m/sec
y = oxygen used, mg/L
z = oxygen required for decomposition, mg/L

Chapter 4

Measurement of
Water Quality

Quantitative measurements of pollutants are obviously necessary before water pollution can be controlled. Measurement of these pollutants is, however, fraught with difficulties.

The first problem is that the specific materials responsible for the pollution are sometimes not known. The second difficulty is that these pollutants are generally at low concentrations, and very accurate methods of detection are therefore required.

Only a few of the many analytical tests available to measure water pollution are discussed in this chapter. A complete volume of analytical techniques used in water and wastewater engineering is compiled as *Standard Methods*.[1] This volume, now in its 16th edition, is the result of a need for standardizing test techniques. It is considered definitive in its field and has the weight of legal authority.

Many of the pollutants are measured in terms of milligrams of the substance per liter of water (mg/L). This is a weight/volume measurement. In many older publications pollutants are measured as parts per million (ppm), a weight/weight parameter.* If the liquid involved is water, these two units are identical, since 1 milliliter (mL) of water weighs 1 gram (g). For pollutants present in very low concentrations (< 10 mg/L), ppm is approximately equal to mg/L. However, because of the possibility of some wastes not having the specific gravity of water, the ppm measure has been scrapped in favor of mg/L.

A third commonly used parameter is percent, a weight/weight relationship. Obviously 10,000 ppm = 1 percent and this is equal to 10,000 mg/L only if 1 mL = 1 g.

---

* In air pollution, however, ppm is in volume/volume.

SAMPLING

Some tests require the measurement to be conducted in the stream since the process of obtaining a sample may change the measurement. For example, if it is necessary to measure the dissolved oxygen (DO) in a stream, the measurement should be conducted right in the stream, or the sample must be extracted with great care to ensure that no transfer of oxygen between the air and water (in or out) has occurred.

Most tests may be performed on a water sample taken from the stream. The process by which the sample is obtained, however, may greatly influence the result.

There are basically three types of samples:

1. grab
2. composite
3. flow weighed composite

The grab, as the name implies, simply measures a point. Its value is that it accurately represents the water quality at the moment of sampling, but it obviously says nothing about the quality before or after the sampling.

The composite sample is obtained by taking a series of grab samples and mixing them together. The flow weighed composite is obtained by taking each sample so that the volume of the sample is proportional to the flow at that time. The last method is especially useful when daily loadings to wastewater treatment plants are calculated.

Whatever the technique or method, however, it is necessary to recognize that the analysis can only be as accurate as the sample, and often the sampling methodology is far more sloppy than the analytical determination.

DISSOLVED OXYGEN

Probably the most important measure of water quality is the DO. Oxygen, although poorly soluble in water, is fundamental to aquatic life. Without free DO, streams and lakes become uninhabitable to gill-breathing aquatic organisms. Dissolved oxygen is inversely proportional to temperature and the maximum oxygen that can be dissolved in water at most ambient temperatures is about 9 mg/L.

This saturation value decreases rapidly with increasing water temperature, as shown in Table 4–1. The balance between saturation and depletion is therefore tenuous.

The amount of oxygen dissolved in water is usually measured either with an oxygen probe or by iodometric titration: Winkler DO Test. The Winkler test for DO, developed almost 100 years ago, is the standard to which all other methods are compared.

Table 4–1. Solubility of Oxygen

Temperature of Water (°C)	Saturation Concentration of Oxygen in Water (mg/L)
0	14.6
2	13.8
4	13.1
6	12.5
8	11.9
10	11.3
12	10.8
14	10.4
16	10.0
18	9.5
20	9.2
22	8.8
24	8.5
26	8.2
28	8.0
30	7.6

Chemically simplified reactions in the Winkler test are as follows:

1. Manganese ions added to the samples combine with the available oxygen

$$Mn^{2+} + O_2 \rightarrow MnO_2$$

forming a precipitate.

2. Iodide ions are added, and the manganous oxide reacts with the iodide ions to form iodine

$$MnO_2 + 2I^- + 4H^+ \rightarrow Mn^{2+} + I_2 + 2H_2O$$

3. The quantity of iodine is measured by titrating with sodium thiosulfate, the reaction being

$$I_2 + 2S_2O_3^{2-} \rightarrow S_4O_6^{2-} + 2I^-$$

Note that all of the DO combines with Mn^{2+}, so that the quantity of MnO_2 is directly proportional to the oxygen in solution. Similarly, the amount of iodine is directly proportional to the manganous oxide available to oxidize the iodide. Although the titration measures iodine, the quantity of iodine is thereby directly related to the original concentration of oxygen.

The Winkler test has obvious disadvantages, including chemical interferences and the difficulty of performing a wet chemical test in the field. The reagents needed to fix the DO as MnO_2 are available in powdered form for field use, but the remainder of the procedure must be performed in a laboratory. These disadvantages are overcome by using a DO electrode, often called a probe.

The simplest (and historically the first) probe is shown in Figure 4–1. The principle of operation is that of a galvanic cell. If lead and silver electrodes are put in an electrolyte solution with a microammeter between, the reaction at the lead electrode would be

$$Pb + 2OH^- \rightarrow PbO + H_2O + 2e^-$$

At the lead electrode, electrons are liberated that travel through the microammeter to the silver electrode where the following reaction takes place:

$$2e^- + \tfrac{1}{2}O_2 + H_2O \rightarrow 2OH^-$$

The reaction would not go unless free DO is available, and the microammeter would not register any current. The trick is to construct and calibrate a meter in such a manner that the electricity recorded is proportional to the concentration of oxygen in the electrolyte solution.

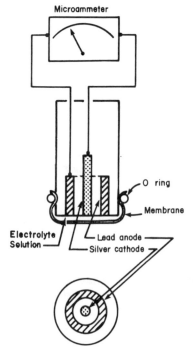

Figure 4–1.   Schematic of a galvanic cell oxygen probe.

In the commercial models, the electrodes are insulated from each other with nonconducting plastic and are covered with a permeable membrane with a few drops of an electrolyte between the membrane and electrodes. The amount of oxygen that travels through the membrane is proportional to the DO concentration. A high DO concentration in the water creates a strong push to get through the membrane, whereas a low DO concentration would force only limited O_2 through to participate in the reaction and thereby create electrical current. Thus the current registered is proportional to the oxygen level in solution.

BIOCHEMICAL OXYGEN DEMAND

Perhaps even more important than the determination of DO is the measurement of the rate at which the oxygen is used. A very low rate of use would indicate either clean water or that the available microorganisms are uninterested in consuming the available organics. A third possibility is that the microorganisms are dead or dying. (Nothing decreases oxygen consumption by aquatic microorganisms quite so well as a healthy slug of arsenic.)

The rate of oxygen use is commonly referred to as *biochemical oxygen demand* (BOD). It is important to understand that BOD is not a measure of some specific pollutant. Rather, it is a measure of the amount of oxygen required by bacteria and other microorganisms while stabilizing decomposable organic matter.

The BOD test was first used for measuring the oxygen consumption in a stream by filling two bottles with stream water, measuring the DO in one, and placing the other in the stream. In a few days the second bottle was retrieved, and the DO was measured. The difference in the oxygen levels was the BOD, or oxygen demand, in milligrams of oxygen used per liter of sample.

This test had the advantage of being very specific for the stream in question since the water in the bottle is subjected to the same environmental factors as the water in the stream, and thus the result was an accurate measure of DO usage in that stream. It was impossible, however, to compare the results in different streams, since three very important variables were not constant: temperature, time, and light.

Temperature has a pronounced effect on oxygen uptake (usage), with metabolic activity increasing significantly at higher temperatures. The time allotted for the test is also important, since the amount of oxygen used increases with time. Light is also an important variable since most natural waters contain algae and oxygen can be replenished in the bottle if light is available. Different amounts of light would thus affect the final oxygen concentration.

The BOD test was finally standardized by requiring the test to be run in the dark at 20°C for 5 days. This is defined as the 5-day BOD, or BOD_5, or the oxygen used in the first 5 days. Although there appear to be some substantial scientific reasons why 5 days was chosen, it has been suggested that the possibility of preparing the samples on a Monday and taking them out on

Friday, thus leaving the weekend free, was not the most important of these reasons.

The BOD test is almost universally run with a standard BOD bottle (about 300-mL volume) as shown in Figure 4–2. It is of course also possible to have a 2-day, 10-day, or any other day BOD. One measure used in some cases is *ultimate BOD*, or the O_2 demand after a very long time.

If we measure the oxygen in several samples every day for 5 days, we may obtain curves such as those shown in Figure 4–3. Sample A had an initial DO of 8 mg/L, and in 5 days this has dropped to 2 mg/L. The BOD therefore is $8 - 2 = 6$ mg/L.

Sample B also had an initial DO of 8 mg/L, but the oxygen was used up so fast that it dropped to zero. If after 5 days we measure zero DO, we know that the BOD of sample B was more than $8 - 0 = 8$ mg/L, but we don't know how much more since the organisms might have used more DO if it were available. For samples containing more than about 8 mg/L, dilution of the sample is therefore necessary.

Figure 4–2. A biochemical oxygen demand (BOD) bottle.

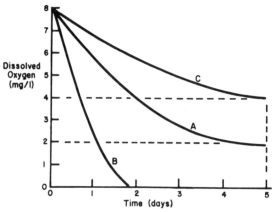

Figure 4–3. Typical oxygen uptake (use) curves in a BOD test.

Suppose sample C shown on the graph is really sample B diluted by 1:10. The BOD of sample B is therefore

$$\frac{8-4}{0.1} = 40 \text{ mg/L}$$

It is possible to measure the BOD of any organic material (e.g., sugar) and thus estimate its influence on a stream, even though the material in its original state might not contain the necessary organisms. *Seeding* is a process in which the microorganisms that create the oxygen intake are added to the BOD bottle.

Suppose we used the water previously described by curve A as seed water since it obviously contains microorganisms (it has a 5-day BOD of 6 mg/L). We now put 100 mL of an unknown solution into a bottle and add 200 mL of seed water, thus filling the 300-mL bottle. Assuming that the initial DO of this mixture was 8 mg/L and the final DO was 1 mg/L, the total oxygen used was 7 mg/L. But some of this was due to the seed water, since it also has a BOD, and only a portion was due to the decomposition of the unknown material. The DO uptake due to the seed water was

$$6 \times \left(\frac{2}{3}\right) = 4 \text{ mg/L}$$

since only two-thirds of the bottle was the seed water, which has a BOD of 6 mg/L. The remaining oxygen uptake ($7 - 4 = 3$ mg/L) must have been due to the unknown material.

If the seeding and dilution methods are combined, the following general formula is used to calculate the BOD:

$$\text{BOD (mg/L)} = \frac{(I - F) - (I' - F')(X/Y)}{D}$$

where   I = initial DO of bottle with sample and seeded dilution water
F = final DO of bottle with sample and seeded dilution water
I' = initial DO of bottle with seeded dilution water
F' = final DO of bottle with seeded dilution water
X = mL of seeded dilution water in sample bottle
Y = mL in bottle with only seeded dilution water
D = dilution of sample

Example 4.1
Calculate the BOD$_5$, given the following data:
Temperature of sample, 20°C
Initial DO is saturation
Dilution is 1:30, with seeded dilution water
Final DO of seeded dilution water is 8 mg/L
Final DO bottle with sample and seeded dilution water is 2 mg/L
Volume of BOD bottle is 300 mL

From Table 4–1, at 20° saturation is 9.2 mg/L; hence, this is the initial DO. Since the BOD bottle contains 300 mL, a 1:30 solution would have 10 mL of sample and 290 mL of seeded dilution water, and

$$BOD_s = \frac{(9.2 - 2) - (9.2 - 8)(290/300)}{0.033} = 18.3 \text{ mg/L}$$

If, instead of stopping the test after 5 days, we allowed the reactions to proceed and measured the DO each day, we might get a curve like that shown in Figure 4–4. Note that some time after 5 days the curve takes a sudden jump. This is due to the exertion of oxygen demand by the microorganisms that decompose nitrogenous organics and that are converting these to the stable nitrate, NO_3^-. The curve is thus divided into nitrogenous and carbonaceous BOD areas. Note also the definition of the ultimate BOD.*

It is important to remember that BOD is a measure of oxygen use, or potential use. An effluent with a high BOD may be harmful to a stream if the oxygen consumption is great enough to cause anaerobic conditions. Obviously, a small trickle going into a great river will have negligible effect, regardless of the milligrams per liter of BOD involved. Similarly, a large flow into a small stream may seriously affect the stream even though the BOD might be low. Accordingly, engineers often talk of "pounds of BOD," a value calculated by multiplying the concentration by the flow rate, with a conversion factor, so that

$$\text{lb of BOD/day} = [\text{mg/L of BOD}] \times \left[\begin{array}{c} \text{flow in million} \\ \text{gallons per day} \end{array} \right] \times 8.34$$

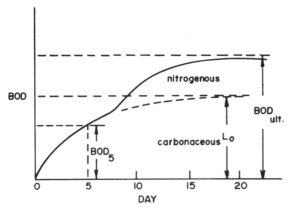

Figure 4–4. Long-term BOD. Note that BOD_{ult} includes the nitrogenous as well as the ultimate carbonaceous BOD (L_o).

---

* This, however, is different from the definition of ultimate BOD on p. 32. The latter would more accurately be termed *ultimate carbonaceous BOD*.

The BOD of most domestic sewage is about 250 mg/L. Many industrial wastes run as high as 30,000 mg/L. The potential detrimental effect of an untreated dairy waste that might have a BOD of 20,000 mg/L is quite obvious.

As discussed in Chapter 3, the carbonaceous part of the BOD curve can be modeled using the equation

$$y = L_o(1 - e^{-k_1't})$$

where $y = BOD_t$, or the amount of oxygen demanded by the microorganisms at some time t, mg/L

L_o = the ultimate carbonaceous demand for oxygen, mg/L

k_1' = rate constant, previously termed the deoxygenation constant, days^{-1}

t = time, days

It should be again emphasized that this equation is applicable only to the carbonaceous BOD curve, and L_o is defined as the ultimate carbonaceous BOD.

It is often necessary, such as when modeling the DO profile in a stream, to know both k_1' and L_o. This must be done by using laboratory BOD tests.

There are a number of techniques for calculating k_1' and L_o. One of the simplest is a method devised by Thomas.[2] Starting with the equation

$$y = L_o(1 - 10^{-k_1't})$$

we can rearrange it to read

$$\left(\frac{t}{y}\right)^{1/3} = (2.3 \, k_1 L_o)^{-1/3} + \left(\frac{k_1^{2/3}}{3.43 \, L_o^{1/3}}\right) t$$

This equation is in the form of a straight line

$$x = a + bt$$

where $x = (t/y)^{1/3}$

$a = (2.3 \, k \, L_o)^{-1/3}$

$b = k_1^{2/3}/(3.43 \, L_o^{1/3})$

Thus plotting x versus t, the slope (b) and intercept (a) can be obtained, and

$$k_1 = 2.61 \, (b/a)$$

$$L_o = 1/(2.3 \, k_1 a^3)$$

Example 4.2

The BOD versus time data for the first 5 days of a BOD test were obtained as follows:

Time (days)	BOD, y (mg/L)
2	10
4	16
6	20

Calculate the k_1 and L_o.

The $(t/y)^{1/3}$ values 0.585, 0.630, and 0.669 are plotted as shown in Figure 4–5. The intercept $a = 0.545$ and the slope $b = 0.021$. Thus

$$k_1 = 2.61 \left(\frac{0.021}{0.545} \right) = 0.10 \text{ day}^{-1}$$

$$(\text{note: } k_1' = 0.23 \text{ day}^{-1})$$

$$L_o = \frac{1}{2.3(0.10)(0.545)^3} = 26.8 \text{ mg/L}$$

For streams and rivers with travel times greater than about 5 days, the ultimate demand for oxygen must include the nitrogenous demand. Although the use of BOD_{ult} (carbonaceous plus nitrogenous) in DO sag calculations is not strictly accurate, it is often assumed that the ultimate BOD may be calculated as

$$BOD_{ult} = a(BOD_5) + b(KN)$$

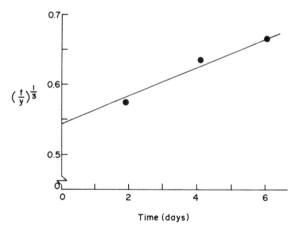

Figure 4–5. Calculation of k_1 and L_o, for Example 4.2.

where   KN = Kjeldahl nitrogen (organic plus ammonia mg/L)
        a and b = constants

The state of North Carolina, for example, uses a = 1.2 and b = 4.0 for calculating the ultimate BOD, which is then substituted for L_o in the DO sag equation.

CHEMICAL OXYGEN DEMAND

Among the many drawbacks of the BOD test, the most important is that it takes 5 days to run. If the organics were oxidized chemically instead of biologically, the test could be shortened considerably. Such oxidation is accomplished with the *chemical oxygen demand* (COD) test. Because nearly all organics are oxidized in the COD test and only some are decomposed during the BOD test, COD values are always higher than BOD values. One example of this is wood pulping waste, in which compounds such as cellulose are easily oxidized chemically (high COD) but are very slow to decompose biologically (low BOD).

Potassium dichromate is generally used as an oxidizing agent. It is an inexpensive compound that is available in very pure form. A known amount of this compound is added to a measured amount of sample, and the mixture is boiled. The reaction, in unbalanced form, is

$$C_X H_Y O_Z + Cr_2 O_7^{=} + H^+ \overset{\Delta}{\rightarrow} CO_2 + H_2 O + Cr^{3+}$$

(organic) (dichromate)

After boiling with an acid, the excess dichromate (not used for oxidizing) is measured by adding a reducing agent, usually ferrous ammonium sulfate. The difference between the chromate originally added and the chromate remaining is the chromate used for oxidizing the organics. The more chromate used, the more organics were in the sample, and hence the higher the COD.

TOTAL ORGANIC CARBON

Since the ultimate oxidation of organic carbon is to CO_2, the total combustion of a sample will yield some significant information on the amount of organic carbon present in a wastewater sample. Without elaboration, this is done by allowing a small portion of the sample to be burned in a combustion tube and measuring the amount of CO_2 emitted. This test is not widely used at present, mainly because of the expensive instrumentation required. It has significant advantages over the BOD and COD tests, however, and will undoubtedly be more widely used in the future.

TURBIDITY

Water that is not clear but is "dirty," in the sense that light transmission is inhibited, is known as turbid water. Turbidity may be caused by many materials, some of which are discussed in Chapter 3. In the treatment of water for drinking purposes, turbidity is of great importance first because of the aesthetic considerations and second because pathogenic organisms can hide on (or in) the tiny colloidal particles.

The standard method of measuring tubidity is with the Jackson candle turbidimeter first developed in 1900. It consists of a long flat-bottomed glass tube under which a candle is placed. The turbid water is poured into the glass tube until the outline of the flame is no longer visible. Obviously, with very turbid water, only a little is required. The centimeters of water in the tube are then measured and compared with the standard turbidity unit, which is

$$1 \text{ mg/L of } SiO_2 = 1 \text{ unit of turbidity}$$

Recent years have seen the development of nephelometers and turbidimeters: photometers that measure the intensity of scattered light. The opaque particles that cause turbidity scatter light, so that the scattered light measured at right angles to a beam of incident light is proportional to the turbidity. These instruments have completely replaced the antiquated Jackson candle. The standard unit of turbidity is now the nephelometric turbidity unit (NTU) instead of the Jackson turbidity unit (JTU).

COLOR AND ODOR

Color and odor are both important measurements in water treatment. Along with turbidity they are called the *physical* parameters of drinking water quality. Color and odor are important from the standpoint of aesthetics. If a water looks colored or smells bad, people will instinctively avoid using it, even though it might be perfectly safe from the public health aspect. Both color and odor may be and often are caused by organic substances such as algae or humic compounds.

Color is measured by comparison with standards. Colored water made with potassium chloroplatinate when tinted with cobalt chloride closely resembles the color of many natural waters. When multicolored industrial wastes are involved, such color measurement is meaningless.

Odor is measured by successive dilutions of the sample with odor-free water until the odor is no longer detectable. This test is obviously subjective and depends entirely on the olfactory senses of the tester.

pH

The pH of a solution is a measure of hydrogen ion concentration. An abundance of hydrogen ions makes a solution acid, whereas a dearth of H^+ ions makes it basic. A basic solution has, instead, an abundance of hydroxide ions, OH^-.

Water molecules, HOH (commonly written H_2O), have the ability to dissociate, or ionize, very slightly. In a perfectly neutral water (neither acidic nor basic) equal concentrations of H^+ and OH^- are present.

If the concentration of both H^+ and OH^- ions is measured in moles (molecular weight in grams) per liter of water, the product of H^+ and OH^- concentrations is always constant.

$$[H^+] \times [OH^-] = K_w = 1 \times 10^{-14}$$

where K_w is the *ion-product constant of water*.* If the hydrogen ion concentration is 10^{-4}, the OH^- concentration must be $10^{-14}/10^{-4} = 10^{-10}$. Since 10^{-10} is smaller than 10^{-4}, the solution is acidic.

The hydrogen ion concentration is so important in aqueous solutions that an easier method of expressing it has been devised. Instead of speaking in terms of moles per liter, we may take the negative logarithm of $[H^+]$ so that

$$pH = -\log[H^+] = \log \frac{1}{[H^+]}$$

The pH value is the negative power to which 10 must be raised to equal the hydrogen ion concentration, or

$$[H^+] = 10^{-pH}$$

For a neutral solution $[H^+]$ is 10^{-7}, or pH $= 7$. Obviously, for greater hydrogen ion concentrations the pH is lower, and for greater hydroxide ion concentrations (or lower hydrogen ion concentrations since the products are constant) the pH drops. The pH range of dilute solutions is from 0 (very acidic; 1 mole of H^+ ions per liter) to 14 (very alkaline). Solutions containing more than 1 mole of H^+ ions per liter have pH < 0.

The measurement of pH is now almost universally by electronic means. Electrodes that are sensitive to hydrogen ion concentration (strictly speaking, the hydrogen ion activity) convert the signal to electrical current.

pH is important in almost all phases of water and wastewater treatment. Aquatic organisms are sensitive to pH changes, and biological treatment

---

* The use of square brackets around a chemical symbol is shorthand for denoting the concentration of the chemical.

requires either pH control or monitoring. In water treatment pH is important in ensuring proper chemical treatment as well as in disinfection and corrosion control. Mine drainage often involves the formation of sulfuric acid (high H^+ concentration), which is extremely deterimental to aquatic life. Continuous acid deposition from the atmosphere (see Chapter 18) may lower the pH of a lake substantially.

SOLIDS

One of the main problems with wastewater treatment is that so much of the wastewater is actually solid. The separation of these solids from the water is in fact one of the primary objectives of treatment.

Strictly speaking, in wastewater anything other than water would be classified as a solid. The usual definition of solids, however, is the residue on evaporation at 103°C (slightly higher than the boiling point of water). The solids thus measured are known as *total solids*. Total solids may be divided into two fractions: the *dissolved solids* and the *suspended solids*. If we put a teaspoonful of common table salt in a glass of water, the salt will dissolve. The water will not look any different, but the salt will remain behind if we evaporate the water. A spoonful of sand, however, will not dissolve and will remain as sand grains in the water. The salt is an example of a dissolved solid, whereas the sand would be measured as a suspended solid.

The separation of suspended solids from dissolved solids is by means of a special crucible, called a *Gooch crucible*. As shown in Figure 4–6, the Gooch crucible has holes on the bottom on which a glass fiber filter is placed. The sample is then drawn through the crucible with the aid of a vacuum. The suspended material is retained on the filter, while the dissolved fraction passes through. If the initial dry weight of the crucible and filter are known, the subtraction of this from the total weight of crucible, filter, and the dried solids caught on the filter yields the weight of suspended solids, expressed as milligrams per liter.

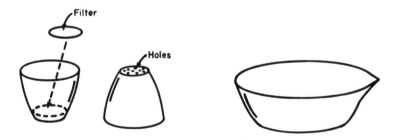

Figure 4–6. The Gooch crucible for suspended solids (with filter) and the evaporating dish for total solids.

Solids may be classified in another way: those that are volatilized at a high temperature and those that are not. The former are known as *volatile solids*, the latter as *fixed solids*. Usually, volatile solids are organic. Obviously, at 600°C, the temperature at which the combustion takes place, some of the inorganics are decomposed and volatilized, but this is not considered as a serious drawback.

The relationship between the total solids and the total volatile solids can best be illustrated by an example.

Example 4.3
  Given the following data:

- weight of dish (such as shown in Figure 4–6) = 48.6212 g
- 100 mL of sample is placed in dish and evaporated. Weight of dish and dry solids = 48.6432 g
- Dish is placed in 600°C furnace, then cooled. Weight = 48.6300 g.

Find the total, the volatile, and the fixed solids.

$$\text{Total Solids} = \frac{(\text{dish} + \text{dry solids}) - (\text{dish})}{\text{volume of sample}}$$

$$= \frac{48.6432 - 48.6212}{100}$$

$$= 220 \times 10^{-6} \text{ g/mL}$$

$$= 220 \times 10^{-3} \text{ mg/mL}$$

$$= 220 \text{ mg/L}$$

$$\text{Fixed Solids} = \frac{(\text{dish} + \text{unburned solids}) - (\text{dish})}{\text{volume of sample}}$$

$$= \frac{48.6300 - 48.6212}{100}$$

$$= 88 \text{ mg/L}$$

$$\text{Volatile Solids} = \text{Total Solids} - \text{Total Fixed Solids}$$

$$= 220 - 88 = 132 \text{ mg/L}$$

It is often necessary to measure the volatile fraction of suspended material, since this is a quick (if gross) measure of the amount of microorganisms present. The *volatile suspended solids* are determined by simply burning the Gooch crucible and weighing it again. The loss in weight is interpreted as volatile suspended solids.

NITROGEN

Recall from Chapter 3 that nitrogen is an important element in biological reactions. Nitrogen may be tied up in high-energy compounds such as amino acids and amines. In this form the nitrogen is known as organic nitrogen. One of the intermediate compounds formed during biological metabolism is ammonia. Together with organic nitrogen, ammonia is considered an indicator of recent pollution. These two forms of nitrogen are often combined in one measure, known as *Kjeldahl nitrogen*, after the scientist who first suggested the analytical procedure.

Aerobic decomposition eventually leads to the nitrite (NO_2^-) and finally nitrate (NO_3^-) nitrogen forms. A high-nitrate and low-ammonia nitrogen therefore suggests that pollution has occurred, but quite some time ago.

All of the above forms of nitrogen can be measured analytically by *colorimetric techniques*. The basic idea of colorimetry is that the ion in question will combine with some compound and form a color and that the intensity of this color is proportional to the original concentration of the ion. For example, ammonia can be measured by adding a compound called Nessler reagent to the unknown sample. This reagent is a solution of potassium mercuric iodide, K_2HgI_4, and combines with ammonium ions to form a yellow-brown colloid. Since Nessler reagent is in excess, the amount of colloid formed is proportional to the concentration of ammonium ions in the sample.

The color is measured either by visual comparison (in long tubes called Nessler tubes) or photometrically. The basic workings of a photometer, illustrated in Figure 4–7, consist of a light source, a filter, the sample, and a photocell. The filter allows only certain wavelengths of light to pass through, thus lessening interferences and increasing the sensitivity of the photocell, which converts light energy to electric current. An intense color will allow only a limited amount of light to pass through and thus create little current. On the other hand, a sample containing very little of the chemical in question will be clear, allow almost all of the light to pass through, and set up a substantial current. If the color intensity (and hence light absorbance) is directly propor-

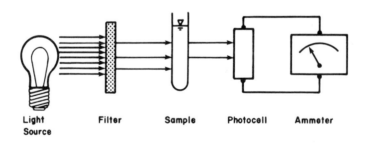

Figure 4–7. Elements of a filter photometer.

tional to the concentration of the unknown ion, the color formed is said to obey Beer's law:

$$I/I_0 = e^{-acx} \quad \text{or}$$

$$\ln I_0/I = acx$$

where I = the intensity of light after it has passed through the sample
I_0 = the intensity of light incident on the sample
a = the "absorption coefficient" of the colored compound
x = the path length of light through the sample
c = the concentration

A photometer may be used to measure ammonia concentration by measuring the absorbance of light by samples containing a known ammonia concentration and comparing the absorbance of the unknown sample with that of these standards.

Example 4.4

Several known samples and an unknown sample were treated with Nessler reagent, and the color was measured with a photometer. Find the ammonia concentration of the unknown sample.

Sample	% Absorbance
Standards: 0 mg/L of ammonia (Distilled water)	0
1 mg/L of ammonia	6
2 mg/L of ammonia	12
3 mg/L of ammonia	18
4 mg/L of ammonia	24
Unknown sample	15

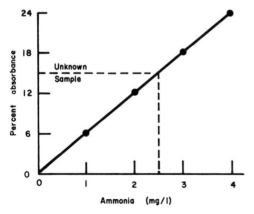

Figure 4–8. Calculation using colorimetric standards.

A plot (Figure 4–8) of ammonia concentration of the standards versus percent absorbance results in a straight line (Beer's Law is adhered to). We may then enter this chart at 15 percent absorbance (the unknown) and read the concentration of ammonia in our unknown as 2.5 mg/L.

PHOSPHATES

The importance of phosphorous compounds in the aquatic environment is discussed in Chapter 3. Phosphorous in wastewater may be either inorganic or organic. Although the greatest single source of inorganic phosphorous is synthetic detergents, organic phosphorus is found in food and human waste as well. All phosphates in nature will, by biological action, eventually revert to inorganic forms to be again used by the plants in making high-energy material.

Ever since phosphates were indicted as one culprit in lake eutrophication, the measurement of total phosphates has assumed considerable importance. Total phosphates may be measured by first boiling the sample in acid solution, which converts all the phosphates to the inorganic forms. From that point the test is colorimertic, using a chemical that when combined with phosphates produces a color directly proportional to the phosphate concentration.

BACTERIOLOGICAL MEASUREMENTS

From the public health standpoint the bacteriological quality of water is as important as the chemical quality. A number of diseases may be transmitted by water, among them typhoid and cholera. However, it is one thing to declare that water must not be contaminated by pathogens (disease-causing organisms) and another to determine the existence of these organisms. First, there are many pathogens. Each has a specific detection procedure and must be screened individually. Second, the concentration of these organisms may be so small as to make their detection impossible. It is a perfect example of the proverbial needle in a haystack. And yet only one or two organisms in the water might be sufficient to cause an infection.

How then can we measure for bacteriological quality? The answer lies in the concept of indicator organisms. The indicator most often used is a group of microbes of the family *Escherichia coli* (often called coliform bacteria), which are organisms normal to the digestive tracts of warm-blooded animals. In addition to that attribute, coliforms are:

- plentiful, and hence not difficult to find
- easily detected with a simple test
- generally harmless except in unusual circumstances
- hardy, surviving longer than most known pathogens

Coliforms have thus become universal indicator organisms. But the presence of coliforms does not prove the presence of pathogens. If a large number of coliforms are present, there is a good chance of recent pollution by wastes from warm-blooded animals, and therefore the water *may* contain pathogenic organisms.

This last point should be emphasized. The presence of coliforms does not mean that there are pathogens in the water. It simply means that there *might* be. A high coliform count is thus suspicious, and the water should not be consumed (although it may be perfectly safe).

There are three ways of measuring for coliforms. The simplest is to filter a sample through a sterile filter, thus capturing any coliforms. The filter is then placed in a petri dish containing a sterile agar that soaks into the filter and promotes the growth of coliforms while inhibiting other organisms. After 24 or 48 hours of incubation the number of shiny black dots, indicating coliform colonies, is counted. If we know how many milliliters were poured through the filter, the concentration of coliforms may be expressed as coliforms/per milliliter. The equipment used for such filter tests is shown in Figure 4–9.

The second method of measuring for coliforms is called the most probable number (MPN), a test based on the fact that in a lactose broth coliforms will produce gas and make the broth cloudy. The production of gas is detected by placing a small tube upside down inside a larger tube (Figure 4–10) so as not to have air bubbles in the smaller tube. After incubation, if gas is produced, some of it will become trapped in the smaller tube and this, along with a cloudy broth, will indicate that tube had been inoculated with at least one coliform.

And here is the trouble. One coliform can cause a positive tube just as easily as a million coliforms can. Hence, it is not possible to ascertain the concentration from just one tube. We get around this problem by inoculating a series of tubes with various quantities of sample, the reasoning being that a 10-mL sample would be more likely to contain a coliform than a 1-mL sample.

For example, if we take three different inoculation amounts, 10, 1, and 0.1 mL of sample, and inoculate three tubes with each amount, after incubation we might have an array such as follows. The plus signs indicate a positive test (cloudy broth with gas formation), and the minus signs represent tubes in which no coliforms were found.

Amount of sample, mL put in test tube	Tube Number 1	2	3
10	+	+	+
1	−	+	−
0.1	−	−	+

On the basis of these data we would suspect that there is at least 1 coliform per 10 mL, but we still have no firm number.

The solution to this dilemma lies in statistics. It can be proven statistically that such an array of positive and negative results will occur most probably if

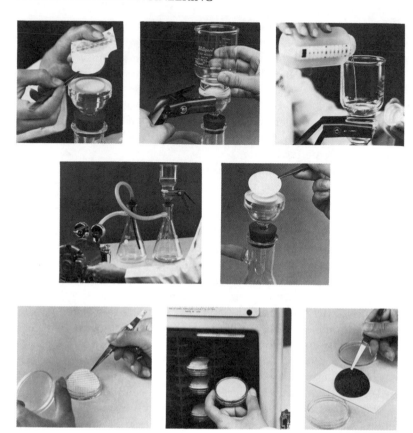

Figure 4–9. Millipore Filter apparatus for the measurement of coliforms. Procedure: (A) Put membrane filter on filter stand; (B) replace funnel; (C) pour in measured amount of sample; (D) turn on vacuum; (E) remove membrane filter; (F) center membrane filter on pad in petri dish containing agar; (G) place petri dish in oven at 35°C; (H) count the coliform colonies after 8 hours.

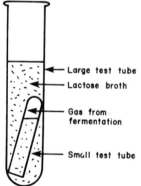

Large test tube

Lactose broth

Gas from fermentation

Small test tube

Figure 4–10. Test tubes used for most probable number (MPN) coliform test.

the coliform concentration was 75 coliforms/100 mL. A higher concentration would most probably have resulted in more positive tubes, whereas a lower concentration would most probably have resulted in more negative tubes. This is how MPN is established.

A third way of measuring coliforms is by a proprietary device called a Coli-Count. A sterile pad with all the necessary nutrients is dipped into the water sample and incubated, and the colonies are counted. The pad is designed to adsorb exactly 1 mL of sample water so that the colonies counted give a coliform concentration per milliliter.

VIRUSES

Because of their minute size and extremely low concentration and the need to culture them on living tissues, pathogenic (or animal) viruses are fiendishly difficult to measure. Because of this problem, there are as yet no standards for viral quality of water supplies as there are for pathogenic bacteria.

One possible method of overcoming this difficulty is to use an indicator organism, much like the coliform group is used as an indicator for bacterial contamination. This can be done by using a bacteriophage, or a virus that attacks only a certain type of bacterium. For example, coliphages attack coliform organisms and, because of their association with wastes from warm-blooded animals, seem to be an ideal indicator. The test for coliphages is performed by inoculating a petri dish containing an ample supply of a specific type of coliform with the wastewater sample. Coliphages will attack the coliforms, leaving visible spots or plaques that can be counted, and an estimate can be made of the number of coliphages per known volume.

HEAVY METALS

The increasing problem of heavy metals in industrial effluent is introduced in Chapter 3. Heavy metals such as arsenic and mercury can harm fish even at low concentrations. Consequently, the method of measuring these ions in water must be very sensitive.

The method of choice for measuring heavy metals is atomic absorption spectrophotometry. In this method, a solution of lanthanum chloride is added to the sample, and the treated sample is sprayed into the flame with an atomizer. Each metallic element imparts a characteristic color to the flame, whose intensity is then measured spectrophotometrically.

TRACE TOXIC ORGANICS

Very low concentrations of chlorinated hydrocarbons and other agrichemical residues in water can be assayed by gas chromatography. Oil residues in water are generally measured by extracting the water sample with Freon and then evaporating the Freon and weighing the residue from this evaporation.

CONCLUSION

Discussed above are only some of the most important tests used in water pollution control. *Standard Methods*, for example, contain no fewer than 239 analytical procedures, many of which can be performed only with special equipment and skilled technicians.

Understanding this, and realizing the complexity, variation, and objectives of some of the measurements of water pollution, how would you answer someone who brings a jug of water to your office, sets it on your desk, and asks, "Can you tell me how much pollution is in this water?"

PROBLEMS

4.1   Given the following BOD test results:
Initial DO   8 mg/L
Final DO   0 mg/L
Dilution   1/10
what can you say about
   a. BOD_5?
   b.   BOD ultimate?
   c.   COD?

4.2   If you had two bottles full of lake water and kept one dark and the other in daylight, which one would have a higher DO after a few days? Why?

4.3   Name three substances you would need to seed if you wanted to measure their BOD.

4.4   The following data were obtained for a sample:
total solids = 4,000 mg/L
suspended solids = 5,000 mg/L
volatile suspended solids = 2,000 mg/L
fixed suspended solids = 1,000 mg/L
Which of these numbers is questionable and why?

4.5   A water has a BOD_5 of 10 mg/L. The initial DO in the BOD bottle was 8 mg/L, and the dilution was 1/10. What was the final DO in the BOD bottle?

4.6   If the BOD_5 of a waste is 100 mg/L, draw a curve showing the effect of adding progressively higher doses of chromium (a toxic chemical) on the BOD_5.

4.7   Consider the following data from a BOD test:

Day	DO (mg/L)	Day	DO (mg/L)
0	9	5	6
1	9	6	6
2	9	7	4
3	8	8	3
4	7	9	3

What is the: (a) BOD_5, (b) carbonaceous BOD, and (c) ultimate BOD? Why was there no oxygen used until the third day? If the sample were "seeded," would the final DO have been higher or lower and why?

4.8   An industry discharges 10 million gallons a day of a waste that has a BOD_5 of 2,000 mg/L. How many pounds of BOD_5 are discharged?

4.9   If you dumped a half gallon of milk every day into a stream, what would be your discharge in pounds of BOD_5 per day? (See p. 57.)

4.10   Given the same standard ammonia samples as in Example 4.3, if your unknown measured 20 percent absorbance, what is the ammonia concentration in the unknown?

4.11   Suppose you ran a multiple-tube coliform test and got the following results: 10-mL samples, all 5 positive; 1-mL samples, all 5 positive; 0.1 mL samples, all 5 negative. Use the table in *Standard Methods* to estimate the concentration of coliforms.

4.12   If coliform bacteria are to be used as an indicator of viral pollution as well as an indicator of bacterial pollution, what attributes must the coliform organisms have (relative to viruses)?

4.13   Draw a typical BOD curve. Label the (a) carbonaceous BOD, (b) nitrogenous BOD, (c) ultimate BOD, and (d) 5-day BOD. On the same graph, plot the BOD curve if the test has been run at 30°C instead of at the usual temperature. Also plot the BOD curve if a substantial amount of toxic materials were added to the sample.

4.14   Consider the BOD data below (no dilution, no seed).
   a.   What is the ultimate carbonaceous BOD?
   b.   What might have caused the lag in the beginning of the test?
   c.   Calculate k_1, the reaction constant.

Day	DO
0	8
1	8
3	7
5	6.5
7	6
15	4
20	4

4.15   Given that the BOD test for a waste diluted 1/10 produced the following curve:

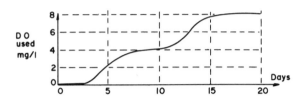

a.   What is the BOD_5 of the waste?
b.   What is the ultimate carbonaceous BOD?
c.   What is the ultimate BOD from nitrification alone?
d.   The curve dosen't follow the "typical" first-order reaction curve. Comment on what happened during the first few days.

4.16   A wastewater is estimated to have a BOD of 200 mg/L.
a.   What dilution is necessary for this BOD to be measured by the usual technique?
b.   If the initial and final DO of the test thus conducted was 9.0 and 4.0, and the dilution water has a BOD of 1 mg/L, what is the dilution?

LIST OF SYMBOLS

$$BOD = \text{biochemical oxygen demand, mg/L}$$
$$BOD_5 = \text{5-day BOD}$$
$$D = \text{dilution (volume of sample/total volume)}$$
$$F = \text{final BOD of sample, mg/L}$$
$$F' = \text{final BOD of seeded dilution water, mg/L}$$
$$I = \text{initial BOD of sample, mg/L}$$
$$I' = \text{initial BOD of seeded dilution water, mg/L}$$
$$X = \text{seeded dilution water in sample bottle, mL}$$
$$Y = \text{volume of BOD bottle, mL}$$

REFERENCES

1. American Public Health Association. *Standard Methods for the Examination of Water and Wastewater*, 16th ed. (Water Pollution Control Federation, American Water Works Association, 1984).
2. Thomas, H.A., Jr. "Graphical Determination of BOD Curve Constants," *Water and Sewer Works* 97:123(1950).

Chapter 5
Water Supply

A supply of water is critical to the survival of life as we know it. People need water to drink, animals need water to drink, and plants need water to drink. Additionally, basic functions of a society require water: cleaning for public health, consumption for industrial processes, and cooling for electricity generation. In this chapter, we discuss the supply of water in terms of:

- the hydrologic cycle and water availability
- groundwater supplies
- surface water supplies
- water transmission

The direction of our discussion is that sufficient water supplies exist for the world, and the nation as a whole, but that many areas are water poor while others are water rich. The trick in water supply is to engineer the supply and transmission from one area to another. In the following chapter, we address treatment methods available to clean up the water once it reaches the areas of demand.

THE HYDROLOGIC CYCLE AND WATER AVAILABILITY

The concept of the hydrologic cycle is a useful starting point to begin our study of water supply. Illustrated in Figure 5–1, this cycle is the precipitation of water from clouds, infiltration into the ground or runoff into surface watercourses, followed by evaporation and transpiration of the water back into the atmosphere.

The rates of precipitation and evaporation/transpiration help define the baseline quantity of water that is available for human consumption. *Precipitation* is the term applied to all forms of moisture originating in the clouds and

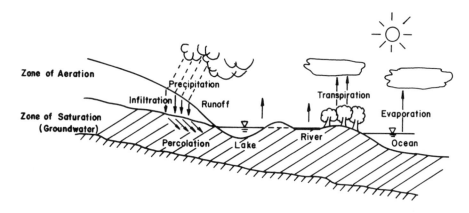

Figure 5–1.   The hydrologic cycle.

falling to the ground, and a range of instruments and technologies have been developed for measuring the amount and intensity of rain, snow, sleet, and hail. The average depth of precipitation over a given region, on a storm, seasonal, or annual basis, is required in many water availability studies. Any open receptacle with vertical sides has application as a rain gauge, but varying wind and splash effects must be considered if amounts collected by different gauges are to be compared.

   Evaporation and *transpiration* help define the baseline quantity of water that is available. The same meteorological factors that influence evaporation are at work in the transpiration process: solar radiation, ambient air temperature, humidity, and wind speed, as well as the amount of soil moisture available to the plants, impact the rate of transpiration. Many measurements for transpiration are made with a *phytometer*, a large vessel filled with soil and potted with selected plants. The surface of the soil is hermetically sealed to prevent evaporation; thus the only escape of moisture is through transpiration, which can be determined by weighing the plant-vessel system at time intervals up to the life of the plant. Because it is impossible to simulate natural conditions, the results of phytometer tests are of limited value. However, they can be used as an index of water demand by a crop under field conditions, and thus relate to the calculations that help an engineer determine the water supply requirements for that crop. Generally speaking, the terms evaporation and transpiration are linked by engineers into *evapotranspiration*, or the total water loss to the atmosphere by the two processes.

GROUNDWATER SUPPLIES

Groundwater is both an important direct source of supply that is tapped by wells and a significant indirect source of supply since a large portion of the flow to streams is derived from subsurface water.

Near the surface of the earth in the *zone of aeration* soil pore spaces contain both air and water. This zone, which may have a zero thickness in swamplands and be several hundred feet thick in mountainous regions, contains three types of moisture. *Gravity water* is in transit after a storm through the larger soil pore spaces. *Capillary water* is drawn through small pore spaces by capillary action and is available for plant uptake. *Hygroscopic moisture* is water held in place by molecular forces during all except the driest climatic conditions. Moisture from the zone of aeration cannot be tapped as a water supply source.

On the other hand, the *zone of saturation* offers water in a quantity that is directly available. In this zone, located below the zone of aeration, the pores are filled with water, and this is what we consider *groundwater*. The stratum that contains a substantial amount of groundwater is called an *aquifer*. At the surface between the two zones, labeled the *water table* or *phreatic surface*, the hydrostatic pressure in the groundwater is equal to atmospheric pressure. An aquifer may extend to great depths, but because the weight of overburden material generally closes pore spaces, little water is found at depths greater than 600 m (2,000 ft). The water readily available from an aquifer is that which will drain by gravity. Each soil type thus has a *specific yield*, defined as the volume of water, expressed as a percentage of the total volume of water in the aquifer, which will drain freely from the aquifer.

Although Hagen and Poiseville were the first to propose that the velocity of flow of water capillary tubes is proportional to the slope of the hydraulic gradient,[1,2] it was Henri Darcy who in 1856 applied their theory to ground-water to obtain:[3]

$$v = ks$$

where v = the velocity of flow
s = the scope of the hydraulic gradient
k = a coefficient having the units of v

The water discharged from an aquifer, q (in cubic meters per second or gallons per minute, gpm), is the product of the cross-sectional area A and velocity. Only the "effective area" is of concern, i.e., the total area of the pores in the ground, so we must correct the total cross-sectional area by the porosity p of the soil. Porosity K_p is defined as the ratio of pore volume to the total volume of a soil formation. After combining terms, we get:

$$q = K_p As$$

Here, q is in gallons per day while all other terms are in feet. K_p is defined as the coefficient of permeability and in common units is expressed as the discharge in gallons of water per day (gpd) through an area one foot square under a gradient of 1 ft/ft at 60°F. In SI units, K_p is expressed as cubic meters per day through an area 1 m^2 under a gradient of 1 m/m. Table 5–1 lists some average porosities, specific yields, and permeabilities of selected materials.

Table 5–1.   Estimate of Average Permeability and Porosity Values for Selected Materials*

Material	Porosity (%)	Specific Yield (%)	Permeability K_p	
			$(gal/day \cdot ft^2)$	$(m^3/day \cdot m^2)$
Clay	45	3	1	0.04
Sand	35	25	800	32
Gravel	25	22	5,000	200
Sandstone	15	8	700	28
Granite	1	0.5	0.1	0.004

*Source: Linsley, R.K. and J.B. Franzini. *Elements of Hydrology.* Copyright © 1958, by McGraw-Hill Book Company. Used with the permission of the McGraw-Hill Book Company.

Values for the coefficient of permeability can be determined for other temperatures, K_{P_T}, by multiplying by the ratio of the kinematic viscosities, ν_{60} and ν_T:

$$K_{P_T} = K_{P_{60}} \frac{\nu_{60}}{\nu_T}$$

Also, it is useful to define the transmissibility T to represent the rate of flow through a section of aquifer of a unit thickness under a unit head (slope = 1 ft/ft). In SI units, this would be cubic meter per day · meter, whereas in common units T is defined in terms of gallons per day through a section 1 foot thick under a unit head (1 ft/1-ft slope). Thus,

$$q = TBS$$

where B is the thickness of the aquifer.

Permeameters are instruments used in the laboratory for measuring the permeability of sample materials. The sample is placed under a water column, and the flows through the sample during given time periods are measured. The applicability of such tests is, however, limited because of the difficulty associated with placing the samples into the permeameter in their natural state; in addition, flows in large-solution cavities (such as would occur in limestone aquifers) and fractures cannot be simulated in this bench-scale test.

Groundwater supplies are often tapped by drilling wells into the ground and pumping the water up and out. If a flow of water is to occur to the well, a gradient must exist. Figure 5–2 illustrates this *cone of depression* in the water table resulting from pumping activities at a sample well site. In 1935, Theis developed a model to relate drawdown to T, the transmissibility of the aquifer:[4]

$$Z_r = \frac{q}{4\pi T} \int_u^\infty \frac{e^{-u}}{u} \, du$$

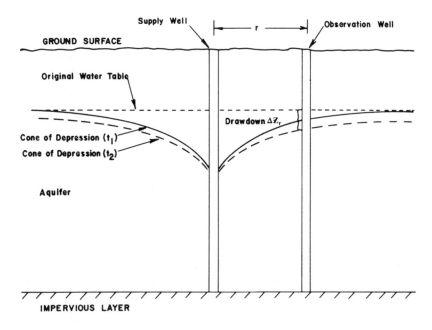

Figure 5–2. Cone of depression.

where   q = the discharge
   Z_r = the drawdown at an observation well a distance r from the water supply well

$$u = \frac{1.87 \, r^2 S_c}{Tt}$$

where   t = the time from pumping start-up at the water supply well
   S_c = the storage constant for the given aquifer (the volume of water removed from a column of aquifer 1 unit dimension square when the water table is lowered 1 unit length, essentially the specific yield of the aquifer)
   r = the distance from the water supply well to the observation well
   u = a constant calculated for each well-aquifer system

When the pumping activity at the water supply well has proceeded for a relatively long time, i.e., as t is large u approaches zero, it is possible to modify the Theis model to:[5]

$$T = \frac{2.3 \, q}{4 \pi \Delta Z} \log_{10} \frac{t_2}{t_1}$$

where ΔZ is the drawdown at the observation well between times t_1 and t_2.

Example 5.1 *

When the drawdown values for a sample well system are plotted versus the time, on log-linear paper, we get results seen in Figure 5–3. We can simplify our calculations when we take ΔZ during a single log cycle. When between 4 hours and 40 hours, $\log_{10}(t_2/t_1) = 1$, and we can easily calculate T from the recorded value of discharged water q:

$$T = \frac{2.3\,q}{4\pi\Delta Z}$$

If q = 200 gpm (288,000 gpd), then:

$$T = \frac{2.3(288,000)}{4(3.14)(5.2)}$$

$$T = 10,142 \text{ gpd per foot width}$$

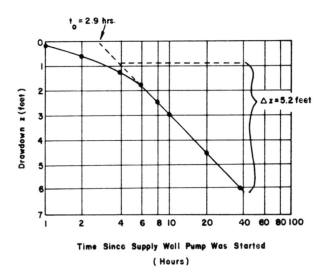

Figure 5–3. Application of the modified Theis method, with sample data.

---

* Adapted from Linsley, R.K. and J.B. Franzini, *Elements of Hydrology* (New York: McGraw-Hill, 1958), p. 138. Copyright © 1958, McGraw-Hill Book Company. Used with permission.

Thus far, we have relied on the existence of perfectly symmetrical cones of depression. In the field, symmetry is not often the case. If several wells are close together, cones of depression may overlap. If a cone of depression intersects a surface stream or lake, the gradient from the stream to the well causes water to move out of the stream, and the cone's surface will also be far from symmetrical. Variation in soil materials also impacts the shape of these drawdown cones.

Many factors help determine the "safe yield" of a given well or a selected aquifer:

- precipitation (P)
- net groundwater flow to the area (G_N)
- evapotranspiration (E_T)
- surface flow out of the area (S_o)
- change in groundwater storage (G_s)
- change in surface water storage (S_s)
- cost of pumping

Any notion of "safe yield" must consider each factor and must reference a given time period. If we focus on the quantity of water available for a year, for example, and neglect cost, we can define safe yield S_F as:

$$S_F = P + G_N - E_T - S_o - S_s - G_s$$

Other factors in the determination of safe yields include the transmissibility of the aquifer (can the aquifer transmit the water to the well(s) at a rate great enough to sustain the demand) and the location of contaminated bodies of water (including salt water intrusion possibilities from the oceans and chemical migrations from disposal or spill sites).

Besides offering a source of well water, aquifers also combine with precipitation to feed surface water courses. Streams, rivers, and lakes also offer options for water supply, and they are addressed next.

SURFACE WATER SUPPLIES

Surface water supplies are not as reliable as groundwater sources since quantities often fluctuate widely during the course of a year or even a week, and qualities are restricted by various sources of pollution. If a river has an average flow of 10 cubic feet per second (cfs), it means often the flow is less than 10 cfs. If a community wishes to use this for a water supply, its demands for the water should be considerably less than 10 cfs to be assured of a reasonably dependable supply.

The variation in the river flow may be so great that even a small demand cannot be met during dry periods in many parts of the nation, and storage facilities must be constructed to hold the water during wet periods so it can be saved for the dry ones. The objective is to build these reservoirs sufficiently large to have dependable supplies.

One method of arriving at the proper reservoir size is by using the *mass curve*. Basically, the total flow in a stream at the point of a proposed reservoir is summed and plotted against the time. On the same curve the water demand is plotted, and the difference between the total water flowing in and the water demanded is the quantity that the reservoir must hold if the demand is to be met. The method is illustrated by the following example problem:

Example 5.2

A reservoir is needed to provide a constant flow of 15 cfs. The monthly stream flow records, in total cubic feet, are:

Month	J	F	M	A	M	J	J	A	S	O	N	D
Cubic feet of water ($\times 10^6$)	50	60	70	40	32	20	50	80	10	50	60	80

We can calculate the storage requirement by plotting the cumulative water as in Figure 5–4. Note that for January, we plot 50 million cubic feet, for February we add 60 to that and plot 110 million cubic feet, and so on.

The demand for water is constant at 15 cfs, or

$$15 \, \frac{\text{cubic feet}}{\text{sec}} \times 60 \, \frac{\text{sec}}{\text{min}} \times 60 \, \frac{\text{min}}{\text{hr}} \times 24 \, \frac{\text{hr}}{\text{day}} \times 30 \, \frac{\text{days}}{\text{month}}$$

$$= 38.8 \times 10^6 \text{ cubic feet/month}$$

This can be represented as a sloped line in Figure 5–4 and plotted on the curved supply line. Note that the stream flow in May was lower than the demand, and this was the start of a drought lasting into June. The demand slope was greater than the supply, and thus the reservoir had to make up the deficit. In July the rains came and the supply increased until the reservoir could be filled up again, late in August. The reservoir capacity needed to get through that particular drought was 60×10^6 cubic feet. A second drought, starting in September, lasted into November and required 35×10^6 cubic feet of capacity. If the municipality therefore had a reservoir with at least 60×10^6 cubic feet capacity, they could have drawn water from it throughout the year.

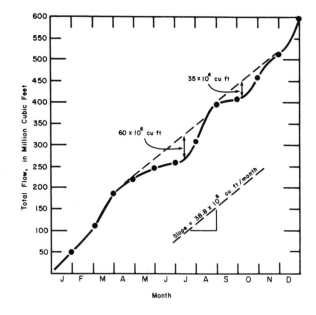

Figure 5–4. Mass curve for determining required reservoir capacity.

A mass curve such as Figure 5–4 is actually of little use if only limited stream flow data are available. One year's data yield very little information about long-term variations. For example, was the drought in the above example that required 60 million cubic feet of storage the worst drought in 20 years, or was the year shown actually a fairly wet year?

To get around this problem, it is necessary to predict statistically the recurrence of events such as droughts and then to design the structures according to a known risk. Water supplies are often designed to meet demands 19 of 20 years. In other words, once in 20 years the drought will be so severe that the reservoir capacity will not be adequate. If running out of water once every 20 years is not acceptable, the people may choose to build a bigger reservoir and thus expect to be dry only once every 50 years. The question really is one of an increasing investment of capital for a steadily smaller added benefit. As in Example 3.2, the probability is best calculated as $m/(n+1)$, where m = ranking (with $m = 1$ for the most severe drought) and n = total number of years.

Example 5.3

Suppose the need is to build a reservoir that could supply the water demand 9 of 10 years. The reservoir capacities, which would be required to prevent running out of water, as calculated from mass curves, follow:

Year	Required Reservoir Capacity, $m^3 \times 10^6$	Year	Required Reservoir Capacity, $m^3 \times 10^6$
1951	60	1961	53
1952	40	1962	62
1953	85	1963	73
1954	30	1964	80
1955	67	1965	50
1956	46	1966	38
1957	60	1967	34
1958	42	1968	28
1959	90	1969	40
1960	51	1970	45

These data must now be ranked, with the highest required capacity getting rank 1, the next highest 2, and so on (n = 20).

Rank	Capacity, $m^3 \times 10^6$	$\dfrac{m}{n+1}$	Rank	Capacity, $m^3 \times 10^6$	$\dfrac{m}{n+1}$
1	90	0.05	11	51	0.52
2	85	0.1	12	46	0.57
3	80	0.14	13	45	0.61
4	73	0.19	14	42	0.66
5	67	0.23	15	40	0.71
6	62	0.28	16	40	0.76
7	60	0.33	17	38	0.81
8	60	0.38	18	34	0.85
9	55	0.43	19	30	0.90
10	53	0.48	20	28	0.95

These data are plotted on a graph in Figure 5–5. A semi-log plot often yields an acceptable straight line. The 1 year in 10 drought has a probability of $m/(n + 1) = 2/(20 + 1) = 0.10$ of occurring. Using the curve in Figure 5–5, we can pick the 10 percent probability drought as $82 \times 10^6 \, m^3$.

This procedure is known as a *frequency analysis* of recurring natural events such as droughts. In the above example we selected a "10-year drought," or the drought that on the average occurs once every 10 years. Recognize that there is no guarantee that it would indeed occur once every 10 years. In fact, it could happen 3 years in a row and then not again for 50 years. On the average, however, a 10-year recurrence interval is a reliable estimate for this example.

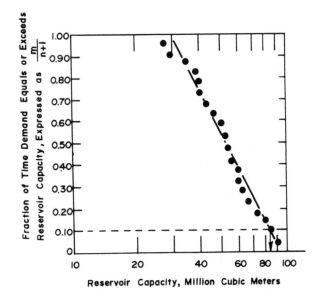

Figure 5–5. Frequency analysis of reservoir capacity.

WATER TRANSMISSION

Water can be transported from either a ground or surface supply source directly to a community or, if water quality considerations indicate, initially to a water treatment facility, by different types of conduits, including:

- Pressure conduits: tunnels, aqueducts, and pipelines
- Gravity-flow conduits: grade tunnels, grade aqueducts, and pipelines

The location of the well field or river reservoir defines the length of the conduits, while the topography indicates whether the conduits are designed to carry the water in open-channel flow or under pressure. The profile of a water supply conduit must generally follow the hydraulic grade line to take advantage of the forces of gravity and thus minimize pumping costs.

Service reservoirs are also necessary in the transmission system to help level out peak demands. In practice, intermediate reservoirs close to the city, or water towers, are sized to meet three design constraints:

- hourly fluctuations in water consumption within the service area
- short-term shutdown of the supply network for servicing
- back-up water requirements to control fires

These distribution reservoirs are most often constructed as open or covered basins, elevated tanks, or, in the past, standpipes. If the service reservoirs are adequately designed to meet these capacity considerations, then the supply conduits leading to them generally must only be designed to carry approximately 50 percent in excess of the average daily demand of the system or subsystem.

Elements of Closed-Conduit Hydraulics

Before discussing flow, it might first be useful to review some important fluid properties.

The *density* of a fluid is its mass per unit volume. In common units, density is expressed as slugs per cubic foot, or as $lb \cdot sec^2/ft^4$. Density in the metric system is in terms of grams per cubic centimeter. The density of water under standard conditions is 1.94 in common units and unity in the metric system.

Specific weight represents the force exerted by gravity on a unit volume of fluid and therefore must be in terms of force per unit volume, such as pounds per cubic foot. The specific weight is related to density as

$$w = \rho g$$

where w = specific weight, lb/ft^3
  ρ = density, $lb \cdot sec^2/ft^4$
  g = gravitational constant, ft/sec^2

The specific weight of water is 62.4 lb/ft^3.

Specific gravity of a liquid is the ratio of its density to that of pure water at a standard temperature. In the metric system the density of water is 1 g/cm and hence the specific gravity has the same numerical value as the density.

The *viscosity* of a fluid is a measure of its resistance to shear or angular deformation and is defined as the proportionality constant relating the shear stress τ to the rate of deformation du/dy. This proportionality constant is usually written as

$$\tau = \mu \, \frac{du}{dy}$$

The assumption inherent in this definition is that the shear stress is directly proportional to the rate of deformation. Such a definition holds for many fluids, which are known as Newtonian fluids. Fluids for which the shear stress is not proportional to the shear rate are known as non-Newtonian fluids. An example of a non-Newtonian fluid is biological sludge. The term viscosity, when applied to biological sludge, is therefore significant only if either the rate of deformation or the shear stress is also specified.

In the metric system, a unit of viscosity is a poise, with units of grams per centimeter-second. Most fluids have low viscosity, and a more convenient unit is the centipoise or 0.01 poise. The viscosity of water at 20°C (68.4°F) is 1 centipoise. In the English system the unit of viscosity is pound seconds per square foot. One $lb \cdot sec/ft^2$ equals 479 poise.

Kinematic viscosity is defined as the absolute viscosity, μ, divided by the density of the fluid, or $\nu = \mu/\rho$. The dimensions of kinematic viscosity are square centimeters per second.

Fluid viscosity is a function of temperature. Some representative values of the viscosity of water are shown in Table 5–2.

Table 5–2. Viscosity of Water

Temperature		Viscosity	
°F	°C	$lb \cdot sec/ft^2$	Centipoise
40	4.4	3.1×10^{-5}	1.5
50	10.0	2.7×10^{-5}	1.3
60	15.5	2.3×10^{-5}	1.1
68.4	20.0	2.1×10^{-5}	1.0
70	21.0	2.0×10^{-5}	0.96
80	26.6	1.8×10^{-5}	0.86
90	32.2	1.6×10^{-5}	0.77

Closed-Conduit Flow

One of the fundamental principles of hydraulics states that in a system, the total energy of a perfect liquid under ideal conditions does not change as it flows from point to point. Total energy is the sum of the position energy, pressure energy, and velocity energy. These energies are usually stated in terms of meters or feet of fluid, so that position energy is static head Z, pressure energy is fluid pressure, P/w, called pressure head, and the velocity energy is velocity head, $v^2/2g$. These terms are defined as:

Z = elevation, m (ft)
P = pressure, kg/m^2 (lb/ft^2)
w = specific weight of the fluid, kg/m^3 (lb/ft^3)
v = velocity, m/s (ft/sec)
g = acceleration due to gravity, m/sec^2 (ft/sec^2)

Consider a system such as is shown in Figure 5–6. Water is flowing through this pipe from a reservoir with constant surface elevation, and assuming there are no losses in this system, it is obvious that the total energy or the total head of water remains constant, although the form of the energy or head of water is converted from one type of energy to another. For example, at point 1, at the surface of the reservoir, all the energy is static head, while at points 3, 4, and 5, the energy is distributed among static, pressure, and velocity head. At point 6,

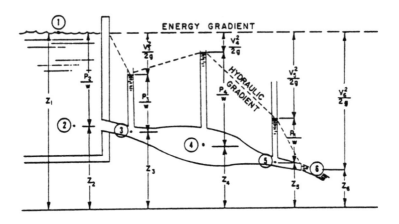

Figure 5–6. Hydraulics for ideal closed-conduit flow.

the energy of the jet, as it enters the atmosphere, is made up of only the velocity head and the static head. The total energy, however, is constant at all points in the system.

This is the constant energy principle, commonly known as Bernoulli's Theorem, which if written between any two points in a system is:

$$Z_1 + \frac{P_1}{w} + \frac{v_1^2}{2g} = Z_2 + \frac{P_2}{w} + \frac{v_2^2}{2g}$$

This equation holds for an ideal fluid, in which no losses occur. In real systems, however, the flow of a fluid is always accompanied by a loss of energy. Taking into account losses occurring in any system between two points, the equation is:

$$Z_1 + \frac{P_1}{w} + \frac{v_1^2}{2g} = Z_2 + \frac{P_2}{w} + \frac{v_2^2}{2g} + h_L$$

in which h_L represents the energy losses within the system.

The energy loss may occur in many places in a system, such as valves, bends, and sudden changes in pipe diameter. One of the major losses of energy is in the friction between the moving fluid and the pipe wall.

Because of the practical problems involved in the application of more elegant and theoretically sound friction loss equations, engineers have resorted to fitted exponential equations for flow calculations. Among these, the Hazen–Williams formula is most widely used in the United States for flow in pressure pipes and the Manning formula for flow in open channels or pipes not flowing full. These formulas are limited to turbulent flow of water and to common ambient temperatures.

The Hazen–Williams formula is

$$v = 1.318 \, Cr^{0.63} s^{0.54}$$

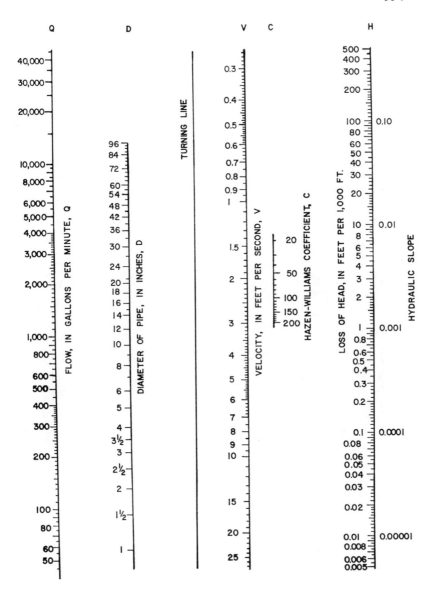

Figure 5–7. Hazen–Williams equation nomograph, common units.

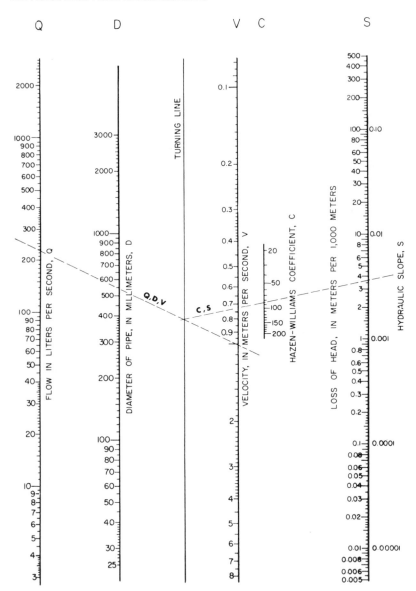

Figure 5–8.   Hazen–Williams equation nomograph, metric units.

where   v = the mean velocity of flow, ft/sec.
r = the hydraulic radius (area divided by wetted perimeter), ft
s = the slope of the hydraulic grade line
C = the Hazen–Williams friction coefficient

Combined with the continuity equation, Q = Av, the formula for discharge, Q, in gallons per minute for circular pipe of diameter D, in inches, is

$$Q = 0.285CD^{2.63}\, s^{0.54}$$

In metric units, these two equations are

$$v = 0.849Cr^{0.63}\, s^{0.54}$$
$$Q = 0.278CD^{2.63}\, s^{0.54}$$

Table 5–3. Values of C for Hazen–Williams Formula

Type of Pipe	C
Asbestos cement	140
Brass	130–140
Brick sewer	100
Cast iron	
New, unlined	130
Old, unlined	40–120
Cement-lined	130–150
Bitumastic enamel-lined	140–150
Tar-coated	115–135
Concrete or concrete-lined	
Steel forms	140
Wooden forms	120
Centrifugally spun	135
Copper	130–140
Fire Hose (rubber-lined)	135
Galvanized iron	120
Glass	140
Lead	130–140
Masonry conduit	120–140
Plastic	140–150
Steel	
Coal-tar enamel-lined	145–150
New, unlined	140–150
Riveted	110
Tin	130
Vitrified	100–140
Wood stave	120

where D and r are in meters and v and Q are meters per second and cubic meters per second, respectively. The nomographs shown as Figures 5–7 and 5–8 are solutions of the above equation in both common and metric units. Table 5–3 summarizes Hazen–Williams coefficients for various pipe materials.

Example 5.4
    The pressure drop through a 6-in. asbestos-cement pipe, 3,000 feet in length is 20 psi. What is the flow rate? 20 psi × 2.31 feet of water/psi = 46.2 feet of head loss (h_L). 46.2 feet ÷ 3,000 feet = 15.4 ft/1,000 ft(s). From Table 5–3 the Hazen–Williams coefficient is 140 and from the nomograph, Figure 5–7, the discharge is 460 gpm. If the flow had been given, the head loss could be found in the reverse manner.

    Energy or head is lost at the entrance to a pipe or conduit; at valves, meters, fittings, and other irregular features; at enlargements, and at flow contractions. These losses, called *minor losses*, are in excess of friction losses over the same length of straight pipe or conduit and may be expressed as:

$$h_L = K \frac{v^2}{2g}$$

for which values of K may be estimated by using Table 5–4.

Example 5.5
    The loss for a flow of 1.0 cfs through a given 6-in. main with a gate valve wide open is 20 feet. Find the head loss with the gate valve 75 percent closed.
    From Table 5–4, the increase in K value when the valve is partially closed is 24.0–0.2 = 23.8 (difference in K for 75 percent closed and fully open).

$$v = \frac{Q}{A} = \frac{1.0}{0.2} = 5 \text{ ft/sec}$$

$$h_L = h_o + K(v^2/2g)$$

$$= 20 + 23.8(5^2/64.4)$$

$$= 29.2 \text{ ft}$$

    Two pipes, two systems of pipes, or a single pipe and a system of pipes are said to be equivalent when their losses of head for equal rates of flow are equal (or flow is equal for equal loss of head). Compound pipes, whether in parallel or

Table 5–4. Minor Losses of Head

Nature of Special Resistance	Loss (K)	Nature of Special Resistance	Loss (K)
Angle valve		45° Elbow	
Wide open	2–5	Use $\frac{3}{4}$ of loss for 90° bend of same radius	
Butterfly valve			
$\theta = 10°$	1	$22\frac{1}{2}°$ Elbow	
$\theta = 40°$	10	Use $\frac{1}{2}$ of loss for 90°	
$\theta = 70°$	320	bend of same radius	
Check valves		Entrance losses	
Horizontal lifts	8–12	Pipe projecting into tank	0.8–1.0
Ball	65–70	End of pipe flush with tank	0.5
Swing	0.6–2.5	Slightly rounded	0.23
		Bell-mouthed	0.04
Gate valves			
Wide open	0.2	Outlet losses	
$\frac{1}{4}$ closed	1.2	From pipe into still water	
$\frac{1}{2}$ closed	5.6	or atmosphere	1.0
$\frac{3}{4}$ closed	24.0	Sudden contraction	
Globe valves		$d/D = \frac{1}{4}$	0.42
Wide open	10	$d/D = \frac{1}{2}$	0.33
		$d/D = \frac{3}{4}$	0.19
90° Elbow		Sudden enlargement	
Regular flanged	0.21–0.30	$d/D = \frac{1}{4}$	0.92
Long radius flanged	0.14–0.23	$d/D = \frac{1}{2}$	0.56
Short radius screwed	0.9	$d/D = \frac{3}{4}$	0.91
Medium radius screwed	0.75		
Long radius screwed	0.60		

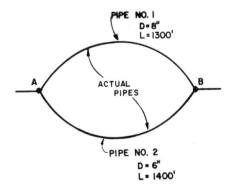

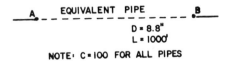

NOTE: C = 100 FOR ALL PIPES

Figure 5–9. Equivalent pipes, parallel.

PIPE NO. 3   PIPE NO. 4

D= 8"   D= 6"

C   L= 400' D   L=600' E

ACTUAL PIPES

C   EQUIVALENT PIPE   E

D= 6.5"
L= 1000'

NOTE: C = 100 FOR ALL PIPES   **Figure 5–10.** Equivalent pipes, series.

in series (Figures 5–9 and 5–10), may be reduced to single, equivalent pipes. The following examples are illustrative:

Example 5.6

For the parallel pipes as shown in Figure 5–9, find the diameter of equivalent pipe (length = 1,000 feet assumed).

1. Loss of head through pipe 1 must always equal loss of head through pipe 2 between points A and B.
2. Assume any arbitrary head loss, say 10 feet.
3. Calculate head loss in feet per 1,000 feet for pipes 1 and 2.
   Pipe 1: $(10/1,300) \times 1,000 = 7.7$ ft/1,000 ft
   Pipe 2: $(10/1,400) \times 1,000 = 7.1$ ft/1,000 ft
4. Use Figure 5–7 to find flow in gallons per minute (gpm).
   Pipe 1: D = 8 in., s = 0.0077,
      Q = 495 gpm
   Pipe 2: D = 6 in., s = 0.0071,
      Q = 220 gpm
   Total Q through both pipes = 715 gpm.
5. Using Figure 5–7 with s = 0.010 and Q = 715 gpm, equivalent pipe size is found to be 8.8 in. in diameter.

Example 5.7

For the pipes in series as shown in Figure 5–10, find the diameter of equivalent pipe (length = 1,000 feet assumed).

1. Quantity of water flowing through pipes 3 and 4 is the same.
2. Assume any arbitrary flow through pipes 3 and 4, say 500 gpm.
3. Using Figure 5–7, find head loss for pipes 3 and 4.
   Pipe 3: D = 8 in., L = 400 feet,
      Q = 500 gpm
      $h_1 = s_1 \times L_1 = 0.008 \times 400 = 3.2$ feet

Pipe 4: D = 6 in., L = 600 feet;
  Q = 500 gpm
  $h_1 = 0.028 \times 600 = 16.8$ feet
  Total head loss in both pipes = 20 feet.
4. Using Figure 5–7, with head loss = 20 feet, s = 20/1,000 and Q = 500 gpm, the equivalent pipe size is found to be 6.5 in. in diameter.

More complex systems may be reduced to a single equivalent pipe by piecework conversion of real and equivalent pipes.

Although the calculations become tedious, solution of network flow problems is dependent on the same basic physical principles as for single pipes, that is, the principles of energy conservation and continuity must be satisfied throughout the network. Examples of complex pipe networks are beyond the scope of this text.

Pumps and Pumping*

Pumps are mechanical devices for converting other forms of energy to hydraulic energy. When interposed in a pipe, they add energy to the liquid passing through the pipes. Nearly always, it is pressure energy. Pumps, like motor vehicles, are not individually designed for public works projects, except for very large or unusual installations. Rather, they are selected from predesigned and manufactured units readily available for a wide range of applications.

Economical selection requires that attention be given to: (1) the normal pumping rate, and also the minimum and maximum rate that the pump will ever be called on to deliver; (2) the total head capacity to meet flow requirements; (3) suction head (or lift); (4) pump characteristics including speed, number, power source, and spatial, environmental, and other special requirements; and (5) the nature of the liquid to be pumped.

There are many types of pumps, but the two most often encountered in environmental engineering are *roto-dynamic* and *displacement* pumps. Roto-dynamic pumps impart energy to liquids with a rotating element, or *impeller*, shaped to force water outward at right angles to its axis (*radial flow*), to give the liquid an axial as well as a radial velocity (*mixed flow*), or to force the liquid in the axial direction alone (*axial flow*). Radial flow and mixed flow machines are commonly referred to as *centrifugal pumps*, and axial flow machines are called *propeller pumps*. Displacement pumps include the *reciprocating* type, in which a piston draws water into a cylinder on one stroke

---

* This discussion on pumps and pumping is adapted from McJunkin, F.E. and P.A. Vesilind, *Practical Hydraulics for Public Works Engineers*, published as a separate issue by *Public Works Magazine* (1968).

and forces it out on the next, and the *rotary* type, in which two cams or gears mesh together and rotate in opposite directions to force water continuously past them. In addition there are *jet pumps (ejectors), air-lift pumps, hydraulic rams, diaphragm pumps,* and others that may be useful under special circumstances.

Pump hydraulics are depicted in Figure 5–11. The *static suction head* on a pump is the vertical distance, i.e., evaluation difference, from a free liquid surface* on the intake to the pump centerline. This static head may be positive or negative (the latter sometimes called *suction lift*). The *net suction head,* h_s, is the static suction head minus friction head losses, including entrance loss, for the capacity under consideration. The *static discharge head* is the vertical distance from the pump centerline to the free surface (or discharge) on the discharge line. Similarly, the *net discharge head,* h_d, is the static discharge head plus the friction head losses. The total head, H, is the difference between the net discharge and suction heads, or,

$$H = h_d - h_s$$

In pumping liquids, the pressure at any point in the suction line should not be reduced to the vapor pressure of the liquid, if air binding, loss of prime, and

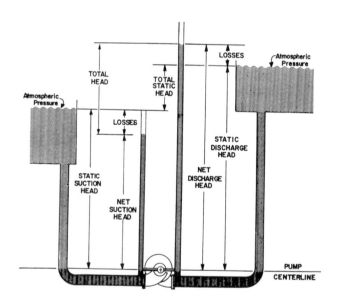

Figure 5–11.   Pump hydraulics defined.

---

* If no free surface exists, the gauge pressure at the pump flange (suction or discharge) for zero discharge, corrected to centerline pump elevation, should be used in lieu of the static head (suction and/or discharge as appropriate).

cavitation are to be avoided. The available energy that can be utilized to move liquid through the suction line to the impeller is thus the net suction head, h_s, plus any pressure (a vacuum being a negative pressure) existing in the suction supply line, and less the vapor pressure of the liquid at the pumping temperature. This energy or head is known as the *net positive suction head* (NPSH).

The useful work done by a pump is the product of the weight of liquid pumped multiplied by the head developed by the pump. The power or work per unit time required is the *water horsepower* (WHP). By definition:

$$WHP \propto QHw$$

where Q = pump discharge
  H = total head
  w = specific weight

For water at 68°F, Q in gallons per minute (gpm), and H in feet:

$$WHP = \frac{QH}{3,960}$$

The total power required to drive a pump is the *brake horsepower* (BHP). The ratio of the water horsepower to the brake horsepower is the pump efficiency, η:

$$\eta = \frac{WHP}{BHP} \times 100$$

Therefore:

$$BHP = \frac{100\,QH}{3,960\eta}$$

for water at 68°F. BHP for other water temperatures and other liquids can be determined by correcting for the change in specific weight. For natural waters, correction of specific weight for temperature is usually negligible.

Pump Characteristic Curves

For a given pump, the total head developed, the power required to drive it, and the resulting efficiency vary with discharge. These interrelations of head, power, efficiency, and capacity are commonly known as the *characteristics of the pump*. They are best shown graphically, as *pump characteristic curves*. The usual practice is to plot head, power, and efficiency against capacity at constant impeller speed, as shown in Figure 5–12.

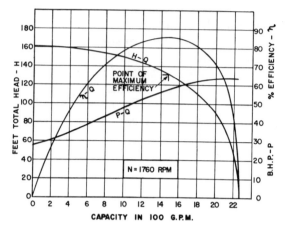

Figure 5–12. Pump characteristic curves.

The curve H–Q, showing the relation between capacity and total head, is called the *head-capacity curve*, and pumps are often classified on the basis of the shape of their head-capacity curves.

The curve η–Q shows the relation between efficiency and capacity and is called simply the *efficiency curve*. The efficiency curve usually exhibits a maximum or peak efficiency.

The curve P–Q showing the relation between power input and pump capacity is generally termed the *power curve*.

System Head Curves

The head loss in a pumping system increases with increasing flow through the system. This can be shown graphically as a *system head curve*. Figure 5–13 is an example. The system head loss for any flow rate is the sum of friction head loss at the rate plus the total static head in the system.

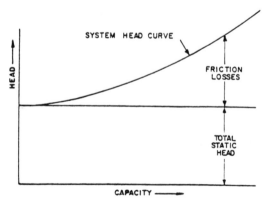

Figure 5–13. System head curve.

The total static head is the difference in elevation between the discharge liquid level and the suction liquid level (see Figure 5–11). Inasmuch as the static head is present, whether the pump is operating or not, it is plotted as the lower portion of the system head curve.

Friction losses, including minor losses, are determined as outlined in the previous section of this chapter. The system piping and fittings may be converted to one equivalent pipe, and head losses for several flow rates may be determined readily from the Hazen–Williams nomograph. Alternatively, the head loss through all pipe and fittings may be computed for a single flow rate, and losses for other flow rates may be determined from the relationship:

$$\frac{Q_1}{Q_2} = \left(\frac{H_1}{H_2}\right)^{0.54}$$

Total head for zero discharge will be equal to the total static head. This point plus several computed points will suffice to plot the curve. Quite commonly, the static head in a system will vary as tanks and reservoirs are filled and lowered. In such cases, system curves may readily be constructed for minimum and maximum heads, thereby enabling prediction of system pumping capacity for the entire range of possible static head conditions.

By separation of static and friction head in preparing the system head curve, the reasonableness of friction losses can be checked. If piping to be used is too small, then the pump and motor capital and operating costs owing to the higher head capacity required will be excessive. As a rule of thumb, the friction losses should not be more than 10 to 20% of the static head for comparatively short pipelines. For elaborate and costly installations, an economic analysis of the tradeoff between pumping and piping costs may be justified.

Operating Head and Discharge

The usual design condition is that a system will be given and the proper pump must be selected. If on the system head curve (such as Figure 5–13), a pump head capacity curve (such as in Figure 5–12) is superimposed, then their intersection will locate the *operating point*. Figure 5–14 is an example. The operating point is the discharge and head at which the given system and given pump will operate. This also locates the efficiency of operation and the power requirements. A pump should be selected that has an operating point at or near its peak efficiency. Many other factors, some previously outlined, must also be considered in selecting from the multitude of options available, but meeting the system discharge requirement (for a given system, satisfying discharge automatically meets the head requirement) at reasonable efficiency is paramount.

The pumping head developed is doubled if two identical pumps are placed in series. Conversely, two pumps in parallel will double the pumping capacity (see Figure 5–15). This does not mean, however, that the head or capacity of a

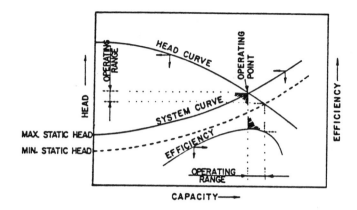

Figure 5–14. Sample determination of pump operating point.

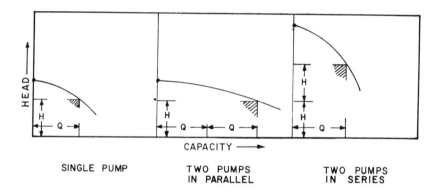

Figure 5–15. Pump characteristic curves with two pumps in parallel and in series.

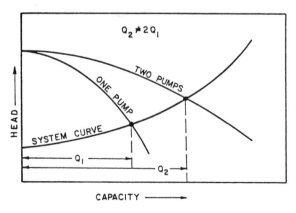

Figure 5–16. System head curve with two pumps in parallel.

system is doubled. Superimposing a system curve on the head-capacity curves for two pumps in parallel, it is clear (from Figure 5–16) that the added capacity results in a greater friction head loss, the relationship being:

$$h_L = Q^2$$

and that the system capacity with two pumps in parallel is not double its capacity with one pump. Similarly, pumping in series will double neither head nor discharge.

CONCLUSION

As the hydrologic cycle indicates, water is a renewable resource because of the driving force of the sun. Thus, the earth is not running out of water; it may just be running out of clean water in general and sufficient water in isolated areas as climatic changes take place.

Both groundwater and surface water supplies are available, although to varying degrees, across the face of the earth. The trick is to develop these supplies by using sound engineering design procedures for wells and dams, and to protect these supplies by using sound engineering and ethical judgments. In the next chapter, we address methods of preparing and treating water for distribution and consumption once the supply has been provided.

PROBLEMS

5.1   Given the following data for monitoring well drawdown, calculate the transmissibility of the sample aquifer using the modified Theis method. Assume the observation well is 300 feet from a well pumped at 350 gpm.

Time of pumping (hr)	1.9	2.8	5.5	9.1	19.0	57.0
Drawdown at observation well (ft)	1.9	2.3	3.4	4.2	5.6	7.5

5.2   A storage reservoir is needed to ensure a constant flow of 20 cfs to a city. The monthly stream flow records are:

Month	J	F	M	A	M	J	J	A	S	O	N	D
ft^3 of water $\times$ 10^6	60	70	85	50	40	25	55	85	20	55	70	90

Calculate the storage requirement.

5.3   If a faucet drips at a rate of 2 drops per second and it takes 25,000 drops to equal 1 gallon of water, how much water is lost each day? Each year?

If water costs \$1.20/1,000 gal, how long would the lost water take to equal the cost of fixing the leak if you did it yourself (free labor!) for 10¢ in parts.

5.4   A 24-in. pipe with centrifugally spun concrete lining carries a flow of 10 mgd. Find: (a) head loss per unit length; (b) velocity.

5.5   A 6-in. main is carrying a flow of 1.0 cfs. What head loss is produced by two regular flanged 45° elbows and a gate valve one-half closed?

5.6   Three pipes, each 1,500 feet long, are arranged in parallel. Two are 6 in. in diameter; the other is 10 in. All three have a Hazen–Williams C value of 100. Find the diameter of the equivalent pipe that is also 1,500 feet long.

5.7   A pump delivers 900 gpm against a total head of 145 feet. Assuming a pump efficiency of 90 percent, what is the BHP?

5.8   Given two identical pumps with the characteristics shown in Figure 5–12, find the discharge for a head of 100 feet if: (a) both pumps are operated in series and (b) both pumps are operated in parallel.

LIST OF SYMBOLS

γ = kinematic viscosity, cm^2/sec
ρ = density, g/cm^3 ($lb \cdot sec^2/ft^3$)
μ = viscosity, poise ($lb \cdot sec/ft^2$)
η = pump efficiency
τ = shear stress
A = cross-sectional area, m^2 (ft^2)
B = thickness of aquifer, m (ft)
BHP = brake horsepower
C = Hazen–Williams friction coefficient
cfs = cubic feet per second
E_T = evapotranspiration, m^3 (gal)
G_N = net groundwater flow, m^3 (gal)
G_s = change in groundwater storage, m^3 (gal)
g = acceleration due to gravity, m/sec^2 (ft/sec^2)
H = total head, m (ft)
h_d = net discharge head, m (ft)
h_L = head loss, m (ft)
h_s = net suction head, m (ft)
k = proportionality constant
K_p = coefficient of permeability, m^3/day (gal/day)
NPSH = net positive suction head, m (ft)
P = pressure, kg/m^2 (lb/ft^2)
P' = precipitation, m^3 (gal)
Q = discharge, m^3/sec (gal/min)
q = discharge through a section of aquifer of unit thickness under a unit head, m^3/sec (gal/min)

r = radius, m (ft)
s = hydraulic gradient
S_c = storage constant, m^3 (gal)
S_F = safe yield, m^3 (gal)
S_o = surface flow, m^3 (gal)
S_s = change in surface water storage, m^3 (gal)
T = transmissibility, m^3/day
t = time
V_d = vapor pressure at drew point, inches of Hg
V_p = vapor pressure, inches of Hg
v = velocity of flow, m/sec (ft/sec)
w = specific weight, kg/m^3 (lb/ft^3)
WHP = water horsepower
Z = elevation, m (ft)
Z_r = drawdown, m (ft)

REFERENCES

1. Hagen, G. "Uber die Bewegung des Wassers in engen cylindrischen Rohren," *Annalen Physik und Chemie* 46:423(1839).
2. Poiseville, G.L.M. "Recherches experimentales sur le mouvement des liquides dans les tubes de tres petit diametre," *Roy. Acad. Sci. Inst. France, Math Phys. Sci. Man.* 9:433(1846).
3. Darcy, H. Les Fontanines Publiques de la Ville de Dijon (Paris, France, 1856).
4. Theis, C.V. "The Relation between the Lowering of the Piezometric Surface and the Rate and Duration of Discharge of a Well Using Groundwater Storage," *Transactions of the American Geophysical Union* 16:519(1935).
5. Jacob, C.E. "Drawdown Test to Determine the Effective Radius of Artesian Well," *Transactions of the American Society of Civil Engineers* 112:1047(1947).

Chapter 6
Water Treatment

Many aquifers and isolated surface waters are high in water quality and may be pumped from the supply and transmission network directly to any number of end uses, including human consumption, irrigation, industrial processes, or fire control. However, such clean water sources are the exception to the rule in many regions of the nation, particularly regions with dense populations or regions that are heavily argicultural. Here, the water supply must receive varying degrees of treatment before distribution.

Impurities enter the water as it moves through the atmosphere, across the earth's surface, and between soil particles in the ground. These background levels of impurities are often supplemented by human activities. Chemicals from industrial discharges and pathogenic organisms of human origin, if allowed to enter the water distribution system, may cause health problems. Excessive silt and other solids may make the water both unsightly and aesthetically unpleasing. Water may be contaminated by many routes. For example, heavy metal pollution, including lead, zinc, and copper, may be caused by corrosion of the very pipes that carry the water from its source to the consumer.

The method and degree of water treatment are important considerations for environmental engineers. Generally speaking, the characteristics of raw water determine the method of treatment. Because most public supply systems are relied on for drinking water, as well as industrial and fire consumption, the highest level of use, human consumption, defines the degree of treatment. Thus, we focus only on treatment technologies that produce potable water.

A typical water treatment plant is diagrammed in Figure 6–1. These plants are designed to remove odors, color, and turbidity as well as bacteria and other contaminants from surface water. Raw surface water entering a water treatment plant usually has significant turbidity caused by tiny colloidal clay and silt particles. These particles have a natural electrostatic charge that keeps them continually in motion and prevents them from colliding and sticking together. Chemicals such as alum (aluminum sulfate) are added to the water, first to neutralize the charge on the particles and then to aid in making the tiny particles

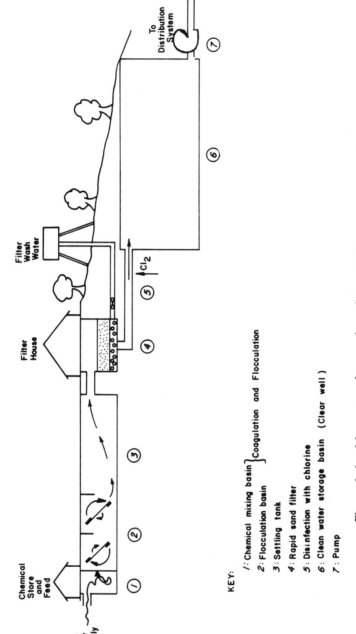

Figure 6–1.   Movement of water through a typical water treatment plant.

"sticky" so they can coalesce and form large particles called *flocs*. This process is called *coagulation* and *flocculation* and is represented by stages 1 and 2 in Figure 6–1.

COAGULATION AND FLOCCULATION

Naturally occurring silt particles suspended in water are difficult to remove because they are very small, often colloidal in size, and they possess negative charges and thus are prevented from coming together to form larger particles that could more readily be settled out. The removal of these particles by settling requires first that their charges be neutralized and second that the particles be encouraged to collide with each other. The charge neutralization is commonly termed coagulation, and the building of larger flocs from the small particles is called flocculation.

A fairly simple but not altogether satisfactory explanation of coagulation is available in the double-layer model. The charge on a particle is represented in Figure 6–2. The solid particle is negatively charged, and it attracts to it counterions that are positively charged. Some of these ions are so strongly attached to the particle that they travel with the particle, thereby forming a slippage plane. Around this inner layer is an outer layer of ions consisting mostly of counterions, but these are loosely attached and can easily slip off. The charge on the particle as it moves through a fluid is the negative charge, diminished in part by the positive ions in the inner layer. This charge is called the *zeta potential*.

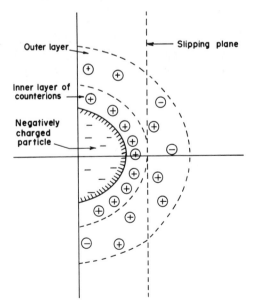

Figure 6–2. Charges on a suspended particle, as explained by the double-layer theory.

If the net negative charge is considered a repulsive charge (since the neighboring particles are also so charged), the charge may be pictured as in Figure 6–3A. In addition to this repulsive charge, however, all particles have an attractive electrostatic charge, known as the *van der Waals force*, which is due to the molecular structure of the particle. This attractive charge is also shown in Figure 6–3A. Combining these two charges, we find a net repulsive charge, or an energy hill, which prevents the particles from coming together. The objective of coagulation is to reduce this energy hill to zero so that particles would not be repulsed by each other.

One means of accomplishing this end is to add trivalent cations to the water. These ions would snuggle up to the negatively charged particle, and because they possess a stronger charge they would displace the monovalent cations. The effect of this would be to reduce the net negative charge and thus lower the repulsive force. This is shown schematically in Figure 6–3B. In this condition, the particles will not repulse each other, and upon colliding, they will stick together. A stable colloidal suspension has thus been made into an unstable colloidal suspension.

The usual source of trivalent cations in water treatment is alum (aluminum sulfate). Alum has an additional advantage in that some fraction of the aluminum may form aluminum oxides/hydroxides, represented simply as:

$$Al^{3+} + 3OH^- \rightarrow AlOH_3 \downarrow$$

These complexes are sticky and heavy and will greatly assist in the clarification of the water in the settling tank if the unstable colloidal particles can be made to come into contact with the floc. This process is enhanced through an operation known as flocculation.

A flocculator introduces velocity gradients into the water so that particles in a fast-moving stream can catch up with and collide with slow-moving particles. Such velocity gradients are commonly introduced by rotating paddles. The power required for moving a paddle through the water is stated as

$$P = \frac{C_D A \rho v^3}{2}$$

where P = power, N/sec (ft · lb/sec)
  A = paddle area, m^2 (ft^2)
  ρ = fluid density, kg/m^3 (lb · sec^2/ft^3)
  v = velocity of the paddle relative to the fluid, m/sec (ft/sec)
  C_D = drag coefficient

The velocity gradient produced as a result of a power input in a given volume of water, $\forall$, is

$$G = \left(\frac{P}{\mu \forall} \right)^{1/2}$$

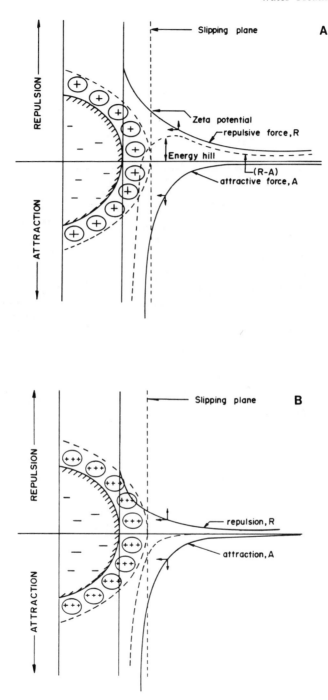

Figure 6–3. The reduction of the net charge on a particle as the result of the addition of trivalent counterions.

where   G = velocity gradient, $\sec^{-1}$
μ = viscosity, dyne · sec/cm^2 (lb · sec/ft^2)
$\forall$ = tank volume, m^3 (ft^3)

Generally accepted design standards require G to be between 30 and 60 $\sec^{-1}$.
Time is also an important variable in flocculation, and the term $G\bar{t}$ is often used in design, where $\bar{t}$ = hydraulic retention time in the flocculation basin. Typically, $G\bar{t}$ values range from 1×10^4 to 1×10^5.

Example 6.1
Assume the Durham, North Carolina, water plant is designed for 30 million gallons per day (mgd). The flocculator dimensions are 100 feet long, 50 feet wide, 16 feet deep. Revolving paddles attached to four horizontal shafts rotate at 1.7 rpm. Each shaft supports four paddles that are 6 inches wide and 48 feet long. Paddles are centered 6 feet from the shaft. Assume $C_D = 1.9$, and mean velocity of water is 35 percent of paddle velocity. Find the velocity differential between the paddles and the water. Also calculate the value of G and the time of flocculation.
The rotational velocity is

$$v_r = \frac{2\pi rn}{60}$$

where   r = radius, ft
n = rpm

$$v_r = \frac{2\pi \times 6 \times 1.7}{60} = 1.07 \text{ ft/sec}$$

The velocity differential between paddles and fluid is assumed to be 65 percent of v_r, so that

$$v = 0.65 \times v_r = 0.65 \times 1.07 = 0.70 \text{ ft/sec}$$

Total power input is

$$P = \frac{C_D A\rho v^3}{2} = \frac{1.9 \times 384 \times 1.94 \times (0.70)^3}{2} = 243 \text{ ft · lb/sec}$$

where C_D = drag coefficient = 1.9, obtained from any fluid mechanics handbook
A = paddle area = $4 \times 4 \times 48 \times (6/12) = 384$ ft^2
v = velocity differential = 0.70 ft/sec
ρ = density of water = 1.94 at 50°F

$$G = \left(\frac{P}{\forall\mu}\right)^{1/2} = \sqrt{\frac{243}{(8 \times 10^4) \times (2.73 \times 10^{-5})}} = 10.5 \frac{\text{ft/sec}}{\text{ft}}$$

where $\forall = 100 \times 50 \times 16 = 80,000 \text{ ft}^3$
$\mu = \text{viscosity} = 2.73 \times 10^{-5} \text{ at } 50°F$

The value of $G = 10.5 \text{ sec}^{-1}$ is a little low.
Time of flocculation

$$\bar{t} = \frac{\forall}{Q} = \frac{80,000 \times 7.48 \times 24 \times 60}{30 \times 10^6} = 28.7 \text{ min}$$

This is within the accepted range.

SETTLING

When the flocs have been formed they must be separated from the water. This is invariably done in gravity settling tanks, which simply allow the heavier-than-water particles to settle to the bottom. Settling tanks are designed to approximate uniform flow and to minimize all turbulence. Hence, the two critical elements of a settling tank are the entrance and exit configurations. Figure 6–4 shows one type of entrance and exit configuration used for distributing the flow entering and leaving the water treatment settling tank.

Alum sludge is not highly biodegradable and thus will not decompose at the bottom of the tank. After some time, usually several weeks, the accumulation of alum sludge at the bottom of the tank is such that it has to be removed. Typically, the sludge exits through a *mud valve* at the bottom and is wasted either into a sewer or to a sludge holding/drying pond. We will see in Chapter 8 that, in contrast to water treatment sludges, sludges collected in wastewater treatment plants can remain in the bottom of the settling tanks only a matter of hours before starting to produce odoriferous gases, thus floating some of the solids.

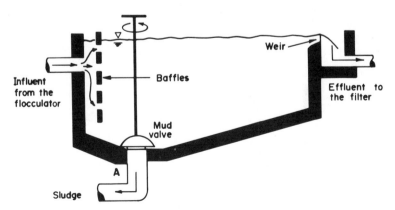

Figure 6–4.   Schematic of a common settling tank.

The water leaving a settling tank is essentially clear. The polishing is performed with a rapid sand filter.

FILTRATION

In Chapter 5 we discussed movement of water into the ground and through soil particles and alluded to the cleansing action the particles have on contaminants in the water. Picture the extremely clear water that bubbles up from "underground streams" in hills and valleys across the nation. The soil particles definitely help filter the groundwater, and through the years environmental engineers have learned to apply this natural process in water treatment and supply systems and have developed what we now know as the rapid sand filter. The actual process of separating impurities from a carrying liquid by rapid sand filtration involves two phases: filtration and backwashing.

A slightly simplified version of the rapid sand filter is illustrated in a cutaway in Figure 6–5. Water from the settling basins enters the filter and seeps through the sand and gravel bed, through a false floor, and out into a clear well that stores the finished water. During filtration valves A and C are open.

Eventually the rapid sand filter becomes clogged and must be cleaned. This cleaning is performed hydraulically. The operator first shuts off the flow of water to the filter (closing valves A and C), then opens valves D and B, which allow wash water (clean water stored in an elevated tank or pumped from the clear well) to enter below the filter bed. This rush of water forces the sand and gravel bed to expand and jolts individual sand particles into motion, rubbing against their neighbors. The light colloidal material trapped within the filter is

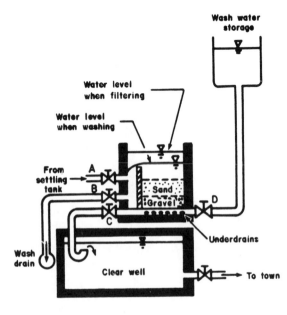

Figure 6–5. Schematic of a rapid sand filter.

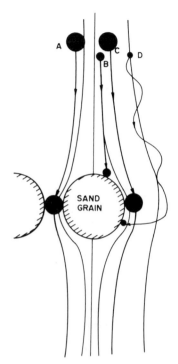

Figure 6–6.   Mechanisms of solids removal in a filter: (A) straining, (B) sedimentation, (C) interception, (D) diffusion.

released and escapes with the wash water. After a few minutes, the wash water is shut off and filtration is resumed.

The solid impurities in the water are removed by numerous processes, the most important of which are straining, sedimentation, interception, and diffusion (see Figure 6–6). Straining, possibly the most important mechanism in general, takes place almost exclusively in the first few centimeters of the filter medium. As the filtering process begins, straining removes only particles in the water that are large enough to get caught in the pores (A in Figure 6–6). After a time, these trapped particles themselves begin to form a screen that has smaller openings than the original filter medium. Smaller particles suspended in the water are trapped by this mat and immediately begin acting as part of the screen. Thus, removal efficiency owing to screening tends to increase in some proportion to the time of the filtration phase.

A second removal mechanism is sedimentation, in which larger and heavier particles do not follow the fluid streamline around the sand grain and settle on the grain (B in Figure 6–6).

Interception occurs with particles that do follow the streamline, but are too large and are caught because they brush up against the sand grain (C in Figure 6–6).

Finally, very small particles are experiencing Brownian motion and may by chance hit the sand grain. This last process is called diffusion (D in Figure 6–6).

The first three mechanisms are most effective for larger particles, whereas diffusion can occur only for subcolloidal-sized particles. This is represented by a

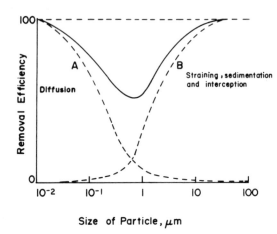

Figure 6–7.   Removal of various-sized particles in a filter.

typical removal efficiency curve for different-sized particles (Figure 6–7) where the efficiency of removal is high for both large and small particles and substantially reduced for midsized (about 1-μm) particles. Unfortunately, many viruses, bacteria, and fine clay particles are about 1 μm in size, and thus the filter is less effective in the removal of these particles.

Filter beds are often classified as single medium, dual media, or tri-media. The latter two are often utilized in the treatment of wastewater because they permit solids to penetrate into the bed, have more storage capacity, and thus increase the required time between backwashings. Also, multimedia filters tend to spread head loss buildup over time and further permit longer filter runs.

One of the primary considerations in filter design is the head loss through the sand. Obviously, as the sand gets progressively dirtier the head loss increases. Figure 6–8 shows a simplified representation of head loss in a filter. Although it is not possible to predict without experimentation what head losses will be experienced in a given application, the head loss in clear sand may be estimated with several different equations. One of the earliest and still most widely used method is called the Carman–Kozeny equation.

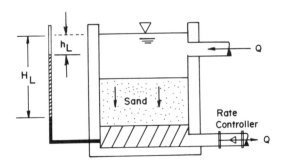

Figure 6–8.   Head loss in a filter as measured by a manometer. Note: h_L is head loss in a clean filter and H_L is head loss in a dirty filter.

A head loss in clean sand while filtering may be estimated by first considering the filter to be a mass of pipes, in which case the Darcy head loss equation holds:

$$h_L = f \frac{L}{D} \frac{v^2}{2g} \tag{6-1a}$$

where   L = depth of filter
D = "pipe" diameter
v = velocity in "pipes"
f = friction factor

But of course the "pipes," or channels through the sand, are not straight, and D varies, so we can substitute D = 4R where

$$R = \frac{\text{area}}{\text{wetted perimeter}} = \frac{\pi D^2/4}{\pi D} = \frac{D}{4} = \text{the hydraulic radius}$$

We thus have

$$h_L = f \frac{L}{4R} \frac{v^2}{2g} \tag{6-1b}$$

The velocity of the water approaching the sand is

$$v_a = Q/A' \tag{6-2a}$$

and the velocity through the bed is

$$v = \frac{v_a}{e} \tag{6-2b}$$

where   e = porosity, fraction of open spaces in the sand
Q = flow rate
A′ = surface area of the sand bed

The total channel volume is the porosity of bed times the total volume, or $e\forall$. The total solids volume is $(1 - e)$ times the total volume, which is also equal to the number of particles times the volume of the particles. Thus total volume =

$$\frac{N\forall_p}{1 - e}$$

where   N = number of particles
$\forall_p$ = volume occupied by each particle

The total channel volume is $e([N\forall_p]/[1 - e])$, and since the total wetted surface is NA_p (where A_p = surface area of each particle), the hydraulic radius

$$R = \frac{Area}{wetted\ perimeter} \propto \frac{Volume}{Area} = \frac{e([N\forall_p]/[1-e])}{NA_p} = \left(\frac{e}{1-e}\right)\left(\frac{\forall_p}{A_p}\right)$$

For spherical particles, $\forall_p/A_p = d/6$, but for particles that are not true spheres, $\forall_p/A_p = \phi(d/6)$, where ϕ = shape factor. For example, $\phi = 0.95$ for Ottawa sand, a common filter sand. We thus have

$$R = \left(\frac{e}{1-e}\right)\left(\phi\frac{d}{6}\right) \tag{6-3}$$

Substituting equations 6-2 and 6-3 into equation 6-1b,

$$h_L = f_1\left(\frac{L}{\phi d}\right)\left(\frac{1-e}{e^3}\right)\left(\frac{v_a^2}{g}\right) \tag{6-4}$$

We may approximate the friction factor by the equation

$$f = 150\left(\frac{1-e}{R}\right) + 1.75$$

where R = Reynolds number = $\phi\left(\frac{\rho v_a d}{\mu}\right)$.

The above discussion is applicable to a filter bed made of only one size particle. For a bed made of nonuniform sand grains,

$$d = \frac{6}{\phi}\left(\frac{\forall}{A}\right)_{avg} \qquad d = \frac{6}{\phi}\left(\frac{\forall_{avg}}{A_{avg}}\right) \tag{6-5}$$

where $\forall_{avg}$ = average volume of all particles
A_{avg} = average area of all particles

Substituting equation 6-5 into 6-4

$$h_L = f\left(\frac{L}{6}\right)\left(\frac{1-e}{e^3}\right)\left(\frac{v_a^2}{g}\right)\left(\frac{A}{\forall}\right)_{avg}$$

and $(A/\forall)_{avg}$ can be approximated as

$$\frac{6}{\phi}\sum\frac{x}{d'}$$

where x = weight fraction of particles retained between any two sieves
d' = geometric mean diameter between these sieves

$$h_L = f\left(\frac{L}{\phi}\right)\left(\frac{1-e}{e^3}\right)\left(\frac{v_a^2}{g}\right)\sum \frac{x}{d'} \tag{6-6}$$

This equation holds for nonstratified beds, such as are found in a slow sand filter, since the fraction factor does not vary with depth. When the particles are stratified, as in a rapid sand filter, we can write equation 6-6 as

$$\frac{dh_L}{dL} = \left[\frac{1}{\phi}\left(\frac{1-e}{e^3}\right)\left(\frac{v_a^2}{g}\right)\right]f\frac{1}{d'}$$

The total head loss is thus

$$h_L = \int_0^{h_L} dh_L = \left[\frac{1}{\phi}\left(\frac{1-e}{e^3}\right)\left(\frac{v_a^2}{g}\right)\right]\int_0^{L}\frac{f}{d'}\,dL$$

Since $dL = Ldx$, where dx is the proportion of particles of size d

$$h_L = \left[\frac{1}{\phi}\left(1-\frac{e}{e^3}\right)\left(\frac{v_a^2}{g}\right)\right]L\int_{x=0}^{x=1} f\frac{dx}{d}$$

and if the particles between adjacent sieve sizes are assumed uniform,

$$h_L = \frac{L}{\phi}\left(\frac{1-e}{e^3}\right)\left(\frac{v_a^2}{g}\right)\sum \frac{fx}{d'}$$

Example 6.2
   A sand consisting of the following sizes is used*:

Sieve Number	Percentage of sand retained on sieve, $\times 10^2$	Geometric mean size of sand, ft $\times 10^{-3}$
14–20	1.10	3.28
20–28	6.60	2.29
28–32	15.94	1.77
32–35	18.60	1.51
35–42	19.10	1.25
42–48	17.60	1.05
48–60	14.30	0.88
60–65	5.10	0.75
65–100	1.66	0.59

---

* Adapted from Clark, J.W., W. Viessman, and M.J. Hammer. *Water Supply and Pollution Control*, 3rd ed. (New York: Thomas Y. Crowell, 1977).

The filter bed measures 20×20 feet and is 2 feet deep. The sand has a porosity of 0.40 and a shape factor of 0.95. The filtration rate is 4 gal/min · ft². Assume the viscosity is 3×10^{-5} lb · sec/ft². Find the head loss through the clean sand. The solution is shown in tabular form:

Reynolds Number	Friction Factor, f	$\dfrac{x}{d}$	$f\dfrac{x}{d}$
1.80	51.7	3.4	174
1.37	67.4	28.8	1,941
1.06	86.6	90.1	7,802
0.91	100.6	123.2	12,394
0.75	121.7	152.8	18,595
0.63	144.6	167.6	24,235
0.53	171.5	162.5	27,868
0.45	201.7	68.0	13,715
0.35	258.8	28.1	7,272

Column 1: The approach velocity is

$$v_a = 4 \frac{\text{gal}}{\text{min} \cdot \text{ft}^2} \times \frac{1}{7.481} \frac{\text{ft}^3}{\text{gal}} \times \frac{1}{60} \frac{\text{min}}{\text{sec}} = 8.9 \times 10^{-3} \text{ ft/sec}$$

For the first particle size, $d = 3.28 \times 10^{-3}$ feet

$$R = \phi \frac{\rho v_a d}{\mu} = \frac{(0.95)(1.94)(89 \times 10^{-3})(3.28 \times 10^{-3})}{3 \times 10^{-5}} = 1.80$$

Column 2:

$$f = 150\left(\frac{1 - e}{R}\right) + 1.75$$

$$= 150\left(\frac{1 - 0.4}{1.8}\right) + 1.75 = 51.75$$

Columns 3 and 4:
for the first size, $x = 1.10\%$ and $d = 3.28 \times 10^{-3}$

$$f\frac{x}{d} = \frac{51.75(0.011)}{3.28 \times 10^{-3}} = 174$$

The last column is summed: $\Sigma\, f(x/d) = 113,977$

$$h_L = \frac{L}{\phi}\left(\frac{1 - e}{e^3}\right)\left(\frac{v_a^2}{g}\right)\Sigma f\frac{x}{d}$$

$$= \frac{2}{0.95}\left(\frac{1 - 0.4}{0.4^3}\right)\left(\frac{(8.9 \times 10^{-3})^2}{32.2}\right) 113,977$$

$$= 5.78 \text{ feet}$$

The deposition of material during the filtering process increases the head loss through the filter. A method of predicting head loss takes advantage of this limitation, predicting head loss

$$H_L = h_L + \sum_{i=1}^{n} (h_{L_i})_t$$

Here, H_L is the total head loss through the filter, h_L is the clear water head loss at time zero, and $(h_{L_i})_t$ is the head loss in the ith layer of medium in the filter at time t. (All values are in meters.) Head loss within an individual layer, $(h_{L_i})_t$, is related to the amount of material caught by the layer:

$$(h_{L_i})_t = x(q_i)_t^y$$

Here, $(q_i)_t$ is the amount of material collected in the ith layer at time t, in milligrams per cubic centimeter, and x and y are experimental constants. Head loss data for sand and anthracite are summarized in Figure 6–9.

A *filter run* is the time a filter operates before it has to be cleaned. The end of a filter run is indicated in one of two ways:

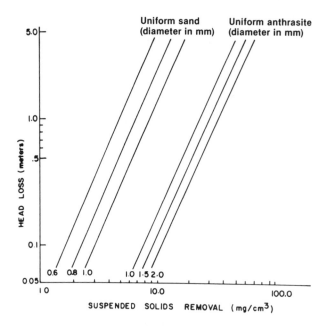

Figure 6–9. Increased head loss in a filter owing to materials deposits. [Adapted from Metcalf and Eddy, Inc., *Wastewater Engineering* (New York: McGraw-Hill, 1979).]

- excessive head loss
- excessive turbidity of the filtered water

If either of these occurs, the filter must be washed.

Example 6.3
The filter described in Example 6.2 is run so that it reduces the suspended solids from 55 mg/L to 3 mg/L during a filter run of 4 hours. Assume this material is captured in the first 6 in. of sand, where the sand grain size is adequately described as 0.8 mm. What is the head loss of the dirty filter?

The total water through the filter is

$$4 \frac{gal}{min \cdot ft^2} \times (20 \text{ ft} \times 20 \text{ ft}) \times 4 \text{ hr} \times 60 \frac{min}{hr} = 384{,}000 \text{ gal}$$

$$384{,}000 \text{ gal} \times 3.785 \frac{L}{gal} = 1.45 \times 10^6 \text{ L}$$

Solids removal is $55 - 3 = 52 \text{ mg/L}$

$$52 \text{ mg/L} \times 1.45 \times 10^6 \text{ L} = 75.4 \times 10^6 \text{ mg}$$

Effective filter volume $= 20 \text{ ft} \times 20 \text{ ft} \times \frac{1}{2} \text{ ft} = 200 \text{ ft}^3$

$$200 \text{ ft}^3 \times 0.02832 \frac{m^3}{ft^3} \times 10^6 \frac{cm^3}{m^3} = 5.65 \times 10^6 \text{ cm}^3$$

Suspended solids removed per unit filter volume is

$$\frac{75.4 \times 10^6 \text{ mg}}{5.65 \times 10^6 \text{ cm}^3} = 13.3 \text{ mg/cm}^3$$

From Figure 6–8, the $h_L \cong 5 \text{ m} = 16.45$ feet, so the total head loss of the dirty filter is

$$5.78 \text{ feet} + 16.45 \text{ feet} = 22.23 \text{ feet}$$

Note that in the preceding example, the head loss of the dirty filter exceeds the total head available (if we assume that the filter is perhaps 10 feet deep). Obviously, a very dirty filter, as in the above example, has a region of negative pressure in the filter bed. This is not recommended since the negative pressure may cause gases to come out of solution and thus further impede the effectiveness of filtration.

The flow rate in most filters is controlled by a value that allows only a given volume of water to pass through, regardless of the pressure. Such *rate con-*

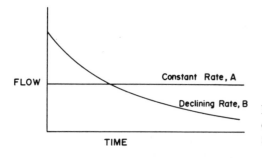

Figure 6–10. Flow through a filter if operated at constant rate or declining rate.

trollers allow the filter to operate at a constant rate, as shown by curve A in Figure 6–10. An alternative method of filter operation is to allow the water to flow through at a rate governed by the head loss, as shown by curve B in Figure 6–10. The relative advantages of these two methods of filter operation are still being debated.

DISINFECTION

After filtration, the finished water is often disinfected with chlorine (Step 5 in Figure 6–1). Disinfection is the process of killing the remaining microorganisms in the water, some of which may be pathogenic. Chlorine from bottles or drums is fed in correct proportions to the water to obtain a desired level of chlorine in the finished water. When chlorine comes in contact with any organic manner, including microorganisms, it oxidizes this material and is in turn reduced to inactive chlorides.

Chlorine gas is very soluble in water and rapidly forms hypochlorous acid through hydrolysis:

$$Cl_2 + H_2O \rightleftharpoons HOCl + H^+ + Cl^-$$

The hypochlorous acid itself ionizes:

$$HOCl \rightleftharpoons OCl^- + H^+$$

At temperatures found in water supply systems, the hydrolysis of chlorine is generally complete in a matter of seconds, whereas the ionization of hypochlorous acid is instantaneous. Both HOCl and OCl$^-$ are effective disinfectants and are called free available chlorine in water. This free available chlorine kills pathogenic bacteria and thus disinfects the water.

Many water plant operators prefer to maintain a residual of chlorine in the water, thus ensuring that if within the distribution system organic matter, such as bacteria, enters the water there is sufficient chlorine present to eliminate this potential health hazard.

It must be noted that the secondary effects of chlorine addition are poorly understood. Chlorine may combine with synthetic organics in the water to possibly produce dangerous halogenated compounds, and the long-term health effects of such compounds as trihalomethane and chloroform are unknown.

In addition to chlorine, a growing number of municipalities now have fluoride added to the water. At the proper concentration fluoride will do much to prevent dental decay in children and young adults.

From the clear well (step 6 in Figure 6–1) the water is pumped to the distribution system. This is a closed network of pipes, all under pressure. In most cases water is pumped to an elevated storage tank that not only serves to equalize pressures but provides storage for fires and other emergencies as well.

CONCLUSION

Water treatment is often necessary if surface water supplies, and sometimes groundwater supplies, are to be available for human utilization. Because the vast majority of cities across the nation use one water distribution system for households, industries, and fire control, large quantities of water often must be available to satisfy the "highest use," which is usually drinking water. Does it make sense to go to the trouble and expense of producing high-quality water, only to use it for lawn sprinkling? This is not an easy question, and the problems of future water supply have prompted the serious consideration of dual water supplies: one of high quality and one of a lower quality, perhaps reclaimed from wastewater. Many engineers are convinced that the next major environmental engineering concern will be the availability and production of potable water. The job, therefore, is far from done.

PROBLEMS

6.1 A water treatment system must be able to deliver 15 mgd of water to a city of 150,000 people. Estimate: (1) the diameter of three equally sized mixing basins 10 feet deep, with detention times of 2 min, (2) length, width, and corresponding surface area of three flocculator basins that are 10 feet deep, (3) surface area of three settling basins that are 10 feet deep with 2-hr detention periods, and (4) the required area of each of 15 rapid sand filters rated at 2 gpm/ft^2.

6.2 An engineer suggests the following design parameters for a city's proposed rapid sand filter: flow rate = 0.6 m^3/sec, loading rate to filter = 125.0 m^3/day · m^2. How much surface area is required for the filter? Select the number of equally sized filters, and size these filters assuming a width-to-length ratio of 1.0 to 2.5 with a maximum surface area of each filter tank of 75 m^2.

6.3 A flocculator is designed to treat 25 mgd. The flocculator basin is 100 feet long, 50 feet wide, and 20 feet deep and is equipped with 12-in. paddles (48 feet long). The paddles are attached to four horizontal shafts, two paddles per shaft, their center line is 8.0 feet from the shaft, and they rotate at 2.5 rpm. Assume the velocity of the water is 25 percent of the velocity of the paddles and the water temperature is 50°F.

Calculate: (1) the time of flocculation, (2) the velocity differential between the water and the paddle, (3) the hydraulic power (and subsequent energy consumption) if the drag coefficient is 1.7 and the velocity differential between water and paddles is 75 percent of the linear velocity of the blades, and (4) the value of G and Ḡt̄.

6.4 Discuss utilization and disposal options for water treatment sludges collected in the settling tanks following flocculation basins.

LIST OF SYMBOLS

ϕ = shape factor
ρ = fluid density, kg/m^3 (lb · sec^2/ft^4)
μ = fluid viscosity (lb · sec/ft^2)
$\forall$ = tank volume, m^3 (ft^3)
$\forall_p$ = volume occupied by each particle, m^3 (ft^3)
A = paddle area, m^2 (ft^2)
A' = surface area of the sand bed, m^2 (ft^2)
A_p = surface area of the sand particles, m^2 (ft^2)
C_D = drag coefficient
D = diameter of "pipe," m (ft)
d = diameter of sand particle, m (ft)
d' = geometric mean diameter between sieve sizes, m (ft)
e = porosity fraction of open spaces in the sand
f = friction factor
G = velocity gradient, sec^{-1}
g = acceleration due to gravity, m/sec^2 (ft/sec^2)
H_L = total head loss through a filter, m (ft)
h'_L = clear water head loss through a filter, m (ft)
L = depth of filter, m (ft)
n = revolutions per minute
$(h_{L_i})_t$ = head loss in ith layer of media at time t
P = power, N/sec (ft · lb/sec)
Q = flow rate, m^3/sec (gal/min)
R = hydraulic radius, m (ft)
$\mathbf{R}$ = Reynolds number
r = radius, m (ft)
t = time of flocculation, min
v = velocity of the paddle relative to the fluid, m/sec (ft/sec)
v = velocity of water through sand bed, m/sec (ft/sec)
v_a = velocity of water approaching sand, m/sec (ft/sec)
x = weight fraction of particles retained between two sieves

Chapter 7
Collection of Wastewater

"The Shambles" is both a street and an area in London, and during the eighteenth and nineteenth centuries it was a highly commercialized area, with meat packing as a major industry. The butchers in those days would throw all of their waste into the street where it was washed away by rainwater into drainage ditches. The condition of the area was so bad that it contributed its name to the English language.

In old cities, drainage ditches like the ones at the Shambles were constructed for the sole purpose of moving stormwater out of the cities. In fact, it was illegal in London to discard human excrement into these ditches. Eventually, these ditches were covered over and became what we now know as *storm sewers*.

As water supplies developed and the use of the indoor water closet increased, the need for transporting domestic wastewaters, called sanitary wastes, became obvious. This was accomplished in one of two ways: (1) discharge of the sanitary waste into the storm sewers, which then carried both sanitary wastes and stormwater and were known as *combined sewers*, and (2) construction of a new system of underground pipes for removing the wastewater, which became known as *sanitary sewers*.

Newer cities and more recently built (post-1900) parts of older cities almost all have separate sewers for sanitary wastes and stormwater. In this chapter, storm sewer design is not covered in detail, since this is discussed in Chapter 10. Emphasis here is on estimating the quantities of domestic and industrial wastewaters, and in the design of the sewerage systems to handle these flows.

ESTIMATING WASTEWATER QUANTITIES

The term *sewage* is used here to mean only domestic wastewater. In addition to sewage, however, sewers also must carry

- industrial wastes
- infiltration
- inflow

123

The quantity of industrial wastes may usually be established by water use records. Alternatively, the flows may be measured in manholes that serve only a specific industry, using a small flow meter in a manhole. Typically, a parshall flume is used, and the flow rate is calculated as a direct proportion of the flow depth. Industrial flows often vary considerably throughout the day, and continuous recording is mandatory.

Infiltration is the flow of groundwater into sanitary sewers. Sewers are often placed under the groundwater table, and any cracks in the pipes will allow water to seep in. Infiltration is the least for new, well-constructed sewers, and can go as high as 500 m³/km-day (200,000 gal/mi-day). Commonly, for older systems, 700 m³/km-day (300,000 gal/mi-day) is used in estimating infiltration. This flow is of course detrimental since the extra volume of water must go through the sewers and the wastewater treatment plant. It thus makes sense to reduce this as much as possible by maintaining and repairing sewers, and keeping sewerage easements clear of large trees that could send roots into the sewers and cause severe damage.

The third source of flow in sanitary sewers is called *inflow*, and represents stormwater that is collected unintentionally by the sanitary sewers. A common source of inflow is a perforated manhole cover placed in a depression, so that stormwater flows into the manhole. Sewers laid next to creeks and drainageways that rise up higher than the manhole elevation, or where the manhole is broken, are also a major source. Last, illegal connections to sanitary sewers, such as roofdrains, may substantially increase the wet weather flow over the dry weather flow. Commonly, the ratio of dry weather flow to wet weather flow is between 1:1.2 and 1:4.

Domestic wastewater flows vary with season, day of the week, and hour of the day. Figure 7–1 shows a typical daily flow for a residential area.

The three flows of concern when designing sewers are the average flow, the peak or maximum flow, and the extreme minimum. The ratio of average flow to both the maximum and minimum flows is a function of the total flow, since a higher average daily discharge implies a larger community in which the extremes

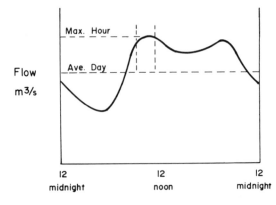

Figure 7–1. Typical dry-weather wastewater flow for a residential area.

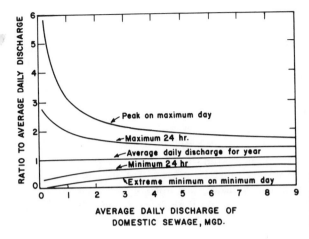

Figure 7–2. Relation of the average daily flow of domestic wastewater to the extremes of flow.

are evened out. Figure 7–2 is a plot showing commonly experienced ratios of average to the extremes as a function of the average daily discharge.

SYSTEM LAYOUT

Sewers that collect wastewater from residences and industrial establishments almost always operate as open channels or gravity flow conduits. Pressure sewers are used in a few places, but these are expensive to maintain and are

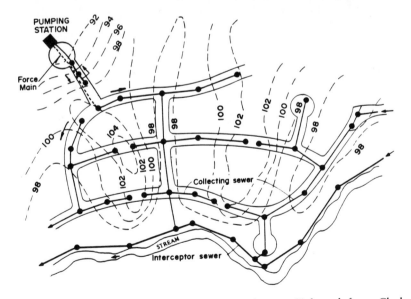

Figure 7–3. Typical wastewater collection system layout. [Adapted from Clark, J., W. Viessman, and M. Hammer, *Water Supply and Sewerage* (New York: IEP, 1977).]

useful only when either there are severe restrictions on water use or the terrain is such that gravity flow conduits cannot be efficiently constructed.

A typical system for a residential area is shown in Figure 7–3. Building connections are usually made with clay or plastic pipe, 6 in. in diameter, to the *collecting sewers* that commonly run under the street. Collecting sewers are sized to carry the maximum anticipated peak flows without surcharging (filling up) and are commonly made of clay, asbestos, cement, concrete or cast iron pipe. They discharge in turn into intercepting sewers, known colloquially as *interceptors*, which collect large areas and discharge finally into the wastewater treatment plant.

Collecting and intercepting sewers must be placed at a sufficient grade to allow for adequate velocity during low flows, but not so great as to promote excessively high velocities when the flows are at their maximum. In addition, sewers must have manholes, commonly every 120 to 180 m (400 to 600 feet) to

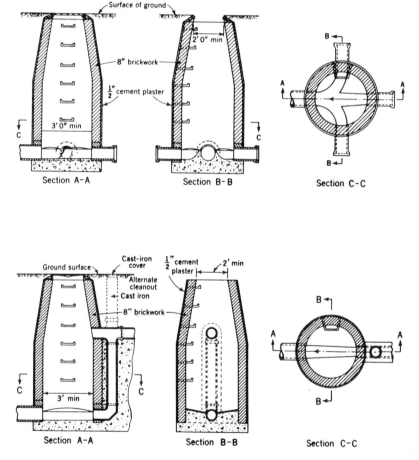

Figure 7–4. Typical manholes used for collecting sewers. [Courtesy of ASCE.]

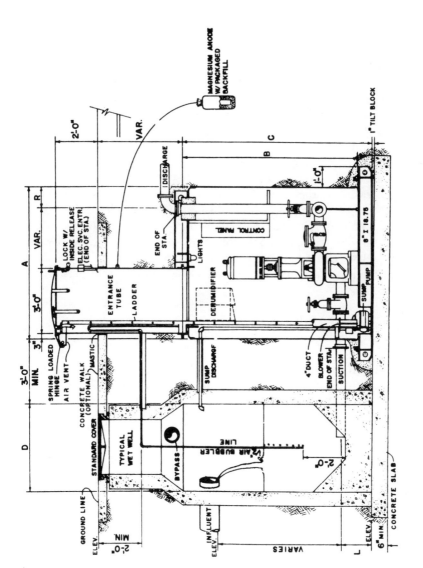

Figure 7–5.   Typical pumping station for domestic wastewater. [Courtesy of Gorman-Rupp.]

facilitate cleaning and repair. Manholes are also necessary whenever the sewer changes grade (slope), size, or direction. Typical manholes are shown in Figure 7–4.

In some cases it becomes either impossible or uneconomical to use gravity flow, and the wastewater must be pumped. A typical packaged pumping station is shown in Figure 7–5, and its use is indicated on the typical system layout in Figure 7–3.

SEWER HYDRAULICS

The design of sewers begins by selecting a reasonable layout and establishing the expected flows within each pipe linking the manholes. For large systems, this often involves the use of economic analysis to determine exactly what routing provides the optimal system. For most smaller systems, this is an unnecessary refinement, and a reasonable system may be drawn by eye.

The average discharge is estimated on the basis of the population served in the drainage area, and the maximum and minimum flows are calculated as shown in Figure 7–2. Once this is done, the design is a search for the right pipe diameter and grade (slope) that will allow the minimum flow to exceed a velocity necessary for the conveyance of solids, while keeping the velocity at maximum flow less than a limit at which undue erosion and structural damage can occur to the pipes. Commonly, the velocity should be held between these limits:

- minimum—0.6 /sec (~2 ft/sec)
- maximum—3.0 m/sec (~10 ft/sec).

The velocity in sewers is usually calculated by using the Manning equation, derived originally from the Chezy open channel flow equation

$$v = c\sqrt{rs}$$

where v = velocity of the flow, m/sec
   r = hydraulic radius of the pipe, or the area divided by the wetted perimeter, m
   s = slope of the pipe
   c = the Chezy coefficient

For circular pipes, the Manning equation is derived by setting

$$c = \frac{k}{n} r^{1/6}$$

where n = roughness factor
   k = constant

Table 7–1.   Manning Roughness Coefficient, n

Type of Channel, Closed Conduits	Roughness Coefficient, n
Cast iron	0.013
Concrete	
Straight	0.011
With bends	0.013
Unfinished	0.014
Clay, vitrified	0.012
Corrugated metal	0.024
Brickwork	0.013
Sanitary sewers, coated with slime	0.013

so that

$$v = \frac{k}{n} r^{2/3} s^{1/2}$$

If v is in feet per second and r is in feet, k = 1.486. In metric units with v as meters per second and r as meters, k = 1.0. The slope s is, of course, dimensionless, calculated as the fall over distance. The n term is dependent on the pipe material, with n increasing as pipes get rougher. Table 7–1 is a listing of n values for some types of sewer pipe.

Example 7.1
An 8-in-diameter cast iron sewer is to be set at a grade (slope) of 1-m fall per 500-m length. What will be the flow in this sewer when flowing full?
Using common units, and noting that n = 0.013 from Table 7–1,

$$V = \frac{1.486}{0.013} \left[\frac{\pi(8/12)^2}{4} \div (\pi 8/12) \right]^{2/3} \left[\frac{1}{500} \right]^{1/2}$$

$$V = 1.54 \text{ ft/sec}$$

and since the area is $[\pi(8/12)^2]/4 = 0.35 \text{ ft}^2$,

$$Q = AV = (0.35)(1.54) = 0.54 \text{ cfs}$$

For circular conduits flowing full, the Manning formula may be rearranged as

$$Q = \frac{0.00061}{n} D^{8/3} s^{1/2}$$

where   Q = discharge, cfs
D = pipe diameter, in.

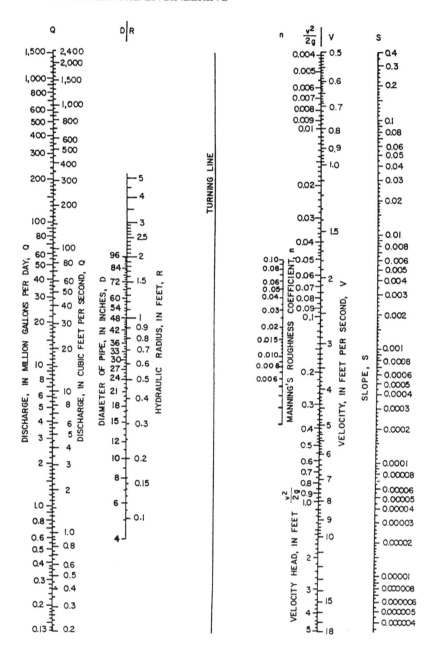

Figure 7–6. Nomograph for the solution of Manning's equation, common units.

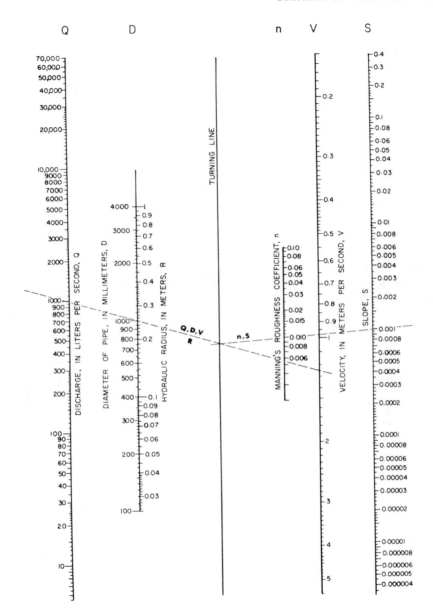

Figure 7–7. Nomograph for the solution of Manning's equation, metric units.

This calculation may be simplified by using a nomograph that represents a graphical solution of the Manning equation. Figures 7–6 and 7–7 are such nomographs in both common and metric units.

It should be emphasized that the Manning equation is used to calculate the velocities in sewers flowing full, hence the maximum expected flow in the sewer is used. For flows less than the maximum, the velocity is calculated using the *hydraulic elements* graph shown as Figure 7–8. In this chart, the hydraulic radius (area per wetted perimeter) is calculated for various ratios of the depth partially full (d) over the depth full (D), the latter of course being the inside diameter of the pipe. The velocity is calculated by using the Manning equation for any depth d, and the discharge is obtained from multiplying the velocity times the area. From experimental evidence, it has been found that the friction factor n also varies slightly with the depth (as noted in Figure 7–8), but this is often ignored in the calculations.

It should be noted that the maximum flow in a given sewer occurs not when it is full, but when d/D is about 0.95. This is because the additional flow area gained by increasing the d/D from 0.95 to 1.00 is very small compared with the large additional pipe surface area that the flow experiences.

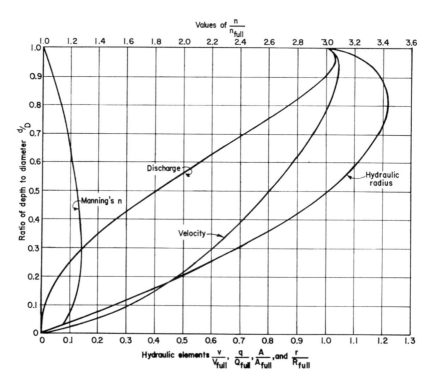

Figure 7–8. Hydraulic elements chart for open channel flow.

Example 7.2

From Example 7.1, it was found that the flow carried by an 8-in. cast iron pipe flowing full at a grade of 1/500 is 0.54 cfs with a velocity of 1.54 ft/sec. What will be the velocity in the pipe when the flow is only 0.1 cfs?

$$\frac{q}{Q} = \frac{0.1}{0.54} = 0.185$$

From the hydraulic elements chart

$$\frac{d}{D} = 0.33$$

and d = 0.33 (8 in.) = 2.64 in.

$$\frac{v}{V} = 0.64$$

and v = 0.64 (1.54 ft/sec) = 0.98 ft/sec.

In cases when the desired slope between manholes is not sufficient to maintain adequate velocities at minimum flow, additional slope must be provided, or a larger pipe diameter used. On the other hand, if the slope of the pipe is too steep, it must be reduced, and often a drop manhole is used, which is in effect a method of wasting energy (see Figure 7–4). Standard American convention requires that a head loss of 0.1 foot (0.06 m) be allowed for each lateral. The invert elevation of the pipe leaving the manhole is thus at least 0.1 foot lower than the invert elevation of the pipe coming in.

Example 7.3

The system shown in Figure 7–9 is to be designed given the following flows: maximum flow = 3.2 mgd, minimum flow = 0.2 mgd, and the minimum and maximum allowable velocities are 2 and 12 ft/sec, respectively. All manholes should be about 10 feet deep, and there is no additional flow between manhole 1 and manhole 4. Design acceptable invert elevations for this system. [Note: The nomograph (Figure 7–6) is used in this solution.]

From 1 to 2: The street slopes at 2/100, so choose the slope of the sewer, s = 0.02. Assume n = 0.013 and try D = 12 in. From the nomograph, connecting n = 0.013 and s = 0.02, hitting the turning line, then connecting the point or the turning line with D = 12 in., read Q = 3/2 mgd and V = 6.1 ft/sec. This is acceptable as the sewer flowing full. To check for minimum velocity, q/Q = 0.2/3.2 = 0.063, and from the hydraulic elements chart (Figure 7–8), read v/V = 0.48, and v = 0.48 (6.1 ft/sec) = 2.9 ft/sec. The downstream invert elevation of manhole 1 is ground elevation minus 10 feet, or 62.0. The upstream

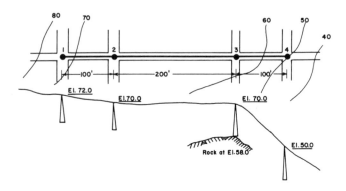

Figure 7–9.   Sewer layout, for Example 7.3.

invert elevation of manhole 2 is thus $62.0 - 2.0 = 60.0$ feet. Allowing 0.1 foot for head loss in the manhole, the downstream invert elevation is 59.9 feet.

From 2 to 3: The slope here will be a problem because of rock. Try a larger pipe, $D = 18$ in. From the nomograph, $V = 2.75$ ft/sec for $Q = 3.2$ mgd. From the hydraulic elements chart, $v/V = 0.48$ and $v = 0.48$ (2.75 ft/sec) = 1.32 ft/sec. Too low. So even with a slope of 0.002 (from the nomograph), the velocity at minimum flow is too low. Try $s = 0.005$ and $D = 18$ in., reading $Q = 4.7$ mgd and $V = 4.1$ ft/sec. From hydraulic elements, $v = 1.9$ ft/sec, which is close enough. Use $D = 18$ in., and $s = 0.005$. Thus the upstream invert elevation of manhole 3 is $59.9 - (0.005)(200) = 58.9$ feet and the downstream is at 58.8 feet, still well above the rock.

From 3 to 4: This street obviously has too much slope. Try using $D = 12$ in., and $s = (58.8 - 40)/100 = 0.188$ (40 feet is the desired invert elevation of manhole 4). Read $Q = 8.5$ mgd and $V = 19$ ft/sec. But only 3.2 mgd is required at maximum flow, hence $q/Q = 3.2/8.5 = 0.38$ and $v = 0.78 \times 18$ ft/sec = 14.8 ft/sec. This is too high. Use a drop manhole, with the invert at, say,

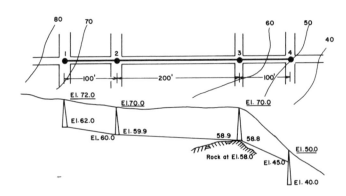

Figure 7–10.   Final design, for Example 7.3.

elevation 45.0. In this case s = 0.138, and Q = 7.3 mgd, V = 14.5 ft/sec, q/Q = 3.2/7.3 = 0.44, and v = 0.85 × 14.5 ft/sec = 12.3 ft/sec. Close enough! The upstream invert elevation of manhole 4 is thus at 45.0 feet and the downstream invert elevation can be at, say, 40.0 feet. There is, of course, little need to check the minimum velocity in this case. The final design is shown in Figure 7–10.

Often it is convenient to organize sewer calculations according to a table such as Figure 7–11. This is a slightly simplified version of a table suggested by the American Society of Civil Engineers and the Water Pollution Control Federation *Manual of Practice on Sanitary Sewer Design.*[1] This volume should be consulted for the details of accepted engineering practice in the design of sewers.

LINE NO.	LOCATION	MAN-HOLE NO.		AREA ACRES		MAX. DISCHARGE TOTAL			MIN. DISCHARGE TOTAL			SLOPE OF SEWER %	DIAMETER in.	CAPACITY FULL cfs	VELOCITY FULL ft/sec	MIN. VELOCITY ft/sec	MAX. VELOCITY ft/sec	MAX. DEPTH ft	MAX. VEL. HEAD ft	MAX. ENERGY ft	MANHOLE LOSS ft	M.HOLE INVERT DROP ft.	FALL IN SEWER ft	SEWER INVERT ELEVATION		GROUND SURFACE ELEVATION	
		FROM	TO	LENGTH ft.	INCREMENT	TOTAL	INFILTRATION	SEWAGE mgd	SEWAGE INFIL-TRATION mgd		cfs	INFILTRATION	SEWAGE mgd	SEWAGE INFIL-TRATION mgd		cfs																UPPER END	LOWER END	UPPER END	LOWER END

Figure 7–11.   Work sheet for the design of sewers.

CONCLUSION

The methods discussed in this chapter are crude hand calculations, and few modern consulting engineers who design sewerage systems will use such cumbersome methods. Computer programs that not only solve for the hydraulic parameters but evaluate the economic alternatives (e.g., a deeper cut or bigger pipe) are widely used. Elegant programs, however, are only useful if the design process is well understood.

PROBLEMS

7.1 A vitrified clay pipe sewer is to discharge 4 cfs when laid on a grade of 0.0016 and flowing half full. What is the required diameter?

7.2 An 18-in. sewer, n = 0.013, is 15,000 feet long and laid on a uniform grade. Difference in elevation of the two ends is 4.8 feet. Find velocity and discharge when sewer is flowing 0.6 full.

7.3 A 48-in. circular sewer is required to discharge 100 cfs when full. What is the required grade, according to the Manning formula, if n = 0.015?

7.4 A 12-in. concrete sewer, flowing just full, is laid on a grade of 0.00405 ft/ft. Find the velocity and rate of discharge.

7.5 A 12-in. sewer, n = 0.013, is laid on a grade of 3.0 ft/1,000 ft. Find the velocity and discharge when the sewer is flowing 0.4 full.

LIST OF SYMBOLS

c = Chezy coefficient
D = diameter of pipe, m (in.)
d = depth of flow in a pipe, m (in.)
n = Manning roughness coefficient
Q = flow rate, full pipe, m^3/sec (gal/min)
q = flow rate, partially full pipe, m^3/sec (gal/min)
r = hydraulic radius, m (ft)
s = slope
V = velocity, full pipe, m/sec (ft/sec)
v = velocity in a partially full pipe, m/sec (ft/sec)

REFERENCE

1. American Society of Civil Engineers and Water Pollution Control Federation. *Design and Construction of Sanitary and Storm Sewers* (New York, 1969).

Chapter 8

Wastewater Treatment

As civilization developed and cities grew, domestic sewage and industrial wastes were eventually discharged into drainage ditches and sewers, and the entire contents were commonly emptied into the nearest watercourse. For major cities, this discharge was often sufficient to destroy even a large body of water. As Samuel Taylor Coleridge observed:

> In Köln, a town of monks and bones
> And pavements fanged with murderous stones
> And rags, and bags, and hideous wenches;
> I counted two and seventy stenches,
> All well defined, and several stinks!
>
> Ye Nymphs that reign o'er sewers and sinks,
> The river Rhine, it is well known,
> Doth wash your city of Köln;
> But tell me Nymphs! What power divine
> Shall hence forth wash the river Rhine?

During the nineteenth century, the River Thames was so grossly polluted that the House of Commons had to have rags soaked in lye stuffed into the cracks in the windows of Parliament to reduce the stench.

Beginning with the pioneering work in the United States and England, sanitary engineering technology eventually developed to the point where it became economically, socially, and politically feasible to treat the wastewater so as to reduce its adverse impact on watercourses. In this chapter, this technology is reviewed, beginning with the simplest (and earliest) treatment systems, and concluding with a description of the most advanced systems in use today. The discussion begins, however, by reviewing the characteristics of wastewaters that make their disposal difficult, and showing why wastewater disposal cannot always be done onsite, and sewers and centralized treatment plants become necessary.

WASTEWATER CHARACTERISTICS

The discharges in a sanitary sewerage system are composed of domestic wastewater, industrial discharge, and infiltration. Infiltration, of course, is only a problem in that it adds to the total volume of wastewater and seldom is it directly a cause of concern in wastewater disposal. Industrial wastes are of course another matter. These discharges vary widely with the size and type of industry, and the amount of treatment applied by the industry before discharge into public sewers. In the United States, the trend has been toward a higher degree of pretreatment (see Chapter 11), prompted in part by regulations limiting discharges and by the imposition of local sewer surcharges. The latter are charges levied by a community to help pay for the extra treatment that an unusual discharge would require. The operation of sewer surcharges is illustrated by the example below.

Example 8.1

A community has a sewer use charge, based on water use, of $0.20/m^3$. In addition, for discharges that exceed the limits of BOD = 250 mg/L and suspended solids (SS) = 300 mg/L, the discharger is to pay $0.50/kg of BOD and $1.00/kg of SS. A chicken-packing plant uses 2,000 m^3 of water per day and discharges a wastewater with a BOD = 480 mg/L and SS = 1,530 mg/L. What is its daily wastewater disposal bill?

The sewer use charge is 2,000 m^3/day × $0.20/m^3$ = $400/day. The BOD discharged is 480 mg/L × 2,000 m^3/day × 1,000 L/m^3 = 960 × 10^6 mg/day = 960 kg/day. The allowable BOD is 250 mg/L × 2,000 m^3/day × 1,000 L/m^3 × 10^{-6} kg/mg = 500 kg/day. Thus, the extra BOD is 960 − 500 = 460 kg/day, and the surcharge is 460 kg/day × $0.5/kg = $230/day. Similarly, the allowable SS = 300 × 2,000 × 1,000 × 10^{-6} = 600 kg/day, while the SS discharged is 1,530 × 2,000 × 1,000 × 10^{-6} = 3,060 kg/day. The surcharge for SS is thus (3060 − 600) × 1.00 = $2,460/day. The total charge to the industry is $400 + $230 + $2,460 = $3,090/day.

Commonly, the problems with industrial discharges are not BOD and SS, both of which may be readily reduced in a wastewater treatment plant, but chemicals such as toxic metals, radioactive materials, refractory organics, etc. Typically, local communities place tight restrictions on such discharges and thus force the industries to pretreat the wastewater before discharging it to the public sewers.

The third component of municipal wastewaters, domestic sewage, tends to vary substantially over time and from one community to the next. For example, the BOD and SS for a small community vary as shown in Figure 8–1. During early morning hours, little domestic wastewater is discharged, and a substantial part of the dry weather flow is infiltration, thus resulting in weak sewage.

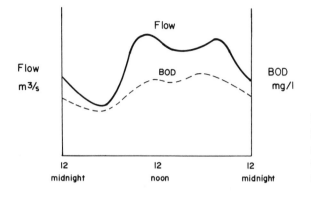

Figure 8–1. Daily variations in flow and biochemical oxygen demand (BOD) in a domestic wastewater from a small community.

Table 8–1. Characteristics of a Typical Domestic Wastewater

Parameter	Typical Value for Domestic Sewage
BOD	250 mg/L
SS	220 mg/L
Phosphorus	8 mg/L
Organic and ammonia nitrogen	40 mg/L
pH	6.8
Chemical oxygen demand	500 mg/L
Total solids	720 mg/L

On the average, however, it is instructive to consider "typical" values for the characteristics of domestic wastewater. Some of these are shown in Table 8–1.

ONSITE WASTEWATER DISPOSAL

Environmental engineers have been severely (and sometimes rightly) criticized for having a "sewer syndrome"—they want to collect all wastewater and provide treatment at a central location. Often this approach does not make much sense.

Consider the situation depicted in Figure 8–2, in which two wastewater treatment options are shown—a centralized treatment plant, and several smaller plants—all discharging their effluents into the same river. The single large plant obviously must provide extremely good treatment to attain acceptable dissolved oxygen (DO) levels downstream. On the other hand, the smaller plants could take advantage of the assimilative capacity of the river, and would not necessarily have to provide the same high degree of treatment. The logical extension of this idea is to not have *any* treatment plants at all, but to dispose of the wastewater onsite, with each house or building having its own treatment system.

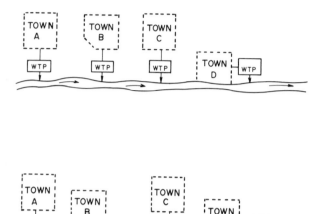

Figure 8–2.   Two wastewater treatment options for several small communities.

The original onsite system of course is the pit privy, glorified in song and fable.* The privy, still used in camps and other temporary residences, consists simply of a deep (perhaps 2 m or 6 feet deep) pit into which human excrement is deposited. When a pit fills up, it is covered and a new pit is dug.

A logical extension of the pit privy idea is a composting toilet that accepts not only human wastes, but food waste as well and produces a useful compost. With such a system, wastewaters from other sources such as washing and bathing are discharged separately.

By far the greatest number of households with onsite disposal systems use a form of the *septic tank* and *tile field*.

As shown in Figure 8–3, a septic tank consists of a concrete box that removes the solids and promotes partial decomposition. The solid particles settle out and eventually fill the tank, thus necessitating periodic cleaning. The water overflows into a tile drain field that promotes the seepage of water discharged.

---

* Probably the most literary work on the theme of the outhouse was by James Whitcomb Riley, who penned "The Passing of the Backhouse." A few lines from this epic:

> "But when the crust was on the snow
>   and the sullen skies were grey,
> In sooth the building was no place where
>   one could wish to stay.
> We did our duties promptly,
>   there one purpose swayed the mind.
> We tarried not nor lingered long
>   on what we left behind,
> The torture of that icy seat
>   would make a Spartan sob. . ."

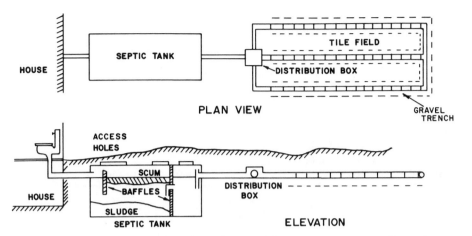

Figure 8–3. Septic tank and tile field used for onsite wastewater disposal.

A tile field consists of pipe laid in about a 1-m (3-feet)-deep trench, end on end but with short gaps between the pipe. The effluent from the septic tank flows into the tile field pipes and seeps into ground through these gaps. Alternatively, seepage pits consisting of gravel and sand may be used for promoting the adsorption of the effluent into the ground.

The most important consideration in designing a septic tank and tile field system is the ability of the ground to absorb the effluent. Percolation tests, used to measure the suitability of ground for the fields, are conducted in the following way:

1. Dig a hole about 6–12 in.^{2} and as deep as the proposed tile field trench.
2. Scratch the soil to remove smeared surfaces and provide a more natural soil interface and put some gravel in the bottom of the pit.
3. Fill the pit with water and let it stand overnight.
4. Next day fill the pit with water to 6 in. above the gravel, and measure the drop in water level in 30 minutes.
5. Calculate the percolation rate as inches per minute.

The U.S. Public Health Service and all county and local departments of health have established guidelines for sizing the tile fields or seepage pits. Typical standards are shown in Table 8–2.

Many areas in the United States have soils that percolate poorly, and septic tank/tile field systems are inappropriate. Several options are now available for onsite wasterwater disposal, one of which is shown in Figure 8–4. Since 1970 new onsite disposal has been discouraged and, in some regions, prohibited. Planning for future growth of communities and the legislative demands for assessment of environmental impact are making onsite disposal obsolete.

Table 8–2.   Adsorption Area Requirements for Private Residences

Percolation Rate (in./min)	Required Area of Adsorption Field, per Bedroom (ft^2)
Greater than 1	70
Between 1 and 0.5	85
Between 0.5 and 0.2	125
Between 0.2 and 0.07	190
Between 0.07 and 0.03	250
Less than 0.03	unsuitable ground

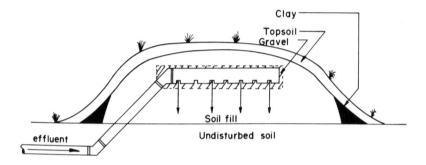

Figure 8–4.   Alternative onsite disposal system.

In urbanized areas, it has been several centuries since there was enough land available for onsite treatment and percolation. Up until the nineteenth century, this problem was solved by constructing large cesspools or holding basins for the wastewater, which had to be pumped out as they filled up. The public health problems with this system were enormous, as discussed previously (see the Broad Street pump incident, Chapter 3). An obviously better way to move the human waste out of a congested community was to use water as a carrier.

The *water closet*, as it is known in Europe, thus became a standard trapping of our urban society. Actually, the invention of the flushed water closet is mired in controversy. Some credit John Bramah[1] as being the inventor in 1778, while others present convincing arguments that Sir John Harington[2] was the inventor, in 1596!* The latter argument is strengthened by Sir John's original description

---

* Sir John, a courtier-poet, installed his invention in his country house at Kelson, near Bath. Queen Elizabeth had one fitted soon afterward at Richmond Palace. The two books that were written about this innovation bear the strange titles, *A New Discourse on a Stale Subject. Called the Metamorphosis of Ajax, & An Anatomie of the Metamorphased Ajax.* Ajax is a play on words; 'a jakes' is a closet.

of the device, although there is no record of him donating his name to the invention. The first recorded use of that euphemism is found in the regulation at Harvard University, where in 1735 it was decreed that "No Freshman shall go to the Fellows' John."

The wide use of waterborne wastewater disposal, however, caused another problem. Now the wastes of a community were all concentrated in one place, and a major effort was necessary to clean up the mess. This demand fostered what is known as centralized treatment.

CENTRAL WASTEWATER TREATMENT

The objective of wastewater treatment is to reduce the concentrations of specific pollutants to the level at which the discharge of the effluent will not adversely affect the environment. Note that there are two important aspects of this objective. First, wastewater is treated only to reduce the concentrations of selected constituents that would cause harm to the environment or pose a health hazard. Not everything in wastewater is troublesome, and thus is not removed. Second, the reduction of these constituents is only to some required level. It is obviously technically possible to produce distilled and deionized H_2O from wastewater, but this is not necessary, and can in fact be detrimental to the watercourse. Fish and other aquatic organisms cannot survive in distilled water.

For any given wastewater in a specific location, the *degree* and *type* of treatment therefore are variables that require engineering decisions.

Often, the degree of treatment is dictated by the assimilative capacity of the recipient. The procedure by which DO sag curves are drawn is reviewed in Chapter 3. As noted there, the amount of oxygen-demanding materials (BOD) discharged determines how far the DO level will be depressed. If this depression (deficit) is too large, some BOD must be removed in the treatment plant. Thus a certain plant on a given watercourse is required to produce a given quality of effluent. Such an *effluent standard* (discussed more fully in Chapter 11) dictates in large part the type of treatment required.

To facilitate the discussion of wastewater treatment, a "typical wastewater" (Table 8–1) will be assumed, and it will be further assumed that the effluent from this wastewater treatment must meet the following effluent standards:

BOD ≤ 15 mg

SS ≤ 15 mg

P ≤ 1 mg

Obviously, other criteria might, in given situations, be important. For example, nitrogen is thought to be the limiting nutrient in estuarine waters, and if the discharge was to be a brackish estuary, the total nitrogen would be an important parameter. In our simplified case, however, we are concerned only with these three constituents.

To further facilitate discussion, the treatment system selected to achieve these effluent levels consists of four major components:

- pretreatment—screening and removal of large objects such as animal carcasses and chunks of debris.
- primary treatment—the major objectives are removal of nonhomogenizable solids and homogenization. Primary treatment systems are always physical processes, as opposed to biological or chemical.
- secondary treatment—which is designed to remove the demand for oxygen. These processes are commonly biological in nature.
- tertiary treatment—a name applied to any number of polishing or cleanup processes, one of which is the removal of nutrients such as phosphorus. These processes may be physical (e.g., filters), biological (e.g., oxidation ponds), or chemical (precipitation of phosphorus).

PRIMARY TREATMENT

The most objectionable aspect of discharging raw sewage into watercourses is the floating material. It is only logical, therefore, that *screens* were the first form of wastewater treatment used by communities, and even today, screens are used as the first step in treatment plants. Typical screens, shown in Figure 8–5, consist of a series of steel bars that might be about 2.5 cm (1 in.) apart. The purpose of a screen in modern treatment plants is the removal of materials that might damage equipment or hinder further treatment. In some older treatment plants screens are cleaned by hand, but mechanical cleaning equipment is used in almost all new plants. The cleaning rakes are automatically activated when the screens get sufficiently clogged to raise the water level in front of the bars.

In many plants, the next treatment step is a *comminutor*, a circular grinder designed to grind the solids coming through the screen into pieces about 0.3 cm (1/8 in.) or smaller. Many designs are in use; one common design is shown in Figure 8–6.

The third treatment step involves the removal of grit or sand. This is necessary because grit can wear out and damage such equipment as pumps and flow meters. The most common *grit chamber* is simply a wide place in the channel where the flow is slowed down sufficiently to allow the heavy grit to settle out. Sand is about 2.5 times as heavy as most organic solids and thus settles much faster than the light solids. The objective of a grit chamber is to

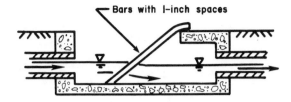

Figure 8–5. Bar screen used in wastewater treatment. The top picture shows a manually cleaned screen, the bottom picture represents a mechanically cleaned screen. [Photo courtesy of Envirex.]

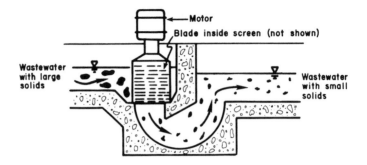

Figure 8–6. A comminutor used to grind up large solids.

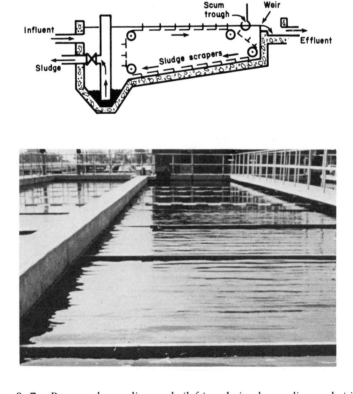

Figure 8–7. Rectangular settling tank (left) and circular settling tank (right).

remove sand and grit without removing the organic material. The latter must be further treated in the plant, but the sand may be dumped as fill without undue odor or other problems.

After the grit chamber most wastewater treatment plants have a *settling tank* (Figure 8–7) to settle out as much of the solid matter as possible. Accordingly, the retention time* is kept long and turbulence is kept to a minimum. The solids settle to the bottom and are removed through a pipe, while the clarified liquid escapes over a V-notch weir, a notched steel plate over which the water flows, promoting equal distribution of liquid discharge all the way around a tank. Settling tanks are also known as *sedimentation tanks* and often as *clarifiers*. The settling tank that follows preliminary treatment such as

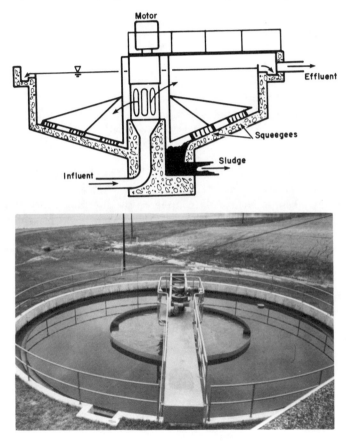

Figure 8–7. (*Continued*)

---

* Retention time is the total time an average slug of water will spend in the tank. This theoretical time is calculated as the time required to fill up a tank. For example, if the volume is 100 m³, and the flow rate is 2 m³/min, the retention time is $100/2 = 50$ minutes. Some authors use *detention time* synonymously with retention time.

screening and grit removal is known as a *primary clarifier*. The solids that drop to the bottom of a primary clarifier are removed as *raw sludge*, a name that does not do justice to the undesirable nature of this stuff.

Raw sludge is generally odoriferous and full of water, two characteristics that make its disposal difficult. It must be both stabilized to retard further decomposition and dewatered for ease of disposal. In addition to the solids from the primary clarifier, solids from other processes must similarly be treated and disposed of. The treatment and disposal of wastewater solids (sludge) is an important part of wastewater treatment and is discussed further in Chapter 9.

Since settling tanks are both ubiquitous and effective, their operation and design are reviewed in greater detail below.

Settling Tank Design and Operation

As noted above, one of the most efficient means of separating solids from the surrounding liquid is to allow the solids to settle out under the force of gravity. As long as the density of the solids exceeds that of the liquid, it should be possible to achieve solid/liquid separation. Unfortunately, this is not always true, since other forces come into play, as we shall see later.

Particles settle in one of three general ways:

- Class I: as discrete particles, unhindered by the container walls or neighboring particles. This is called *discrete particle settling*.
- Class II: as discrete particles but hindered by their neighbors and changing in size owing to particle contact resulting from close proximity. This is called *flocculent settling*.
- Class III: as a mass of particles, with no interparticle movement so that all particles have the same settling velocity, called *thickening*.

The simplest case is Class I, discrete particle settling. The single particle is an infinite fluid, having attained terminal velocity, and has three forces acting on it: drag, buoyancy, and gravity.

If the particle is settling in a fluid at its terminal velocity, these forces balance as

$$F_g = F_D + F_B$$

$$\rho_s g \forall = \rho g \forall + C_D \left(\frac{A \rho v^2}{2} \right)$$

where
F_g = force due to gravity = mg
F_D = drag force = $(C_D \rho v^2 A)/2$
F_B = buoyancy = $\rho \forall g$
m = mass of particle = $\rho_s \forall$
g = gravitational acceleration
ρ_s = particle density, kg/m^3
ρ = fluid density, kg/m^3
C_D = drag coefficient

v = velocity of particle, m/sec
A = projected area of particle, m^2
$\forall$ = volume of particle, m^3

Solving for the velocity, we get

$$v = \left[\frac{2\forall g(\rho_s - \rho)}{C_D \rho A} \right]^{1/2}$$

If we now assume that the particle is a sphere,

$$v = \left[\frac{4}{3} \frac{dg(\rho_s - \rho)}{C_D \rho} \right]^{1/2}$$

which is the well-known Newtonian equation, where d = particle diameter. If, as is the case in most settling in wastewater treatment, the Reynolds number (R) is sufficiently low and a laminar boundary layer is maintained, the drag coefficient can be expressed as

$$C_D = \frac{24}{R}$$

where $R = vd\rho/\mu$
μ = fluid viscosity

If $C_D > 1$, this no longer holds, and the drag coefficient may be approximated as

$$C_D = \frac{24}{R} + \frac{3}{\sqrt{R}} + 0.34$$

For laminar conditions, substituting $C_D = 24/R$ into the Newtonian equation, we get the well-known Stokes equation,

$$v = \frac{d^2 g(\rho_s - \rho)}{18\mu}$$

The velocity of particles may be related to the expected settling tank performance by idealizing a settling tank, as originally proposed by Camp.[3]

As shown in Figure 8–8, a rectangular tank is first divided into four zones: inlet zone, outlet zone, sludge zone, and settling zone. The first two zones are designated for the dampening of currents caused by the influent and effluent. The sludge zone is simply storage space for the settled solids. Settling, then, takes place only in the fourth zone.

Several initial assumptions are required in this analysis:

- Uniform flow occurs within the settling zone.
- All particles entering the sludge zone are removed.
- Particles are evenly distributed in the flow as they enter the settling zone.
- All particles entering the effluent zone escape the tank.

Consider now a particle entering the settling zone at the water surface. This particle has a settling velocity of v_o and a horizontal velocity V such that the

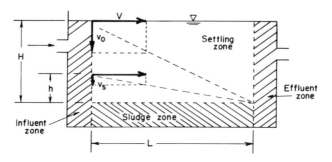

Figure 8–8. Ideal settling tank schematic.

component of the two defines a trajectory as shown in Figure 8–8. In other words, the particle just is removed (it barely squeaks into the sludge zone). Note that had the particle entered the settling zone at any other height, its trajectory would have always carried it into the sludge zone. Particles having this velocity are termed the "critical particles" in that particles with lower settling velocities are not all removed. For example, the particle having velocity v_s, entering the settling zone at the surface, will end up in the effluent zone and escape. If this same particle, however, entered at some height h, it would have just been removed. Any of these particles that happen to enter the settling zone at height h or lower would thus be removed, and those entering above h would not. Since the particles entering the settling zone are equally distributed, the proportion of those particles with a velocity of v_s that are removed is thus equal to h/H, where H is the height of the settling zone. With reference to Figure 8–8, similar triangles yield

$$\frac{v_o}{H} = \frac{v_s}{h} = \frac{V}{L}$$

The time that the critical particle spends in the settling zone is

$$\bar{t} = \frac{L}{V} = \frac{h}{v_s} = \frac{H}{v_o}$$

hence

$$v_o = \frac{H}{\bar{t}}$$

The time $\bar{t}$ is also equal to the hydraulic retention time, or $\forall/Q$, where Q is the flow rate and $\forall$ is the volume of the settling zone, or $\forall = AH$, with A = surface area of settling zone. Thus

$$v_o = \frac{H}{\bar{t}} = \frac{H}{AH/Q} = \frac{Q}{A}$$

This equation represents an important design parameter for settling tanks, and is called the *overflow rate*. Note the units:

$$v_o = \frac{m}{s} = \frac{Q}{A} = \frac{m^3/s}{m^2}$$

Commonly, overflow rate is expressed as "gallons/day/square foot," but it actually is a velocity term, and is in fact equal to the velocity of the critical particle. When the design of a clarifier is specified by overflow rate, what is really defined then is the critical particle.

It should also be noted that when any two of the following—overflow rate, retention time, or depth—are defined, the remaining parameter is also fixed.

Example 8.2

A primary clarifier has an overflow rate of 600 gal/day/ft^2 and a depth of 6 ft. What is its hydraulic retention time?

$$v_o = 600 \; \frac{gal}{d\text{-}ft^2} \times \frac{1}{7.48} \; \frac{ft^3}{gal} = 80.2 \; ft/d$$

$$\bar{t} = \frac{H}{v_o} = \frac{6}{80.2} \; \frac{ft}{ft/d} = 0.0748 \; days = 1.8 \; hours$$

Overflow rate is interesting in that we may obtain a better understanding of settling by looking at individual variables. For example, increasing the flow rate, Q, in a given tank increases the v_o; i.e., the critical velocity increases and thus fewer particles are removed since fewer particles have a $v_s > v_o$.

We would, of course, like to *decrease* v_o, so more particles can be removed. This is done by either reducing Q or increasing A. The latter term may be increased by changing the dimensions of the tank so that the depth is shallow and the length and width are very large. For example, the area may be doubled by taking a 3-m-deep tank, slicing it in half (two 1.5-m slices), and placing them alongside each other. The new shallow tank has the same horizontal velocity, but double the surface area, hence v_o is half of the original value.

Why not then make *very* shallow tanks? Other than the problem of hydraulics, even distribution of flow, and the great expense in concrete and steel,* this seems to be a reasonable conclusion.

We have, however, made a major assumption that isn't really true in wastewater treatment: solids do *not* settle as discrete particles but are indeed influenced by their neighbors, and tend to stick together as they come into contact, thus forming larger particles. This process is commonly called *floc-*

---

* Remember the max-min problems in calculus? What's the shape of a six-sided prism with the smallest surface area?

culent settling, and may be illustrated by setting up a cylinder of dirty water and allowing the particles to settle. If the water is sampled at various elevations, the clarification of the water with time can be observed (Figure 8–9).

If we first introduce particles all with the same settling velocity, we should find that there is perfect clarification as the highest particle drops to the bottom. If the v_s = 120 cm/2 hr, then the curve in Figure 8–9A should result. Everything below the settling curve would be at the original suspended solids concentration (1,000 mg/L), and above the line the water would be clean.

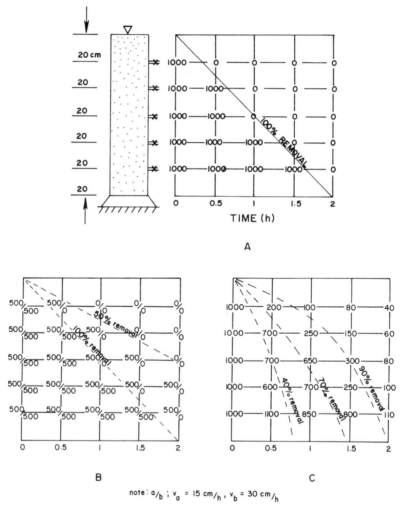

Figure 8–9. Settling test with a column of dirty water and periodic sampling at five ports. The initial suspended solids (SS) concentration (t = 0) is assumed to be 1,000 mg/L.

If now we had two sizes of particles, $v_a = 60/2 = 30$ cm/hr and $v_b = 60$ cm/hr, each at 500 gm/L, the curves would look like Figure 8–9B. The curves are still straight lines, indicating that a mixture of different-sized particles would still produce straight settling lines.

A real-life wastewater, however, produces curves as shown in Figure 8–9C. The curves that show the fraction of solids removed (or reduced) at various sampling ports are curved lines, with their slopes *increasing* with time. This must mean that the increased velocities are due to collisions and the subsequent building of larger particles. It is for this reason (also) that a shallow tank is not as efficient as the ideal tank theory suggests.

Curves such as Figure 8–9C may be used for estimating the removal efficiencies of real slurries. At a retention time of 2 hour, the data in Figure 8–9C show that the entire tank experienced 90 percent removal. The top section, however, is much clearer, since it only has about 40 mg/L of SS, and thus experienced $[(1,000 - 40)/(1,000)]100 = 96$ percent SS removal. The general equation for estimating solids removal, giving credit for the cleaner water on top, is

$$R = P + \sum_{i=1}^{n-1} \left(\frac{h}{H}\right)(P_i - P)$$

where R = overall solids recovery, %
           P = solids recovery at the bottommost section
           P_i = solids recovery in section i
           n = number of sections
           h = height of each section
           H = height of column ($H = nh$)

Example 8.3

A chemical waste at an initial SS concentration of 1,000 mg/L and flow rate of 200 m^3/hr is to be settled in a tank 1.2 m deep, 10 m wide, and 31.4 m long. The results of a laboratory test are shown in Figure 8–9C. Calculate the fraction of solids removed, the overflow rate, and the velocity of the critical particle.

The surface area of the tank is $A = WL = (31.4)(10) = 314$ m^2. The overflow rate is therefore

$$\frac{Q}{A} = \frac{200}{314}$$

$$= 0.614 \text{ m}^3/\text{hr-m}^2$$

The critical velocity is thus $v_o = 0.614$ m/hr, which is not germane to the problem since we have flocculent settling.

The hydraulic retention time is

$$\bar{t} = \frac{\forall}{Q} = \frac{AD}{Q} = \frac{(314)(1.2)}{200} = 1.88 \text{ hour}$$

On Figure 8–9C, the 85 percent removal line approximately intersects $\bar{t} = 1.88$ hr. Thus 85 percent of the solids are clearly removed. In addition to this, however, even better removal is indicated at the top of the water column. At the top 20 cm, assume the SS = 40 mg/L, equal to $[(1,000 - 40) \times 100]/1,000 = 96$ percent removal, or 11 percent better than the entire column. The second shows $[(1,000 - 60) \times 100]/1,000 = 94$ percent removal, and so on. The total removed, ignoring the lowermost section, is

$$R = P + \sum_{i=1}^{n-1} \left(\frac{h}{H}\right)(P_i - P)$$

$$R = 85 + \tfrac{1}{6}(11) + \tfrac{1}{6}(9) + \tfrac{1}{6}(5) + \tfrac{1}{6}(4) = 90.9\%$$

Typically, solids capture efficiency of the primary clarifier for domestic waste may be plotted as a function of the retention time (which, for a given depth, is also related to the overflow rate), as shown in Figure 8–10.

The third type of settling, Class III, *thickening*, in which the particles are so tightly packed together that they settle as a single bed of solids, is covered in Chapter 9.

Primary treatment then is mainly a removal of solids, although some BOD is removed as a consequence of the removal of decomposable solids. Typically, the wastewater that was described earlier might now have these characteristics:

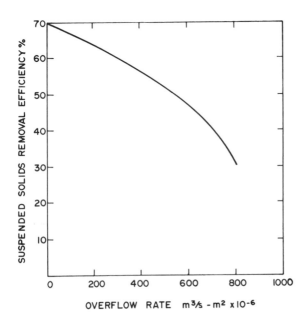

Figure 8–10. Performance of primary clarifiers.

	Raw Wastewater	*After Primary Treatment*
BOD mg/L	250	175
SS mg/L	220	60
P mg/L	8	7

A substantial fraction of the solids has been removed, as well as some BOD and a little P (as a consequence of the removal of raw sludge).

In a typical treatment plant, the wastewater would now move on to secondary treatment.

SECONDARY TREATMENT

The water leaving the primary clarifier has lost much of the solid organic matter but still contains a high demand for oxygen; i.e., it is composed of high-energy molecules that will decompose by microbial action, thus creating a BOD. This demand for oxygen must be reduced (energy wasted) if the discharge is not to create unacceptable conditions in the watercourse. The objective of secondary treatment is thus to remove BOD, whereas the objective of primary treatment is to remove solids.

The trickling filter, shown in Figure 8–11, consists of a filter bed of fist-sized rocks over which the waste is trickled. A very active biological growth forms on the rocks, and the organisms obtain their food from the waste stream dripping through the bed of rocks. Air is either forced through the rocks or, more commonly, air circulation is obtained automatically by a temperature difference between the air in the bed and ambient temperature. In the older filters the waste is sprayed onto the rocks from fixed nozzles. The newer designs utilize a rotating arm that moves under its own power, like a lawn sprinkler, distributing the waste evenly over the entire bed. Often the flow is recirculated, thus obtaining a higher degree of treatment. The name trickling filter is obviously a misnomer since no filtration takes place.

Around the turn of the century when trickling filtration was already firmly established, some researchers began musing about the wasted space in a filter taken up by the rocks. Could the microorganisms not be allowed to float free and could they not be fed oxygen by bubbling in air? Although this concept was quite attractive, it was not until 1914 that the first workable pilot plant was constructed. It took some time before this process became established as what we now call the *activated sludge system*.

The key to the activated sludge system is the reuse of microorganisms. The system, shown as a block diagram in Figure 8–12, consists of a tank full of waste liquid (from the primary clarifier) and a mass of microorganisms. Air is bubbled into this tank (called the *aeration tank*) to provide the necessary oxygen for the survival of the aerobic organisms. The microorganisms come in contact with the dissolved organics and rapidly adsorb these organics on their

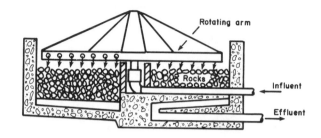

Figure 8–11. Trickling filter.

surface. In time, the microorganisms decomposed this material to CO_2, H_2O, some stable compounds, and more microorganisms. The production of new organisms is relatively slow, and most of the aeration tank volume is in fact used for this purpose.

Once most of the food has been utilized, the microorganisms are separated from the liquid in a settling tank, sometimes called a *secondary* or *final clarifier*. The liquid escapes over a weir and may be discharged into the recipient. The separation of microorganisms is an important part of the system. In the settling tanks, the microorganisms exist without additional food and become hungry. They are thus activated; hence the term *activated sludge*.

The settled microorganisms, now known as *return activated sludge*, are pumped to the head of the aeration tank where they find more food (organics in the effluent from the primary clarifier) and the process starts all over again. The activated sludge process is a continuous operation, with continuous sludge pumping the clean water discharge.

As mentioned earlier, one of the end products of this process is more microorganisms. If none of the microorganisms are removed, their concentration will soon increase to the point at which the system is clogged with solids. It

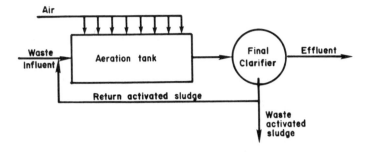

Figure 8–12. Block diagram of the activated sludge system.

is therefore necessary to waste some of the microorganisms, and this *waste activated sludge* must be processed and disposed of. Its disposal is one of the most difficult aspects of waste treatment.

Activated sludge systems are designed on the basis of loading, or the amount of organic matter (food) added relative to the microorganisms available. This ratio is known as the food-to-microorganisms ratio (F/M) and is a major design parameter. Unfortunately it is difficult to measure either F or M accurately, and engineers have approximated these by BOD and the SS in the aeration tank, respectively. The combination of the liquid and microorganisms undergoing aeration is known (for some unknown reason) as *mixed liquor*, and thus the SS are called *mixed liquor suspended solids* (MLSS). The ratio of incoming BOD to MLSS, the F/M ratio, is also known as the *loading* on the system, calculated as pounds of BOD/day per pound of MLSS in the aeration tank.

If this ratio is low (little food for many microorganisms) and the aeration period (retention time in the aeration tank) is long, the microorganisms make maximum use of available food, resulting in a high degree of treatment. Such systems are known as *extended aeration* and are widely used for isolated sources (e.g., motels, small developments). An added advantage of extended aeration is that the ecology within the aeration tank is quite diverse and little excess biomass is created, resulting in little or no waste activated sludge to be disposed of—a significant saving in operating costs and headaches.

At the other extreme is the "high-rate" system in which the aeration periods are very short (thus saving money by building smaller tanks) and the treatment efficiency is lower. The efficiencies and F/M ratios for the three types of activated sludge systems are shown in Table 8–3.

Example 8.4

The BOD_5 of the liquid from the primary clarifier is 120 mg/L at a flow rate of 0.05 mgd. The aeration tank is $20 \times 10 \times 20$ feet, and the MLSS = 2,000 mg/L. Calculate the F/M ratio.

lb of BOD = 120 mg/L × 0.05 mgd × 8.34 = 50 lb/day

(see p. 56 for a discussion of how to convert from flow and concentration to pounds)

lb of MLSS = $(20 \times 10 \times 20)$ ft^3 × 2,000 mg/L × 3.83 L/gal

$$\times 7.481 \text{ gal/ft}^3 \times 2.20 \times 10^{-6} \text{ lb/mg} = 229 \text{ lb}$$

$$F/M = \frac{50}{229} = 0.22 \frac{\text{lb of BOD/day}}{\text{lb of MLSS}}$$

Table 8–3. Loadings and Efficiencies of Activated Sludge Systems

Process	Loading $\frac{F}{M} = \frac{\text{lb of BOD/day}}{\text{lb of MLSS}}$	Aeration Period (hr)	Efficiency of BOD Removal (%)
Extended aeration	0.05–0.2	30	95
Conventional	0.2–0.5	6	90
High rate	1–2	4	85

When the microorganisms first come in contact with the food, the process requires a great deal of oxygen. Accordingly, the DO level in the aeration tank drops immediately after the point at which the waste is introduced. If DO levels are measured over the length of a tank, extremely low concentrations are often found at the influent end of the aeration tank. These low levels of DO may be detrimental to the microbial population. Accordingly, two variations of the activated sludge treatment have found some use: *tapered aeration* and *step aeration* (Figure 8–13). The former method consists of blasting additional air where needed, whereas step aeration involves the introduction of the waste at several locations, thus evening out the initial oxygen demand.

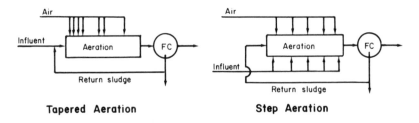

Tapered Aeration   **Step Aeration**

Figure 8–13.   Tapered and step aeration schematics.

The third modification is *contact stabilization*, or *biosorption*, a process in which the sorption and bacterial growth phases are separated by a settling tank. The advantage is that the growth can be achieved at high solids concentrations, thus saving tank space. Many existing activated sludge plants can be converted to biosorption plants when tank volume limits treatment efficiency. Figure 8–14 is a diagram of the biosorption process.

The two principal means of introducing sufficient oxygen into the aeration tank are by bubbling compressed air through porous diffusers or beating air in mechanically. Both diffused air and mechanical aeration are shown in Figure 8–15.

The success of the activated sludge system depends on many factors. Of critical importance is the separation of the microorganisms in the final clarifier. The microorganisms in the system are sometimes very difficult to settle out, and the sludge is said to be a *bulking sludge*. Often this condition is characterized by a biomass composed almost totally of filamentous organisms that form a kind of lattice structure with the filaments and refuse to settle.*

Treatment plant operators should keep a close watch on settling characteristics because a trend toward poor settling may be the forerunner of a badly upset (and hence ineffective) plant. The settleability of activated sludge is most often described by the sludge volume index (SVI), which is determined by measuring the milliliters of volume occupied by a sludge after settling for 30 minutes in a 1-L cylinder, and calculated as

$$\text{SVI} = \frac{(\text{volume of sludge after 30 min, in mL}) \times 1{,}000}{\text{mg/L of suspended solids}}$$

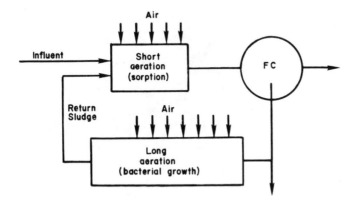

Figure 8–14. The biosorption modification of the activated sludge process.

---

* You may picture this as filling a glass with cotton balls, then pouring water in it. The cotton is simply not heavy enough to settle to the bottom of the glass.

Example 8.5

A sample of mixed liquor was found to have SS = 4,000 mg/L and, after settling for 30 minutes in a 1-L cylinder, occupied 400 mL. Calculate the SVI.

$$SVI = \frac{400 \times 1,000}{4,000} = 100$$

SVI values below 100 are usually considered acceptable, with SVIs greater than 200 defined as badly bulking sludges. Some common loadings, as a function of the SVI, for final clarifiers are shown in Figure 8–16.

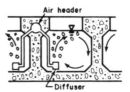

Diffused Aeration

Figure 8–15. Activated systems with diffused aeration and mechanical (surface) aeration. [Photos courtesy of Envirex.]

The causes of poor settling (high SVI) are not always known, and hence the solutions are elusive. Wrong or variable F/M ratios, fluctuations in temperature, high concentrations of heavy metals, and deficiencies in nutrients in the incoming wastewater have all been blamed for bulking. Cures include chlorination, changes in air supply, and dosing with hydrogen peroxide (H_2O_2) to kill the filamentous microorganisms.

When the sludge does not settle, the return activated sludge becomes thin (low SS concentration) and thus the concentration of microorganisms in the

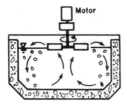

Mechanical Aeration

Figure 8-15. (*Continued*)

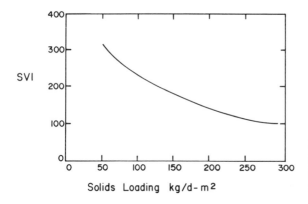

Figure 8–16. Allowable solids loading for final clarifiers increase for better-settling sludge.

aeration tank drops. This results in a higher F/M ratio (same food input, but fewer microorganisms) and a reduced BOD removal efficiency.*

Biological Process Dynamics Applied to the Activated Sludge System

The organics in the influent are degraded and oxidized to CO_2 and H_2O. Some of the high-energy organics are used to build new microorganisms. These organics, known in the language of biological process dynamics as *substrate*, are usually measured indirectly as the BOD, although methods such as organic carbon, for example, may be more accurate measures of substrate concentration. The substrate (S) is biodegraded and used by the microorganisms (X), expressed as SS, at a rate dS/dt. The rate of new cell mass (microorganisms) production as a result of the destruction of the substrate is

$$\frac{dX}{dt} = Y \frac{dS}{dt}$$

where Y = the yield, or mass of microorganisms produced per mass of substrate used; commonly expressed as kilograms of SS produced/kilograms of BOD used, or Y = dX/dS.

---

* You may think of the microorganisms as workers in an industrial plant. If the total number of workers is decreased, the production is cut. Similarly, if fewer microorganisms are available, less work is done. This will become evident in the discussion on process dynamics.

The expression for substrate utilization commonly employed is the Monod model,*

$$\frac{dS}{dt} = \frac{X}{Y} (\mu) = \frac{X}{Y} \left(\frac{\hat{\mu}S}{K_s + S} \right)$$

where μ = is the specific growth rate, in terms of days^{-1}, and equal to $(dX/dt)(1/X)$
$\hat{\mu}$ = maximum growth rate constant, days^{-1}
K_s = saturation constant, mg/L

This expression is an empirical model, based on experimental work with pure cultures. The two constants, $\hat{\mu}$ and K_s, must be evaluated for each substrate and microorganism culture.

The application of biological process dynamics to the activated sludge process is best illustrated by considering a system shown in Figure 8–17. This is a simple continuous biological reactor, of volume V and an effluent of Q. The reactor is *completely mixed*, meaning that the influent is dispersed within the tank immediately upon introduction, thus there are no concentration gradients in the tank and the quality of the effluent is *exactly* that of the tank contents.

In such a continuous reactor, there are two types of retention times: liquid and solids. The liquid, or *hydraulic retention time*, is expressed as

$$\bar{t} = \frac{V}{Q}$$

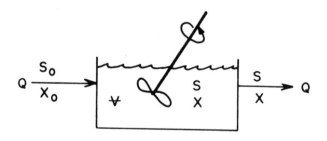

Figure 8–17.   A biological reactor without microorganism recycle.

---

* The argument for the model validity is beyond the scope of this text. Suffice it to say that the model is empirical, but reasonable. If you are interested in the development of this model, see any number of modern textbooks on wastewater processing.[4-6]

where $\bar{t}$ = hydraulic retention time, min. The solids retention time is also known as *sludge age*, or the average time a solid (microorganism) particle stays in the system. What wastewater treatment plant operators call sludge age is known also as the *mean cell residence time* in research, and is expressed as

$$\Theta_c = \frac{\text{mass of SS in the system}}{\text{mass rate of solids wasted}}$$

For the system pictured in Figure 8–17, the mean cell residence time is

$$\Theta_c = \frac{\forall X}{QX}$$

(which of course is equal to the hydraulic retention time). The units of Θ_c commonly are expressed in days.

For the system in Figure 8–17, we now assume that there are no organisms in the influent ($X_0 = 0$) and that the reactor has attained steady state (the growth rate of microorganisms is balanced by the loss rate of microorganisms in the effluent). For the sake of simplicity, we also ignore the death of microorganisms.

We can write a mass balance in terms of the microorganisms

$$\begin{bmatrix} \text{Rate of change} \\ \text{in the} \\ \text{reactor} \end{bmatrix} = \begin{bmatrix} \text{Rate} \\ \text{of} \\ \text{inflow} \end{bmatrix} - \begin{bmatrix} \text{Rate} \\ \text{of} \\ \text{outflow} \end{bmatrix} + \begin{bmatrix} \text{Net rate of} \\ \text{growth of} \\ \text{microorganisms} \end{bmatrix}$$

$$\frac{dX}{dt} \forall \quad = \quad QX_0 \quad - \quad QX \quad + \quad \left(Y \frac{dS}{dt} \right) \forall$$

In a steady-state system, $(dX/dt)\forall = 0$, and since we assumed no cells in the inflow, $QX_0 = 0$. Substituting the Monod substrate utilization model,

$$\frac{dS}{dt} = \frac{\hat{\mu} XS}{Y(K_s + S)}$$

into the above relationship and introducing the mean cell residence time Θ_c,

$$\frac{1}{\Theta_c} = \frac{\hat{\mu} S}{K_s + S}$$

and

$$S = \frac{K_s}{\hat{\mu} \Theta_c - 1}$$

This is an important expression since it implies that the substrate concentration, S, is a function of the kinetic constants (which are beyond our control for a given substrate) and the mean cell residence time. The value of S (and thus the treatment efficiency) is influenced then by the mean cell residence time (or the sludge age as previously defined).

The system pictured in Figure 8–17 is not very efficient, however, since long hydraulic residence times are necessary to prevent the microorganisms from being flushed out. The success of the activated sludge system for wastewater treatment is based on the recycle of the microorganisms. Such a system is shown in Figure 8–18.

We need to make some simplifying assumptions before we can conveniently model this system. First, we again assume that $X_o = 0$, and further that the microorganism separator (the final settling tank or clarifier) is a perfect device, so that there are no microorganisms in the effluent ($X_e = 0$). We also once again assume steady-state conditions and perfect mixing. The excess microorganisms (waste activated sludge) are removed from the system at a flow rate Q_w and a solids concentration X_r, which is the settler underflow concentration and the concentration of the solids being recycled to the aeration tank. And last, we assume that there is no substrate removal in the settling tank and that the settling tank has no volume, so that all of the microorganisms in the system are in the reactor (aeration tank). The volume of the aeration tank is thus the only active volume, and the settling (microorganism separation) is assumed to take place magically in a zero-volume tank, an obviously incorrect assumption. Note that the mean cell residence time in this case is

$$\Theta_c = \frac{\text{microorganisms in the system}}{\text{microorganisms lost or wasted/time}}$$

$$= \frac{XV}{Q_w X_r + (Q - Q_w)X_e}$$

and since $X_e = 0$

$$\Theta_c = \frac{XV}{Q_w X_r}$$

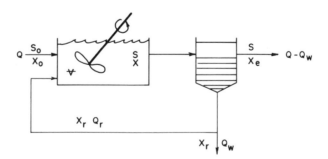

Figure 8–18.   A biological reactor with microorganism recycle.

The removal of substrate is often expressed in terms of a *substrate removal velocity* (q) and defined as

$$q = \frac{\text{mass of substrate removed}}{\text{mass of microorganisms under aeration} \cdot \text{time}}$$

In our notation

$$q = \frac{S_o - S}{X\bar{t}}$$

The substrate removal velocity is a rational measure of the substrate removal activity, or the mass of BOD removed in a given time per mass of microorganisms doing the work. This is also called, in some texts, the process-loading factor, with the clear implication that it is a useful operational and design tool.

The substrate removal velocity may be derived by doing a mass balance in terms of substrate or a continuous system with microorganism recycle (Figure 8–18).

$$\begin{bmatrix} \text{Rate of} \\ \text{change of} \\ \text{substrate} \end{bmatrix} = \begin{bmatrix} \text{Rate of} \\ \text{substrate} \\ \text{inflow} \end{bmatrix} - \begin{bmatrix} \text{Rate of} \\ \text{substrate} \\ \text{outflow} \end{bmatrix} - \begin{bmatrix} \text{Rate of} \\ \text{substrate} \\ \text{utilization} \end{bmatrix}$$

$$\frac{dS}{dt}\forall \quad = \quad QS_o \quad - \quad QS \quad - \quad qXV$$

Note: The rate of substrate utilization is

$$qXV = \left[\frac{\text{mass of substrate removed}}{\text{mass of microorganisms} \cdot \text{time}} \right] \cdot \left[\frac{\text{microorganisms}}{\text{volume of reactor}} \right]$$

$$\times [\text{volume of reactor}] = \frac{\text{mass of substrate removed}}{\text{time}}$$

Solving for q,

$$q = \frac{S_o - S}{X\bar{t}}$$

The Monod rate (μ) was defined previously as

$$\frac{dS}{dt} = \frac{\hat{\mu}SX}{Y(K_s + S)} = \frac{X}{Y}(\mu)$$

$$\mu = \frac{dX}{dt}\frac{1}{X} = \frac{\text{mass of microorganisms produced}}{\text{time} \cdot \text{mass of microorganisms}}$$

and the yield (Y) as

$$Y = \frac{dX}{dS} = \frac{\text{mass of microorganisms produced}}{\text{mass of substrate removed}}$$

Using now a mass balance in terms of the microorganisms:

$$\begin{bmatrix} \text{Rate of change} \\ \text{in microorganism} \\ \text{concentration} \end{bmatrix} = \begin{bmatrix} \text{Rate of} \\ \text{inflow of} \\ \text{microorganisms} \end{bmatrix} - \begin{bmatrix} \text{Rate of} \\ \text{outflow of} \\ \text{microorganisms} \end{bmatrix} + \begin{bmatrix} \text{Rate of} \\ \text{production of} \\ \text{microorganisms} \end{bmatrix}$$

$$\frac{dX}{dt} \forall \quad = \quad QX_o \quad - Q_w X_r - (Q - Q_w)X_e + \quad \mu X \forall$$

Assuming again steady state ($[dX/dt] = 0$) and $X_o = X_e = 0$,

$$\mu = \frac{X_r Q_w}{X \forall}$$

Note that this is the reciprocal of the previously defined sludge age, or mean cell residence time, so that

$$\Theta_c = \frac{XV}{X_f Q_w} = \frac{1}{\mu}$$

The substrate removal velocity may then be expressed as

$$q = \frac{\mu}{Y}$$

$$= \frac{\text{mass of microorganisms produced}}{\text{time} \times \text{mass of microorganisms in the reactor}}$$

$$= \frac{\text{mass of substrate removed}}{\text{mass of microorganisms produced}}$$

and substituting

$$q = \frac{\hat{\mu} S}{Y(K_s + S)}$$

Previously we defined q as $(S_o - S)/Xt$. Equating these two expressions and solving for $S_o - S$, the substrate removal (reduction in BOD),

$$S_o - S = \frac{\hat{\mu} S X \bar{t}}{Y(K_s + S)}$$

Since we defined the substrate removal velocity q, as

$$q = \frac{\mu}{Y}$$

the mean cell residence time (sludge age) is

$$\Theta_c = \frac{1}{qY}$$

and the concentration of microorganisms in the reactor (MLSS) is

$$X = \frac{S_o - S}{\bar{t}q}$$

Example 8.6

An activated sludge system operates at a flow rate (Q) of 4,000 m³/day, with an incoming BOD (S_o) of 300 mg/L. Through pilot plant work the kinetic constants for this system were determined to be Y = 0.5 kg of SS/kg of BOD, K_s = 200 mg/L, μ = 2 day^{-1}. We need to design a treatment system that will produce an effluent BOD of 30 mg/L (90 percent removal). Determine: (1) what should be (a) the volume of the aeration tank, (b) the MLSS, and (c) the sludge age, and (2) how much sludge will be wasted daily?

The MLSS concentration is usually limited by the ability to keep an aeration tank mixed and to transfer sufficient oxygen to the microorganisms. Assume in this case that X = 4,000 mg/L. The hydraulic retention time is then obtained from the equation

$$S_o - S = \frac{\hat{\mu} S X \bar{t}}{Y(K_s + S)}$$

Rearranged,

$$\bar{t} = \frac{Y(S_o - S)(K_s + S)}{\hat{\mu} S X}$$

$$= \frac{0.5(300 - 30)(200 + 30)}{2(30)(4,000)}$$

$$= 0.129 \text{ day} = 3.1 \text{ hours}$$

The volume of the tank is then $V = \bar{t}Q = 4,000(0.129) = 516$ m³. The sludge age is

$$\Theta_c = \frac{1}{qY}$$

where

$$q = \frac{S_o - S}{X\bar{t}} = \frac{300 - 30}{(4,000)(0.129)} = 0.523 \frac{\text{kg of BOD removed/day}}{\text{kg of SS in the reactor}}$$

or, equivalently

$$q = \frac{\hat{\mu}S}{Y(K_s + S)} = \frac{2(30)}{(0.5)(200 + 30)} = 0.522 \text{ day}^{-1}$$

and

$$\Theta_c = \frac{1}{qY} = \frac{1}{(0.522)(0.5)} = 3.8 \text{ days}$$

Since

$$\frac{1}{\Theta_c} = \frac{\text{kg of sludge solids wasted/day}}{\text{kg of sludge in aeration tank}}$$

$$\Theta_c = \frac{XV}{X_r Q_w}$$

$$X_r Q_w = \frac{XV}{\Theta_c} = \frac{(4,000)(516)(10^3 \text{ L/m}^3)(1/10^6 \text{ kg/mg})}{3.8}$$

$$= 543 \text{ kg/day}$$

Example 8.7

Using the same data as in the previous example, we now ask what mixed liquor solids concentration is necessary to attain a 95 percent BOD removal (i.e., $S = 15$ mg).

The substrate removal velocity is

$$q = \frac{\hat{\mu}S}{Y(K_s + S)} = \frac{2(15)}{0.5(200 + 15)} = 0.28 \text{ days}^{-1}$$

and

$$X = \frac{S_o - S}{tq} = \frac{300 - 15}{(0.129)(0.28)} = 7,890 \text{ mg/L}$$

The sludge age would be

$$\Theta_c = \frac{1}{qY} = \frac{1}{(0.28)(0.5)} = 7.1 \text{ days}$$

Note that more microorganisms are required in the aeration tank if higher removal efficiencies are to be attained.

The SVI (see p. 159) can be used to estimate the return sludge pumping rates. After 30 minutes of settling, the solids in the cylinder are at an SS concentration that would be equal to the expected return sludge solids, or

$$X_r \cong \frac{H}{h} (x)$$

where X_r = the expected return suspended solids concentration, mg/L
   X = MLSS, mg/L
   H = height of cylinder, m
   h = height of settled sludge, m

The mixed liquor solids concentration is of course a combination of the return solids diluted by the influent, or

$$X = \frac{Q_r X_r + QX_o}{Q_r + Q}$$

If we again assume no solids in the influent, $(X_o = 0)$

$$X = \frac{Q_r X_r}{Q_r + Q}$$

The success or failure of an activated sludge system often depends on the performance of the final clarifier. If this settling tank is not able to achieve the required return sludge solids, the MLSS will drop and of course the treatment efficiency will be reduced.

Final clarifiers act as both settling tanks (flocculent settling, class II) and thickeners (class III). The design of these tanks therefore requires that both the overflow rate and solids loading be considered. The latter, more fully explained in the next chapter, is expressed in terms of kilograms of solids per day applied to a surface area of square meters. Figure 8–16 shows some commonly used solids loadings for final clarifiers as a function of the SVI.

Secondary treatment of wastewater then usually consists of a biological step such as activated sludge, which removes a substantial part of the BOD and the remaining solids. Looking once again at the typical wastewater, we now have the following approximate water quality:

	Raw Wastewater	After Primary Treatment	After Secondary Treatment
BOD mg/L	250	175	15
SS mg/L	220	60	15
P mg/L	8	7	6

The effluent, in fact, meets our previously established effluent standards for BOD and SS. Only the phosphorus remains high. The removal of inorganic chemicals such as phosphorus is accomplished in tertiary (or advanced) wastewater treatment.

TERTIARY TREATMENT

Primary and biological treatments make up the conventional wastewater treatment plant. However, secondary treatment plant effluents still contain a significant amount of various types of pollutant. Suspended solids, in addition to contributing to BOD, may settle out in streams and form unsightly mud banks. The BOD, if discharged into a stream with low flow, may still cause damage to aquatic life by depressing the DO. Neither primary nor secondary treatment is effective in removing phosphorus and other nutrients or toxic substances. For BOD removal, by far the most popular advanced treatment method is the polishing pond, often called the *oxidation pond*. This is essentially a hole in the ground, a large pond used to confine the plant effluent before it is discharged. Such ponds are designed to be aerobic, hence light penetration for algal growth is important, and a large surface area is needed. The reactions occurring within an oxidation pond are depicted in Figure 8–19. Oxidation ponds are sometimes used as the only treatment step if the waste flow is small and the pond area is large. When the rate of decomposition in an oxidation pond is too great and oxygen availability becomes limiting, the pond may be aerated by either diffused or mechanical aerators. Such ponds are known as *aerated lagoons* and are widely used in treating industrial effluents. A typical aerated lagoon serving a pulp and paper mill is shown in Figure 8–19.

Activated carbon adsorption is another method of BOD removal, and this process has the added advantage that inorganics as well as organics are removed. The mechanism of adsorption on activated carbon is both chemical and physical, with tiny crevices catching and holding colloidal and smaller particles. An activated carbon column is a completely enclosed tube with dirty water pumped up from the bottom and the clear water exiting at the top. As the carbon becomes saturated with various materials, it must be removed from the column and regenerated, or cleaned. Removal is often continuous, with clean carbon being added at the top of the column. The cleaning or regeneration is usually done by heating the carbon in the absence of oxygen. A slight loss in efficiency is noted with regeneration, and some virgin carbon must always be added to ensure effective performance.

*Reverse osmosis** is also finding acceptance as a treatment for various types of trace pollutant, organic as well as inorganic. The wastewater is forced through a semipermeable membrane that acts as a superfilter, rejecting dissolved as well as suspended solids.

Nitrogen removal may be accomplished in two ways. The first method makes use of the fact that even after secondary treatment, most of the nitrogen exists as ammonia. Increasing the pH produces the following reaction:

$$NH_4^+ + OH^- - NH_3\uparrow + H_2O$$

---

* This term is a misnomer. Reverse osmosis has nothing to do with osmotic pressure; this process is better described as "ultrafiltration."

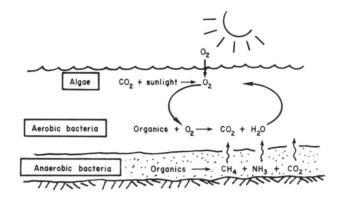

Figure 8–19. Schematic of an oxidation pond and photograph of an aeration lagoon. [Photo courtesy of N.S. Nokkentved.]

Much of the dissolved ammonia gas may then be expelled from the water into the atmosphere. The resulting air pollution problem has not been resolved.

A second method of getting rid of nitrogen is to first treat the waste thoroughly enough to produce nitrate ions. This usually involves longer detention times in secondary treatment, during which bacteria such as *Nitrobacter* and *Nitrosomonas* convert ammonia nitrogen to NO_3^-, a process called *nitrification*. These reactions are

$$2NH_4^+ + 3O_2 \xrightarrow{\textit{Nitrosomonas}} 2NO_2^- + 2H_2O + 4H^+$$

$$2N_2^- + O_2 \xrightarrow{\textit{Nitrobacter}} 2NO_3^-$$

These reactions are slow and thus require long retention times in the aeration tank, as well as sufficient DO. The kinetics constants for these reactions are low, with very low yields, so that the net sludge production is limited, making washout a constant danger.

Once the ammonia has been converted to nitrate, it may be reduced by a broad range of facultative and anaerobic bacteria such as *Pseudomonas*. This reduction, called *denitrification*, requires a source of carbon, and methanol (CH_3OH) is often used for that purpose.

$$6NO_3^- + 2CH_3OH \rightarrow 6NO_2^- + 2CO_2\uparrow + 4H_2O$$

$$6NO_2^- + 3CH_3OH \rightarrow 3N_2\uparrow + 3CO_2\uparrow + 3H_2O + 6OH^-$$

Phosphate removal is usually a chemical process. The most popular chemicals used for phosphorus removal are lime, $Ca(OH)_2$, and alum, $Al_2(SO_4)$. The calcium ion, in the presence of high pH, will combine with phosphate to form a white, insoluble precipitate called calcium hydroxyapatite that is settled out and removed. Insoluble calcium carbonate is also formed and removed and may be recycled by burning in a furnace.

$$CaCO_3 \xrightarrow{\Delta} CO_2 + CaO$$

Quick lime, CaO, is slaked by adding water,

$$CaO + H_2O \rightarrow Ca(OH)_2$$

thus forming lime, which may be reused.

The aluminum ion from alum precipitates out as poorly soluble aluminum phosphate,

$$Al^{3+} + PO_4^{3-} \rightarrow AlPO_4\downarrow$$

and also forms aluminum hydroxides,

$$Al^{3+} + 3OH^- \rightarrow Al(OH_3)\downarrow$$

which are sticky flocs and help to settle out the phosphates. The most common point of alum dosing is in the final clarifier.

The amount of alum required to achieve a given level of phosphorus removal depends on the amount of phosphorus in the water, as well as other constituents. The amount of sludge produced may be calculated by using stoichiometric relationships.

Example 8.8

A wastewater has 6.3 mg/L of P and it is found that an alum dosage of 13 mg/L as Al^{3+} achieves an effluent with a P concentration of 0.9 mg/L.

$$Al^{3+} + PO_4^{3-} \rightarrow AlPO_4$$

Molecular weights:　　　　27　　95　　122
(or 31 as P)

or 31 mg of P removal results in the formation of 122 mg of aluminum phosphate sludge.

$$5.4 \text{ mg/L (P removed)} \times \frac{122}{31} = 21.3 \text{ mg/L (AlPO}_4 \text{ sludge produced)}$$

The amount of Al^{3+} used to produce the $AlPO_4$ is

$$5.4 \text{ mg/L (P removed)} \times \frac{27 \text{ mg of Al}^{3+}}{31 \text{ mg of P}} = 4.7 \text{ mg/L of Al}^{3+}$$

The excess aluminum, $13 - 4.7 = 8.3$ mg/L, probably went into the formation of aluminum hydroxide. The simplest assumption is that

$$Al^{3+} + 3OH \rightarrow Al(OH)_3$$

$$27 \quad 3 \times 17 = 51 \quad 78$$

and so the production of $Al(OH)_3$ is

$$8.3 \text{ mg/L of Al}^{3+} \left(\frac{78 \text{ mg of Al(OH)}_3}{27 \text{ mg of Al}^{3+}} \right) = 24 \text{ mg/L (Al(OH)}_3 \text{ sludge)}$$

The total sludge production is then

$$AlPO_4 + Al(OH)_3 = 21.3 + 24 \cong 45 \text{ mg/L}$$

　　Phosphorus removal may also be accomplished biologically. When the microorganisms in return activated sludge are starved by aeration after their removal in the final clarifier, they show a strong tendency to adsorb phosphorus almost instantly upon introduction of wastewater. Such a "luxury uptake" is at a rate much higher than what they could eventually require in their metabolic activity. If these phosphorus-rich microorganisms are quickly removed and wasted, the excess phosphorus will leave with the waste activated sludge. Thus, it is possible to achieve both nitrogen *and* phosphorus reduction by solely biological means in secondary treatment.

　　Assuming alum precipitation of phosphorus (and a bonus of more efficient SS and BOD removal resulting from the formation of $Al(OH)_3$), we now have attained our effluent goal:

	Raw Wastewater	After Primary Treatment	After Secondary Treatment	After Tertiary Treatment
BOD mg/L	250	175	15	10
SS mg/L	220	60	15	10
P mg/L	8	7	6	0.5

An alternative to high-technology advanced wastewater treatment systems is to spray secondary effluent on land and allow the soil microorganisms to degrade the remaining organics. Such systems, known as *land treatment*, have been used for many years in Europe, but only recently have they been used in North America. They appear to represent a reasonable alternative to complex and expensive systems, especially for smaller communities.

Probably the most promising land treatment method is irrigation. Commonly, from 1,000 to 2,000 hectares of land are required for every $1 \, m^3/sec$ of wastewater flow, depending on the crop and soil. Nutrients such as N and P remaining in secondary effluent are of course beneficial to the crops.

CONCLUSION

A typical wastewater treatment plant is shown schematically in Figure 8–20. We have discussed primary, secondary, and tertiary treatment. The treatment and

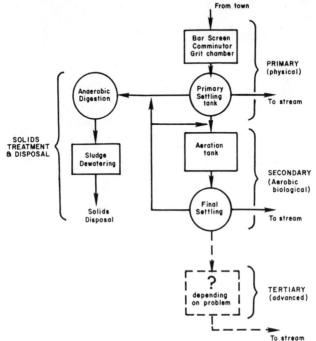

Figure 8–20. Block diagram of a complete wastewater treatment plant.

disposal of the solids removed from the liquid stream deserves special attention, and the topic is covered in the next chapter.

An aerial view of a typical wastewater treatment plant is shown in Figure 8–21. If such plants are well operated, the effluents are often much less polluted than the stream into which they are discharged.

However, not all plants perform that well. The sad fact is that many of the existing wastewater treatment plants are only marginally effective in controlling water pollution, and much of the blame can be placed on plant operation.

The operation of modern wastewater treatment plants is a complex and demanding job.* Unfortunately, operators have historically been considered to be at the bottom of the totem pole, in terms of both pay and community stature, and few qualified people were willing to make plant operation a career. Municipalities were forced to use whatever help was available, which often resulted in poor operation.

Many states now require licensing of treatment plant operators, and their pay and social stature has greatly improved. This is a welcome change, for it

Figure 8–21.   An aerial view of a secondary wastewater treatment plant. [Courtesy of Envirex.]

---

* Although seldom as demanding as the problem faced by an operator finding a pickup truck in the primary clarifier (Figure 8–22).

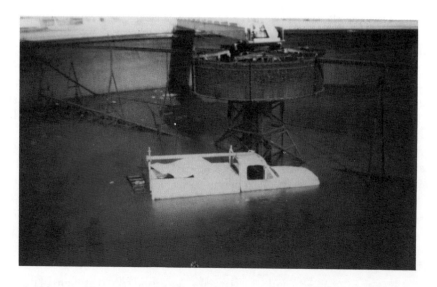

Figure 8–22. An unusual operating problem—a pickup in the primary clarifier. [Courtesy of Phillip Karr.]

makes little sense to entrust unqualified and unreliable workers with the operation of multimillion-dollar facilities.

Wastewater treatment thus requires first the proper design of the plant and second the proper plant operation. One without the other is a waste of money.

PROBLEMS

8.1 The following data were reported on the operation of a wastewater treatment plant:

	Influent (mg/L)	*Effluent* (mg/L)
BOD_5	200	20
SS	220	15
P	10	0.5

(a) What percent removal was experienced for each of these?

(b) What type of treatment plant would produce such an effluent? Draw a block diagram showing the treatment steps.

8.2 Describe the condition of a primary clarifier one day after the raw sludge pumps broke down.

8.3 One operational problem with trickling filters is "ponding," the excessive growth of slime on the rocks and subsequent clogging of the spaces so that the water no longer flows through the filter. Suggest some cures for the ponding problem.

8.4   One problem with sanitary sewers is illegal connections. Suppose a family of four, living in a home with a roof area of 70×40 feet, connects the roof drain to the sewer. For a typical rain of 1 in./hr, what percent increase will there be in the flow from their house over the "dry weather" flow (assumed at 50 gal/capita/day)?

8.5   "Good secondary treatment plants remove 90 percent of measured BOD. That does not mean that 90 percent of the total oxygen-demanding wastes are removed, but only the part that is measured by certain laboratory tests" (Environmental Quality, CEQ, 1970). How could 90 percent of the BOD be removed and not 90 percent of the oxygen-demanding material?

8.6   Suppose an industry decided to build a wastewater treatment plant and hired an engineer to design it for them. One of the first steps would be to sample the waste and run some analyses to determine what its characteristics are. If you had to specify the tests to be run, what would you choose as the five most important wastewater parameters of interest? Name the five and state why you want to know these values.

8.7   The influent and effluent data for a secondary treatment plant are:

	Influent (mg/L)	Effluent (mg/L)
BOD	200	20
SS	200	100
Total phosphorus	10	8

Calculate the removal efficiencies. What is wrong with the plant?

8.8   Draw block diagrams of the unit operations necessary to treat the following wastes to effluent levels of $BOD_5 = 20$ mg/L, SS = 20 mg/L, P = 1 mg/L.

Waste	BOD_5 (mg/L)	SS (mg/L)	P (mg/L)
A. Domestic	200	200	10
B. Chemical industry	40,000	0	0
C. Pickle cannery	0	300	1
D. Fertilizer mfg.	300	300	200

8.9   The success of an activated sludge system depends on the settling of the solids in the final settling tank. Suppose the sludge in a system started to "bulk," not settle very well, and the SS concentration of the return activated sludge dropped from 10,000 mg/L to 4,000 mg/L.
   a. What will this do to the MLSS?
   b. What will this in turn do to the BOD removal? Why?

8.10   If in the above problem the volume of settled sludge (30 min) was 300 mL for both before and during the bulking problem, and the SVI was 100 and 250, respectively, what was the MLSS before and after the bulking problem? Does this agree with your answer to Problem 8.10?

8.11 If you conducted a percolation test and discovered that in 30 minutes the water level dropped 5 in., what size percolation field would you need for a two-bedroom house? What size would you need if it dropped 0.5 in.?

8.12 A 1-mgd conventional activated sludge plant has an influent BOD_5 of 200 mg/L. The primary clarifier removes 30 percent of that BOD. The three aeration tanks are each $20 \times 20 \times 100$ feet. What MLSS are necessary to attain 90 percent BOD removal in the plant?

8.13 An aeration system with a hydraulic retention time of 2.5 hours receives a flow of 0.2 mgd at a BOD of 150 mg/L. The SS in the aeration tank are 4,000 mg/L. The effluent BOD is 20 mg/L, and effluent SS are 30 mg/L. Calculate the F/M ratio (food to microorganism) for this system.

8.14 The MLSS in an aeration tank are 4,000 mg/L. The flow from the primary settling tank is 0.2 m³/sec, and the return sludge flow is 0.1 m³/sec. What must the return sludge SS concentration be to maintain the 4,000 mg/L MLSS?

8.15 A wastewater is composed of SS that are small beads, 0.1-mm diameter with a specific gravity of 2.65. Assume $\mu = 1.31$ cP.
   a. How long would it take for a settling tank 4 m deep to clarify the waste?
   b. If a false floor were put into the settling tank so that it was now only 2 m deep, what fraction of the SS would be removed?
   c. What is the "overflow rate" for this tank?

8.16 A primary clarifier (settling tank) is 80 feet long, 30 feet wide, and 12 feet deep and receives a flow of 200,000 gpd.
   a. Calculate the overflow rate in gpd/ft².
   b. What is the settling velocity in ft/day of the critical particle (assuming entire tank is used for clarification)?

8.17 A rectangular primary clarifier (settling tank) is 10 m wide, 25 m long and 5 m deep. The flow is 0.5 m³/sec.
   a. What is the design overflow rate?
   b. Would all particles having a settling velocity of 0.01 cm/min be removed?

8.18 A community receives 0.5 m³/sec of flow into its wastewater treatment plant. The total surface area of the primary clarifiers is 2,700 m², and the retention time is 3 hours. Find the overflow rate and the depth of the clarifiers.

8.19 A transoceanic flight on a Boeing 747, with 430 persons on board, takes 7 hours. Estimate the weight of the water necessary to flush the toilets if each flush required 2 gallons. Make any assumptions necessary. What fraction of the total payload (people) would the flush water represent? How could you reduce this weight? (The railroad system, for obvious reasons, is illegal.)

8.20   A family of four wants to build a house on a lot for which the percolation test results show 1.0 mm/min. The county requires a septic tank hydraulic retention time of 24 hours. Find the volume of the tank required and the area of the tile field. Sketch the system, including all dimensions.

8.21   A consulting engineer has proposed the following design for a primary clarifier: $Q = 0.150 \, m^3/sec$, influent $SS = 310 \, mg/L$, raw primary sludge concentration = 4 percent solids, SS removal efficiency = 60 percent, length = 40 m, width = 15 m, depth = 2.2 m. Is this a reasonable design and expected performance?

8.22   A final clarifier is to be able to handle a MLSS of 2,800 mg/L at a flow of $0.30 \, m^3/s$. Determine the tank diameter and solids loading. The design overflow rate is 64 m/day, and the expected SVI is 150. Assume a reasonable retention time.

8.23   A wastewater contains soluble phosphate at a concentration of 4 mg/L. How much ferric chloride would theoretically be necessary to precipitate this nutrient?

8.24   A community with a wastewater flow of 10 mgd is required to meet effluent standards of 30 mg/L for both BOD_5 and SS. Pilot plant results with the influent of $BOD_5 = 250 \, mg/L$ estimate the kinetic constants at $K_s = 100 \, mg/L$, $\hat{\mu} = 0.25 \, day^{-1}$, and $Y = 0.5$. It is decided to maintain the MLSS at 2,000 mg/L. What is the hydraulic retention time, the sludge age, and the required tank volume?

LIST OF SYMBOLS

A = area, m^2
BOD = biochemical oxygen demand, mg/L
C_D = drag coefficient
d = diameter, m
F = food (BOD), mg/L
F_B = buoyancy force, N
F_D = drag force, N
F_g = gravitational force, N
g = gravitational acceleration, m/sec^2
H = height of a settling tank or settling cylinder, m
h = height, m
K_s = saturation constant, mg/L
L = length, m
M = microorganisms (SS), mg/L
$MLSS$ = mixed liquor suspended solids, mg/L
m = mass, kg
P = phosphorus, mg/L
Q = flow rate, m^3/sec

Q_w = waste sludge flow rate, m^3/sec
q = substrate removal velocity, sec^{-1}
R = Reynolds number
R = overall recovery of SS in settling tank, %
S = substrate concentration, estimated as BOD, mg/L
S_o = influent substrate concentration, estimated as influent BOD, mg/L
SS = suspended solids, mg/L
t = time, sec
$\bar{t}$ = retention time, sec
v = velocity, m/sec
v_o = settling velocity of a critical particle, m/sec
v_s = settling velocity of any particle, m/sec
X = microorganism concentration, estimated as SS, mg/L
X_e = effluent microorganism concentration, estimated as effluent SS, mg/L
X_r = return sludge microorganism concentration, estimated as return sludge SS, mg/L
X_o = influent microorganism concentration, estimated as influent SS, mg/L
Y = yield, SS produced/BOD utilized
Θ_c = mean cell residence time, or sludge age, days
μ = fluid viscosity, $N \cdot sec/m^2$
μ = growth rate constant, sec^{-1}
$\hat{\mu}$ = maximum growth rate constant, sec^{-1}
ρ = fluid density, kg/m^3
ρ_s = density of a solid, kg/m^3
$\forall$ = volume, m^3

REFERENCES

1. Kirby, R.S., et al. *Engineering in History* (New York: McGraw-Hill, 1956).
2. Reyburn, W. *Flushed with Pride* (London: McDonald, 1969).
3. Camp, T.R. "Sedimentation and the Design of Settling Tanks," *Transactions of the American Society of Civil Engineers* 111: 895(1946).
4. Metcalf and Eddy. *Wastewater Engineering* (New York: McGraw-Hill, 1979).
5. Grady, L., and H.C. Lim. *Biological Wastewater Treatment* (New York: Marcel Dekker, 1980).
6. Benefield, L.D., and C.W. Randall. *Biological Process Design for Wastewater Treatment* (Englewood Cliffs, NJ: Prentice-Hall, 1980).

Chapter 9
Sludge Treatment and Disposal

The field of wastewater treatment engineering is littered with unique and imaginative processes for achieving high degrees of waste stabilization at attractive costs. Few of these "wonder plants" have proven themselves in practice, and quite often the problem has been the inattention to the sludge problem. Drawing a flow diagram with a little arrow labeled "SLUDGE TO DISPOSAL" has often been the total extent of the consideration for solids handling, treatment, and disposal.

In the past few years the fact that sludge treatment and disposal accounts for over fifty percent of the treatment costs in a typical secondary plant has prompted a renewed interest in this none-too-glamorous, but essential aspect of wastewater treatment.

This chapter is devoted to the problem of sludge treatment and disposal. The sources and quantities of sludge from various types of wastewater treatment systems are examined first, followed by a definition of sludge characteristics. Such solids concentration techniques as thickening and dewatering are discussed next, concluding with considerations for ultimate disposal.

SOURCES OF SLUDGE

The first source of sludge is the suspended solids (SS) that enter the treatment plant and are partially removed in the primary settling tank or clarifier. Commonly about 60 percent of the SS become *raw primary sludge*, which is highly putrescible and very wet (about 96 percent water).

The removal of BOD is basically a method of wasting energy, and secondary wastewater treatment plants are designed to reduce this high-energy material to low-energy chemicals. This process is typically accomplished by

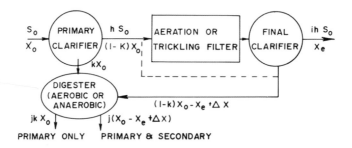

Figure 9–1. Schematic of a generalized secondary treatment plant.

biological means, using microorganisms (the "decomposers" in ecological terms) that use the energy for their own life and procreation. Secondary treatment processes such as the popular activated sludge system are *almost* perfect systems. Their major fault is that the microorganisms convert too little of the high-energy organics to CO_2 and H_2O and too much of it to new organisms. Thus the system operates with an excess of these microorganisms, or *waste activated sludge*. As defined in the previous chapter, the mass of waste activated sludge produced per mass of BOD removed in secondary treatment is known as the *yield*, expressed as kilograms of SS produced per kilogram of BOD removed.

Phosphorus removal processes also invariably end up with excess solids. If lime is used, the calcium carbonates and calcium hydroxyapatites are formed and must be disposed of. Aluminum sulfate similarly produces solids, in the form of aluminum hydroxides and aluminum phosphates. Even so-called "totally biological processes" for phosphorus removal end up with solids. The use of an oxidation pond or marsh for phosphorus removal is possible only if some organics (algae, water hyacinths, fish, etc.) are periodically harvested.

The quantities of sludges obtained from various wastewater treatment processes may be calculated as shown in Figure 9–1.[1] The symbols are defined as follows:

S_o = influent BOD, kg/hr (5 days, 20°C)
X_o = influent SS, kg/hr
h = fraction of BOD not removed in primary clarifier
i = fraction of BOD not removed in aeration or trickling filter
X_e = plant effluent SS, kg/hr
k = fraction of X_o removed in primary clarifier
j = fraction of solids not destroyed in digestion
ΔX = net solids produced by biological action, kg/hr
Y = yield = $\Delta X / \Delta S$, where $\Delta S = hS_o - ihS_o$

Example 9.1

The plant influent has a BOD_5 of 250 mg/L at a flow rate of 1,570 m³/hr (10 mgd), and an influent SS of 225 mg/L. The raw sludge produced would be equal to kX_o, where k is the fraction of SS removed in the primary clarifier and X_o is the influent SS in kilograms/hour. If k is (typically) 0.6, and X_o can be calculated as

$$225 \text{ mg/L} \times 1{,}570 \text{ m}^3/\text{hr} \times 1{,}000 \text{ L/m}^3 \times 10^{-6} \text{ mg/kg} = 353 \text{ kg/hr}$$

The raw sludge produced is then

$$0.6 \times 353 = 212 \text{ kg/hr}$$

Typical values for the constants, for domestic wastewater, are shown below:

$$S_o = \begin{cases} 250 \times 10^{-3} \times Q = \text{kg/h if Q is m}^3/\text{hr} \\ 250 \times 8.34 \times Q = \text{lb/day if Q is mgd} \end{cases}$$

$$X_o = \begin{cases} 220 \times 10^{-3} \times Q = \text{m}^3/\text{h if Q is m}^3/\text{hr} \\ 220 \times 8.34 \times Q = \text{lb/day if Q is mgd} \end{cases}$$

$$k = 0.6$$
$$h = 0.7$$

$$X_e = \begin{cases} 20 \times 10^{-3} \times Q = \text{m}^3/\text{h if Q is m}^3/\text{hr} \\ 20 \times 8.34 \times Q = \text{lb/day if Q is mgd} \end{cases}$$

$j \approx 0.5$ for anaerobic
$j \approx 0.8$ for aerobic
$i = 0.1$ for well-operated activated sludge
$i = 0.2$ for trickling filters
$Y = 0.5$ for activated sludge
$Y = 0.2$ for trickling filters

Quantities of chemical sludges resulting from the precipitation of phosphorus, for example, must be added to the total sludge produced.

CHARACTERISTICS OF SLUDGES

The characteristics of sludges of interest depend entirely on what is to be done with a sludge. For example, if the sludge is to be thickened by gravity, its settling and compaction are important. On the other hand, if the sludge is to be digested anaerobically, the concentrations of volatile solids, heavy metals, etc., are important.

Another fact of immense importance in the design of sludge handling and disposal operations is the variability of the sludges. In fact, this variability may be stated in terms of three "laws":

1. No two wastewater sludges are alike in all respects.
2. Sludge characteristics change with time.
3. There is no "average sludge."

The first statement reflects the fact that no two wastewaters are alike and that if the variable of treatment is added, the sludges produced will have significantly different characteristics.

The second statement is often overlooked by designers. For example, the settling characteristics of chemical sludges from the treatment of plating wastes (e.g., $PB(OH)_2$, $Zn(OH)_2$, or $Cr(OH)_3$) vary with time simply because of uncontrolled pH changes. Biological sludges are of course continually changing, with the greatest change occurring when the sludge changes from aerobic to anaerobic (or vice versa). It is thus quite difficult to design sludge-handling equipment, since the sludge may change in some significant characteristic in only a few hours.

The third "law" is constantly violated. Tables showing "average values" for "average sludges" are useful for illustrative and comparative purposes only, and should not be used for design.

With that caveat, we now proceed to discuss some characteristics of "average sludges," tabulated in Table 9–1.

The first characteristic, solids concentration, is perhaps the most important variable, defining the volume of sludge to be handled and determining whether the sludge behaves as a liquid or a solid. The importance of volatile solids is, of course, in the disposability of the sludge. With high volatiles, a sludge would be difficult to dispose of into the environment. As the volatiles are degraded, gases and odors are produced, and thus a high volatile solids concentration would restrict the methods of disposal.

The rheological characteristics of sludge are of interest in that this is one of only a few truly basic parameters describing the physical nature of a sludge. Two-phase mixtures like sludges, however, are almost without exception non-Newtonian and thixotropic. Sludges tend to act as pseudoplastics, with an apparent yield stress and a plastic viscosity. The rheological behavior of a pseudoplastic fluid is defined by a rheogram shown as Figure 9–2. The term thixotropic relates to the time dependence of the rheological properties.

Sludges tend to act more like plastic fluids as the solids concentration increases. True plastic fluids may be described by the equation

$$\tau = \tau_y + \eta \, \frac{du}{dy}$$

where τ = shear stress
          τ_y = yield stress
          η = plastic viscosity
          du/dy = rate of shear, or the slope of the viscosity (u)-depth (y) profile

Table 9–1. Sludge Characteristics

Type of Sludge	Physical					Chemical		
	Solid Concentration (mg/L)	Volatile Solids (%)	Yield Strength (dyne/cm²)	Plastic Viscosity (g/cm·sec)		Nitrogen (% as N)	Phosphorus (% as P₂O₂)	Potassium (% as K₂O)
Water			0	0.01				
Raw primary	60,000	60	40	0.3		2.5	1.5	0.4
Mixed digested	80,000	40	15	0.9		4.0	1.4	0.2
Waste activated	15,000	70	0.1	0.06		4.0	3.0	0.5
Alum, ppt	20,000	40				2.0	2.0	
Lime, ppt	200,000	18				2.0	3.0	

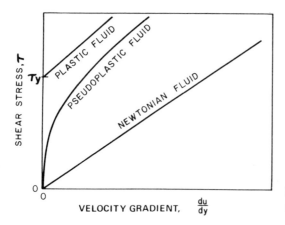

Figure 9–2. Rheograms for three different fluids.

Although sludges are seldom true plastics, the above equation may be used as an approximation of the rheogram.

The yield stress may vary from above 40 dyne/cm^2 for 6 percent raw sludge to only 0.07 dyn/cm^2 for a thickened activated sludge. The large differences suggest that rheological parameters could well be used for scale-up purposes. Unfortunately, few researchers have bothered to measure the rheological characteristics (such analyses are not even included in *Standard Methods*) and many gaps exist in the available data.

The chemical composition is important for several reasons. First, the fertilizer value of the sludge is dependent on the availability of N, P, and K, as well as trace elements. A more important measurement, however, is the concentration of heavy metals and other toxins that would make the sludge toxic to the environment. The ranges of heavy metal concentrations are very large (e.g., cadmium can range from almost zero to over 1,000 mg/kg). Since a major source of such toxins is in industrial discharges, a single poorly operated industrial firm may contribute enough toxins to make the sludge worthless as a fertilizer. Most engineers agree that although it would be most practical to treat a sludge at the plant to achieve the removal of toxic components, there are no effective methods available for removing heavy metals, pesticides, and other potential toxins from the sludge, and that the control must be over the influent (that is, tight sewer ordinances).

In addition to the physical and chemical characteristics, the biological parameters of sludge may also be important. The volatile solids parameter is in fact often interpreted as a biological characteristic, the assumption being that the volatile suspended solids (VSS) is a gross measure of viable biomass. Another important parameter, especially in regard to ultimate disposal, is the concentration of pathogens, both bacteriological and viral. The primary clarifier seems to act as a viral and bacteriological concentrator, with a substantial fraction of these microorganisms existing in the sludge instead of the liquid effluent.

SLUDGE TREATMENT

A great deal of money could be saved, and troubles averted, if sludge could be disposed of as it is drawn off the main process train. Unfortunately, the sludges have three characteristics that make such a simple solution unlikely: they are aesthetically displeasing, they are potentially harmful, and they have too much water.

The first two problems are often solved by stabilization, such as *anaerobic* or *aerobic digestion*. The third problem requires the removal of water by either thickening or dewatering. Accordingly, the next three sections cover the topics of stabilization, thickening, and dewatering, followed finally by considerations of ultimate disposal.

Before we embark on a discussion on sludge treatment, however, we should be reminded that sludge treatment should be used only when necessary. It makes little sense to spend time and money to treat sludges if their ultimate disposal does not require it. The "Get the water out!" syndrome often found in engineering offices is often associated with the "Gotta digest it!" disease. These processes all cost money and should be considered only if direct disposal in undigested liquid form is not practical.

Sludge Stabilization

The objective of sludge stabilization is to reduce the problems associated with two of the detrimental characteristics listed above: sludge odor and putrescibility, and the presence of pathogenic organisms.

There are three primary means of sludge stabilization:

- lime
- aerobic digestion
- anaerobic digestion

Lime stabilization is achieved by adding lime (either as hydrated lime, $Ca(OH)_2$, or as quicklime, CaO) to the sludge and thus raising the pH to about 11 or above. This significantly reduces the odor and helps in the destruction of pathogens. The major disadvantage of lime stabilization is that it is temporary. With time (days) the pH drops and the sludge once again becomes putrescible.

Aerobic stabilization is merely a logical extension of the activated sludge system. Waste activated sludge is placed in dedicated aeration tanks for a very long time, and the concentrated solids are allowed to progress well into the endogenous respiration phase, in which food is obtained only by the destruction of other viable organisms. This results in a net reduction in total and volatile solids. Aerobically digested sludges are, however, more difficult to dewater than anaerobic sludges.

The third commonly used method of sludge stabilization is anaerobic digestion. The biochemistry of anaerobic decomposition of organics is illustrated in Figure 9–3. Note that this is a staged process, with the solution of organics by extracellular enzymes being followed by the production of organic acids by a large and hearty group of anaerobic microorganisms known, appropriately enough, as the *acid formers*. The organic acids are in turn degraded further by a group of strict anaerobes called *methane formers*. These microorganisms are the prima donnas of wastewater treatment, getting upset at the least change in their environment. The success of anaerobic treatment thus boils down to the creation of a suitable condition for the methane formers. Since they are strict anaerobes, they are unable to function in the presence of oxygen and are very sensitive to environmental conditions such as temperature, pH, and toxins. If a digester goes "sour," the methane formers have been inhibited in some way. The acid formers, however, keep chugging away, making more organic acids. This has the effect of further lowering the pH and making conditions even worse for the methane formers. A sick digester is therefore difficult to cure without massive doses of lime or other antacids.

Most treatment plants have two kinds of digesters—primary and secondary (Figure 9–4). The primary digester is covered, heated, and mixed to increase the reaction rate. The temperature of the sludge is usually about 35°C (95°F). Secondary digesters are not mixed or heated and are used for storage of gas and for concentrating the sludge by settling. As the solids settle, the liquid supernat-

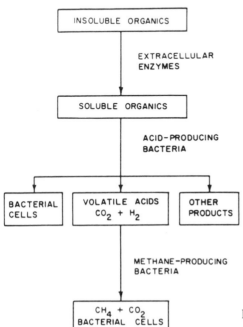

Figure 9–3. Generalized biochemical reactions to anaerobic digestion.

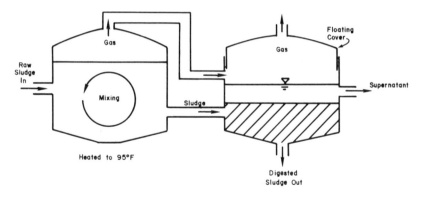

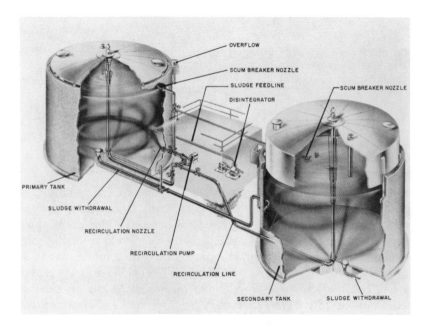

Figure 9–4.   Anaerobic sludge digesters. [Photo courtesy of Dorr Oliver Inc.]

ant is pumped back to the main plant for further treatment. The cover of the secondary digester often floats up and down, depending on the amount of gas stored. The gas is high enough in methane to be used as a fuel, and is in fact usually used to heat the primary digester.

Anaerobic digesters are commonly designed on the basis of solids loading. Experience has shown that domestic wastewaters contain about 120 g (0.27 lb) of SS per day per capita. This may be translated, knowing the population served, into total SS to be handled. To this, of course, must be added the production of solids in secondary treatment. Once the solids production is calculated, the digester volume is estimated by assuming a reasonable loading factor such as 4 kg of dry solids/$m^3 \cdot$ day (0.27 lb/$ft^3 \cdot$ day). This loading factor is decreased if a higher reduction of volatile solids is desired.

Example 9.2

Raw primary and waste activated sludge at 4 percent solids is to be anaerobically digested at a loading of 3 kg/$m^3 \cdot$ day. The total sludge produced in the plants is 1,500 kg of dry solids per day. Calculate the required volume of the primary digester and the hydraulic retention time.

The production of sludge requires

$$\frac{1,500 \text{ kg/day}}{3 \text{ kg/m}^3 \cdot \text{day}} = 500 \text{ m}^3 \text{ digester volume}$$

The total mass of wet sludge pumped to the digester is

$$\frac{1,500 \text{ kg/day}}{0.04} = 37,500 \text{ kg/day}$$

and since 1 L of sludge weighs about 1 kg, the volume of sludge is 37,500 L/day or 37.5 m^3/day and the hydraulic residence time is $\bar{t} = (500 \text{ m}^3)/(37.5 \text{ m}^3/\text{day}) = 13.3$ days.

The production of gas from digestion varies with the temperature, solids loading, solids volatility, and other factors. Typically, about 0.6 m^3 of gas per kg of volatile solids added (10 ft^3/lb) has been observed. This gas is about 60 percent methane and burns readily, usually being used to heat the digester and answer additional energy needs within a plant. It has been found that an active group of methane formers operates at 35°C (95°F) in common practice, and this process has become known as *mesophilic digestion.* As the temperature is increased, to about 45°C (115°F), another group of methane formers predominates, and this process is tagged *thermophilic digestion.* Although the latter process is faster and produces more gas, such elevated temperatures are more difficult and expensive to maintain.

Finally, a word of caution should be expressed about the problem of mixing a primary digester. The assumption is invariably made that the tank is totally mixed, either by mechanical means or by bubbling gases through it. Unfortunately, digester mixing is quite difficult, and some recent studies have shown that on the average, only about 20 percent of the tank volume is well mixed!

All three stabilization processes reduce the concentration of pathogenic organisms, but to varying degrees. Lime stabilization achieves a high degree of sterilization, owing to the high pH. Further, if quicklime (CaO) is used, the reaction is exothermic and the elevated temperatures assist in the destruction of pathogens. Aerobic digestion at ambient temperatures is not very effective in the destruction of pathogens.

Anaerobic digesters have been well studied from the standpoint of pathogen viability since the elevated temperatures should result in substantial sterilization. As early as 1958, however, it was found that *Salmonella typhosa* organisms and many other pathogens can survive digestion. Polio viruses similarly survive with little reduction in virulence. An anaerobic digester cannot, therefore, be considered a method of sterilization.

Sludge Thickening

Sludge thickening is a process in which the solids concentration is increased and the total sludge volume is correspondingly decreased, but the sludge still behaves like a liquid instead of a solid. Typically, for mixed digested sludges, the point at which sludge begins to have the properties of a solid is between 15 and 20 percent solids. Thickening also implies that the process is gravitational, using the difference between particle and fluid densities to achieve greater compacting of solids.

The advantages of sludge thickening in reducing the volume of sludge to be handled are substantial. With reference to Figure 9–5, a sludge with 1 percent solids thickened to 5 percent results in an 80 percent volume reduction. A 20 percent solids concentration, which might be achieved by mechanical dewatering (discussed in the next section), would result in a 95 percent reduction in volume. The savings in treatment, handling, and disposal costs accrued may be substantial.

Two types of nonmechanical thickening operations are presently in use: the gravity thickener and the flotation thickener. These are not very good names, since the latter also uses gravity to separate the solids from the liquid. For the sake of simplicity, however, we will continue to use the two descriptive terms.

A typical gravity thickener is shown in Figure 9–6. The influent, or feed, enters in the middle, and the water moves to the outside, eventually leaving as the clear effluent over the weirs. The sludge solids settle as a blanket and are removed out the bottom.

The thickening characteristics of a sludge have for many years been described by the sludge volume index (SVI). The parameter is defined as the

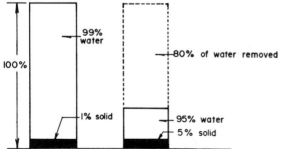

Figure 9–5. Volume reduction owing to sludge thickening.

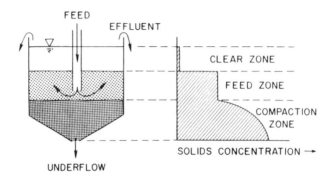

Figure 9–6. Gravity thickener. [Photo courtesy of Dorr Oliver Inc.]

volume occupied by 1 g of sludge after settling for 30 minutes. The sludge is commonly settled in a 1-L cylinder, and the volume of sludge (in milliliters) is measured after 30 minutes. The SVI is calculated as

$$SVI = \frac{mL \text{ of sludge after 30 min} \times 1,000}{\text{sludge SS mg/L}}$$

Treatment plant operators usually consider sludges with an SVI of less than 100 as well-settling sludges and those with an SVI of greater than 200 as potential problems. These numbers are of some use in estimating the settleability of activated sludge, and the SVI is without question a valuable tool in running a secondary treatment plant. But the SVI also has some drawbacks and potential problems.

One of these problems is that the SVI is not necessarily independent of solids concentration. Substantial increases in SS can change the SVI severalfold. Further, for high levels of SS, the SVI has a maximum above which it can never occur. Take, for example, a lime sludge with an SS concentration of 40,000 mg/ L and suppose that it doesn't settle at all (the sludge volume after 30 minutes is 1,000 mL). Calculating the SVI gives us

$$SVI = \frac{1,000 \times 1,000}{40,000} = 25$$

How can a sludge that does not settle at all have an SVI = 25, when we originally noted that a SVI of less than 100 indicated a good settling sludge?

The point is that the maximum SVI for a sludge with an SS of 40,000 is 25. Similar maximum values may be calculated and plotted as shown in Figure 9–7.

A better way to describe the settling characteristics of a sludge is a "flux plot." This graph is developed by first running a series of settling tests at various solids concentrations and recording the sludge-water interface height with time. The velocity of settling is calculated as the slope of the initial straight line portion of the curve and plotted against the solids concentration. Multiplying a velocity by its corresponding solids concentration yields solids flux, which is plotted against the concentration (as shown in Figure 9–8). The flux curve is a "signature" of the sludge thickening characteristics. Further, it may be used to design thickeners. Note that at any concentration C_i, the interface settling velocity is v_i and thus the solids flux at the concentration is $C_i v_i$ (kg/m^3 × m/hr = kg/m$^2 \cdot$ hr).

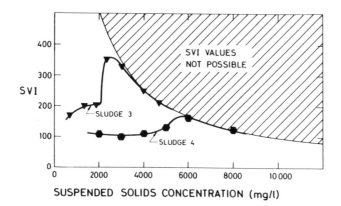

Figure 9–7.   Maximum values of sludge volume index (SVI).

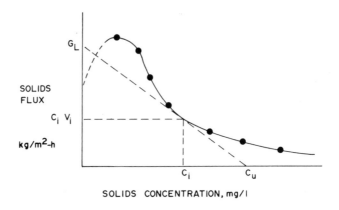

Figure 9–8. Calculation of limiting solids flux in gravitational thickening.

The design involves choosing a desired thickened sludge solids concentration, C_u (estimated in laboratory tests), and drawing a line from this value tangent to the underside of the flux curve. The intercept at the ordinate is the limiting flux, G_L, or that solids flux that controls the thickening operation. In a continuous thickener, it is impossible to pass more solids than that (at the stated concentration) through a unit area.

The units of flux are kg of solids/$m^2 \cdot$ hr (or other comparable units), and are defined as

$$G = \frac{Q_o C_o}{A}$$

where   G = solids flux
        Q_o = interflow rate (m^3/h)
        C_o = inflow solids concentration (mg/L)
        A = surface area of thickener (m^3)

If G is the limiting flux, the necessary thickener area can be calculated as

$$A_L = \frac{Q_o C_o}{G_L}$$

where the subscript L specifies the limiting area and flux.

This graphical procedure may be used to develop an optimal process. If the calculated thickener area is too big, a new (less ambitious) underflow solids concentration may be selected and a new required area calculated.

Example 9.3
If, on Figure 9–8, the C_u was chosen as 25,000 mg/L and the G_L was then read off the graph as 3 kg/$m^2 \cdot$ hr, what is the required area for a thickener if the feed is 60 m^3/hr of sludge with 1 percent solids? Suppose the C_u was next

chosen as 40,000 mg/L and the G_L was read off as 1.8 kg/m² · day. What is the required area in the second case?

The limiting area in the first case is

$$A_L = \frac{Q_o C_o}{G_L} = \frac{60 \text{ m}^3/\text{h} \times 0.01 \times 1,000 \text{ kg/m}^3}{3 \text{ kg/m}^2 \cdot \text{hr}} = 200 \text{ m}^2$$

and in the second case

$$A_L = \frac{60 \times 0.01 \times 1,000}{1.8} = 333 \text{ m}^2$$

Note that the area requirement increases as the desired underflow concentration increases.

In the absence of laboratory data, thickeners are designed on the basis of solids loading, which is another way of expressing the limiting flux. Gravity thickener design loadings for some specific sludges are shown in Table 9–2.

Table 9–2.   Design Loadings for Gravity Thickeners[a]

Sludge	Design Loading (kg of solids/m² · hr)
Raw primary	5.2
Waste activated	1.2
Raw primary + waste activated	2.4
Trickling filter sludge	1.8

[a]Note: $0.204 \times \text{kg/m}^2 \cdot \text{hr} = \text{lb/ft}^2 \cdot \text{hr}$.

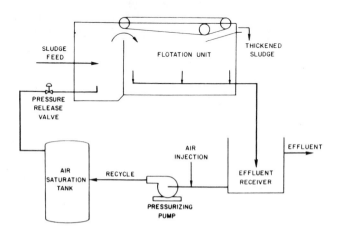

Figure 9–9.   Flotation thickener.

A flotation thickener, shown in Figure 9–9, operates by forcing air under pressure to dissolve in the return flow and releasing the pressure as the return is mixed with the feed. As the air comes out of solution, the tiny bubbles attach themselves to the solids and carry them upward, to be scraped off.

Sludge Dewatering

As defined above, dewatering differs from thickening in that the sludge should behave as a solid after it has been dewatered. Dewatering is seldom used as an intermediate process, unless the sludge is to be incinerated. Most wastewater plants use dewatering as a final method of volume reduction before ultimate disposal.

In the United States, five dewatering techniques have been most popular: sand beds, vacuum filters, pressure filters, belt filters and, centrifuges. Each of these is discussed below.

Sand beds have been in use for a great many years and are still the most cost-effective means of dewatering when land is available and labor is not exorbitant. The beds consist of tile drains in gravel, covered by about 26 cm (10 in.) of sand. The sludge to be dewatered is poured on the beds at about 15 cm (6 in.) deep. Two mechanisms combine to separate the water from the solids: seepage and evaporation. Seepage into the sand and through the tile drains, although important in the total volume of water extracted, lasts for only a few days. The sand pores are quickly clogged, and all drainage into the sand ceases. The mechanism of evaporation takes over, and this process is actually responsible for the conversion of liquid sludge to solid. In some northern areas sand beds are enclosed under greenhouses to promote evaporation as well as prevent rain from falling into the beds.

For mixed digested sludge, the usual design is to allow for 3 months of drying time. Some engineers suggest that this period be extended to allow a sand bed to rest for a month after the sludge has been removed. This seems to be an effective means of increasing the drainage efficiency once the sand beds are again flooded.

Raw sludge will not drain well on sand beds and will usually have an obnoxious odor. Hence raw sludges are seldom dried on beds. Raw secondary sludges have a habit of either seeping through the sand or clogging the pores so quickly that no effective drainage takes place. Aerobically digested sludge may be dried on sand, but usually with some difficulty.

If dewatering by sand beds is considered impractical, mechanical dewatering techniques must be used. The first successful mechanical dewatering device used in sludge treatment was the *vacuum filter*. As shown in Figure 9–10, the vacuum filter consists of a perforated drum covered with a fabric. A vacuum is drawn in the drum, and the covered drum is dipped into sludge. The water moves through the filter cloth, leaving the solids behind, eventually to be scraped off.

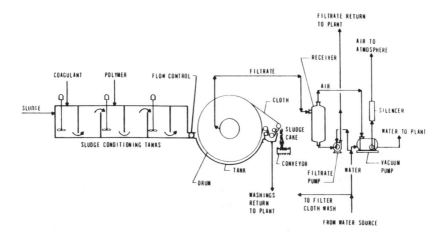

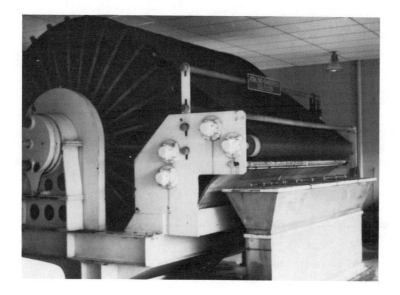

Figure 9–10. Vacuum filter.

The effectiveness of a vacuum filter in dewatering a specific sludge is measured most frequently by the *specific resistance to filtration* test. The resistance of a sludge to filtration can be stated as

$$r = \frac{2PA^2b}{\mu w}$$

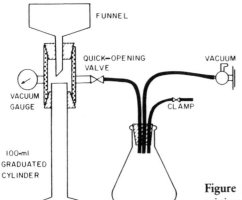

Figure 9–11. Büchner funnel test for determining specific resistance to filtration.

where P = vacuum pressure, N/m^2
   A = area of the filter, m^2
   μ = filtrate viscosity, $N \cdot sec/m^2$
   w = cake deposited per volume of filtrate (for dry cakes this may be approximated as the feed solids concentration), kg/m^3
   b = slope of the filtrate volume versus time/filtrate volume curve

Using the units above, specific resistance is in terms of meters/kilogram.

The factor b may be determined from a simple test with a Büchner funnel, as shown in Figure 9–11. The sludge is poured onto the filter, the vacuum is exerted, and the volume of the filtrate is recorded against time. The plot of these data yields a straight line with a slope b.

Example 9.4
   Results of a Büchner funnel sludge filterability experiment are shown below. Calculate the specific resistance to filtration.

Time (min)	Θ (sec)	V (mL)	Corrected[a] V (mL)	$\frac{\Theta}{V}\left(\frac{sec}{mL}\right)$
−2		0		
0	0	1.5	0	
1	60	2.8	1.3	46.3
2	120	3.8	2.3	52.3
3	180	4.6	3.1	58.0
4	240	5.5	4.0	60.0
5	300	6.1	4.6	65.2

[a]First 2 min are ignored; this is due partly to storage of water in the filter and funnel, and partly because the resistance owing to the filter must be allowed to become negligible.

Other test variables:

$$P = \text{pressure} = 10 \text{ psi} = 703 \text{ g/cm}^2$$
$$= 6.90 \times 10^4 \text{ N/m}^2$$
$$\mu = \text{viscosity} = 0.0011 \text{ poise} = 0.0011 \text{ N} \cdot \text{sec/m}^2$$
$$w = 0.075 \text{ g/mL} = 75 \text{ kg/m}^3$$
$$A = 44.2 \text{ cm}^2 = 0.00442 \text{ m}^2$$

The data are plotted as Figure 9–12. The slope of the line, b from the curve, is $5.73 \text{ sec/cm}^6 = 5.73 \times (100 \text{ cm/m})^6 = 5.73 \times 10^{12} \text{ sec/m}^6$

$$r = \frac{2 \times 6.9 \times 10^4 \times (0.00442)^2 \times 5.73 \times 10^{12}}{0.0011 \times 75} = 1.86 \times 10^{13} \text{ m/kg}$$

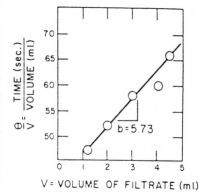

Figure 9–12.   Typical laboratory filtration data, as used in Example 9.4.

Again, although there are no "average sludges," it may be instructive to list some data from experience and compare the filterability of different sludges. Such a listing is shown in Table 9–3.

Table 9.3.   Specific Resistance of Typical Sludges

	Specific Resistance[a] (m/kg)
Raw primary	$10–30 \times 10^{14}$
Mixed digested	$3–30 \times 10^{14}$
Waste activated	$5–20 \times 10^{14}$
Lime and biological sludge	$1–5 \times 10^{14}$
Lime slurry	$5–10 \times 10^{13}$
Alum	$2–10 \times 10^{13}$

[a]As a rule of thumb, a sludge will not filter well if the specific resistance exceeds 1×10^{12} m/kg.

The Büchner funnel test for specific resistance is rather clumsy and time consuming. An indirect method for estimating how well a sludge will filter uses the *capillary suction time* (CST) apparatus, developed in Great Britain. The CST device, as illustrated in Figure 9–13, allows the water to seep out of a sludge (in effect, a filtration process) and onto a blotter. The speed at which the water is taken up by the blotter is measured by timing of the travel over a set distance. This time, in seconds, may be correlated to specific resistance. A short time would indicate a highly filterable sludge, whereas a long time would portend problems in filtration.

A major drawback of the CST test is that the CST, unlike the specific resistance to filtration, is dependent on the solids concentration. Again, using the Darcy Equation, it is possible to derive an expression that appears to be an excellent measure of sludge filterability.[2]

$$\chi = \phi\left(\frac{\mu X}{t}\right)$$

where χ = sludge filterability, $kg^2 m^4/sec^2$
   ϕ = dimensionless instrument constant, specific for each manufacturer
   μ = fluid (not sludge) viscosity, poise
   X = suspended solids concentration, mg/L
   t = capillary suction time, sec

Neither the specific resistance nor the CST sludge filterability tests, although the best parameters for filtration, should be used for design. The *filter leaf* test is by far superior, simply because it simulates the workings of a vacuum filter so well. Figure 9–14 is a diagram showing the filter leaf apparatus. The filter cloth (a sample of the proposed fabric) is wrapped around a hollowed-out disc, the disc (filter leaf) is inserted into the sludge for about 20 seconds (the length of time a typical vacuum filter might be in contact with the sludge) and then raised out of the sludge for maybe 40 seconds (simulating the drying cycle), and the

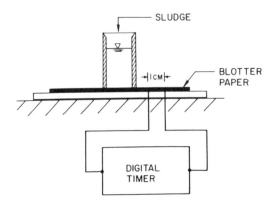

Figure 9–13. Capillary suction time apparatus.

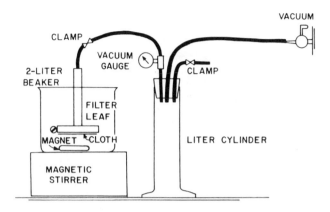

Figure 9–14. Filter leaf apparatus for vacuum filter design.

vacuum is cut off. The cake developed is a very good approximation of a cake formed by a continuous filter.

The design parameter calculated from the filter leaf test is the *filter yield,* in terms of kilograms of dry solids per square meter of filter area per hour. If this value is known, the required filter size may be determined by selecting the hours of operation and establishing how much sludge needs to be removed from the treatment plant.

Typical filter yields for some representative sludges with polymer conditioning are shown in Table 9–4.

The *pressure filter,* shown in Figure 9–15, uses positive pressure to force the water through a filter cloth. Typically, the pressure filters are built as plate-and-frame filters, in which the sludge solids are captured in the spaces between the plates and frames, which are then pulled apart to allow for sludge cleanout.

The *belt filter,* shown in Figure 9–16, operates as both a pressure filter and by a gravity drainage. As the sludge is introduced onto the moving belt, the free water drips through the belt, but the solids are retained. The belt then moves

Table 9–4.   Filter Yields for Typical Sludges

	Filter Yield[a] $(kg/m^2 \cdot hr)$
Raw primary (with polymer)	40
Mixed anaerobically digested (with polymer)	20
Waste activated	(will not filter)
Lime and biological sludge	15

[a]$kg/m^2 \cdot hr \times 0.204 = lb/ft^2 \cdot hr.$

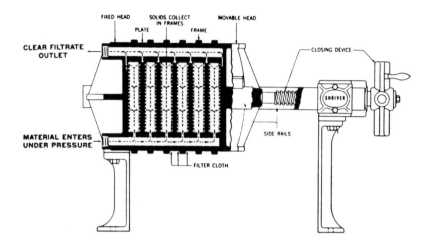

Figure 9–15. Pressure filters. [Photo courtesy of Envirex.]

into the dewatering zone where the sludge is squeezed between two belts. These machines are quite effective in dewatering many different types of sludge and are being widely installed in small wastewater treatment plants.

Centrifugation became popular in wastewater treatment only after organic polymers were available for sludge conditioning. Although the centrifuge will work on any sludge (unlike the vacuum filter, which will not pick up some sludges, resulting in zero filter yield), without good conditioning, most unconditioned sludges cannot be centrifuged with greater than 60 or 70 percent solids recovery.

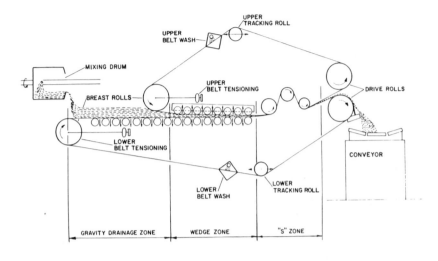

Figure 9–16. Belt filter.

The centrifuge most widely used is the solid bowl decanter, which consists of a bullet-shaped body rotating on its long axis. The sludge is placed into the bowl, and the solids settle out under about 500 to 1,000 gravities (centrifugally applied) and are scraped out of the bowl by a screw conveyor (Figure 9–17).

Although laboratory tests are of some value in estimating centrifuge applicability, tests with continuous models are considerably better and highly recommended whenever possible.

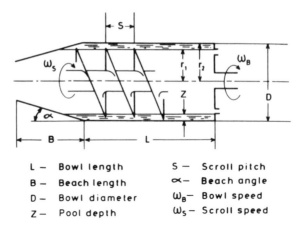

L — Bowl length   S — Scroll pitch
B — Beach length   ∝— Beach angle
D — Bowl diameter ω_B— Bowl speed
Z — Pool depth ω_S— Scroll speed

Figure 9–17. Solid bowl centrifuges. [Photo courtesy of Indersoll Rand and I. Krüger.]

A centrifuge must be able first to settle the solids and then to move them out of the bowl. Accordingly, two parameters have been suggested for scale-up between two geometrically similar machines.

Settling characteristics are measured by the *Sigma equation*.[3] Without going into the derivation of this parameter, it is simply assumed that if two machines (1 and 2) are to have equal effects on the settling within the bowl, the relationship

$$\frac{Q_1}{\Sigma_1} = \frac{Q_2}{\Sigma_2}$$

must hold. The symbol Q represents the liquid flow rate into the machine, and Σ is a parameter composed of machine (not sludge!) characteristics. For solid bowl centrifuges, Sigma is calculated as

$$\Sigma = \frac{\forall \omega}{g \ln(r_2/r_1)}$$

where ω = rotational velocity of bowl, rad/sec
      g = gravitational constant, m/sec^2
      r_2 = radius from centerline to inside bowl wall, m
      r_1 = radius from centerline to surface of sludge, m
      $\forall$ = liquid volume in the pool, m^3

Thus if machine 1, at a certain Σ_1 as calculated from the machine parameters, at the flow rate Q_1 produced a satisfactory result, it is expected that a second (geometrically similar) machine with a (larger) Σ_2 will achieve equal dewatering performance at a flow rate of Q_2.

This analysis does not, however, consider the movement of the solids out of the bowl, a very necessary component of centrifugation. The solids movement may be calculated by the *Beta equation*,[4] in which again, for two machines,

$$\frac{Q_1}{\beta_1} = \frac{Q_2}{\beta_2}$$

where Q now is the terms of solids (pounds per hour or other such dimension) and where $\beta = (\Delta\omega)SN\pi Dz$. These terms are defined as

    $\Delta\omega$ = difference in rotational velocity between the bowl and conveyor,
        $\omega_B - \omega_s$, rad/sec
    S = scroll pitch (distance between blades), m
    N = number of leads
    D = bowl diameter, m
    z = depth of sludge in the bowl, m

The scale-up procedure involves the calculation of Q_2 for liquid as well as solids throughout, and the lowest value would govern the centrifuge capacity.

Example 9.5

    A solid bowl centrifuge (machine 1) was found to perform well if fed at 0.5 m^3/hr of 1 percent solids sludge. It was decided to scale-up to a larger machine (machine 2), and it is necessary to determine the flow rate at which this geometrically similar machine would perform equally well. The machine variables are as follow:

	Machine 1	Machine 2
Bowl diameter (cm)	20	40
Pool depth (cm)	2	4
Bowl speed (rpm)	4,000	3,200
Bowl length (cm)	30	72
Conveyor speed (rpm)	3,950	3,150
Pitch (cm)	4	8
Number of leads	1	1

Looking first at the settling (Sigma) scale-up, the volume of sludge in the pool may be approximated as

$$\forall = 2\pi \left(\frac{r_1 + r_2}{2} \right) (r_2 - r_1) L$$

where the variables are defined in Figure 9–17. The volumes of the two machines are

$$\forall_1 = 2(3.14) \left(\frac{8 + 10}{2} \right) (10 - 8)(30) = 3,400 \text{ cm}^3$$

$$\forall_2 = 2(3.14) \left(\frac{16 + 20}{2} \right) (20 - 16)(72) = 32,600 \text{ cm}^3$$

Since

$$\omega = (\text{rpm}) \left(\frac{1}{60} \frac{\text{min}}{\text{sec}} \right) \left(2\pi \frac{\text{rad}}{\text{rev}} \right)$$

$$\omega_1 = 4,000 \left(\frac{1}{60} \right) (2)(3.14) = 420 \text{ rad/sec}$$

$$\omega_2 = 3,200 \left(\frac{1}{60} \right) (2)(3.14) = 335 \text{ rad/sec}$$

$$\Sigma_1 = \frac{(420)^2}{980} \left(\frac{3,400}{\ln(10/8)} \right) = 2.76 \times 10^6$$

$$\Sigma = \frac{(335)^2}{980} \left(\frac{32,600}{\ln(20/16)} \right) = 16.8 \times 10^6$$

Thus

$$Q_2 = \frac{\Sigma_2}{\Sigma_1} Q_1 = \frac{16.8}{2.78} (0.5) = 3.2 \text{ m}^3/\text{hr}$$

Next looking at the solids loading (Beta), the solids flow rate is

$$Q_1 = 0.5 \frac{\text{m}^3}{\text{h}} \times 0.01 \times 1,000 \frac{\text{kg}}{\text{m}^3} = 5 \text{ kg/hr}$$

$$\beta = (\Delta\omega)(S)(N)(\pi)(D)(z)$$

$$\beta_1 = (4,000 - 3,950)(4)(1)(3.14)(20)(2) = 25,120$$

$$\beta_2 = (3,200 - 3,150)(8)(1)(3.14)(40)(4) = 200,960$$

$$Q_2 = \frac{200,960}{25,120} (5) = 40 \text{ kg/hr}$$

corresponding to a liquid flow rate of

$$\frac{40}{0.01} \times 10^{-3} = 4 \text{ m}^3/\text{hr}$$

The liquid loading thus governs, and the big machine (machine 2) cannot be fed at a flow rate greater than 3.2 m²/hr.

ULTIMATE DISPOSAL

The options for ultimate disposal of sludge are limited to air, water, and land. Strict controls on air pollution complicate incineration, although this certainly is an option. Disposal of sludges in deep water (such as oceans) is decreasing owing to adverse or unknown detrimental effects on aquatic ecology. Land disposal may be either dumping in a landfill or spreading the sludge out over land and allowing natural biodegradation to assimilate the sludge back into the environment. Because of environmental and cost considerations, only incineration and land disposal are presently encountered. There is, however, increasing interest in the use of sludge as a fertilizer.

Incineration is actually not a method of disposal at all, but rather a sludge treatment step in which the organics are converted to H_2O and CO_2 and the inorganics drop out as a nonputrescible residue.

Two types of incinerator have found use in sludge treatment: multiple hearth and fluid bed. The multiple hearth incinerator, as the name implies, has several hearths stacked vertically, with rabble arms pushing the sludge progressively downward through the hottest layers and finally into the ash pit (Figure 9-18).

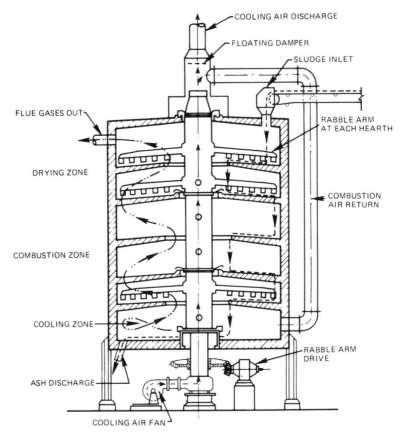

Figure 9–18. Multiple hearth incinerator. [Courtesy of Nichols Engineering and Research Corp.]

The fluid bed incinerator is full of hot sand that is suspended by air injection, and the sludge is incinerated within the moving sand. Owing to the violent motion within the fluid bed, scraper arms are unnecessary. The sand acts as a "thermal flywheel," allowing intermittent operation.

The second method of disposal—land spreading—is a science in its infancy. It has become popular only in the last few years, and design and operating data are only now being developed.

The ability of land to absorb sludge and to assimilate it depends on such variables as soil type, vegetation, rainfall, slope, etc. In addition, the important variable of the sludge itself will influence the capacity of a soil to assimilate sludge.

Generally, sandy soils with lush vegetation, low rainfall, and gentle slopes have proven most successful. Mixed digested sludges have been spread from tank trucks, and activated sludges have been sprayed from both fixed and moving nozzles. The application rate has been variable, but 100 dry tons/acre/ year is not an unreasonable estimate. Most unsuccessful land application systems may be traced to overloading the soil. Given enough time (and absence of toxic materials) any soil will assimilate sprayed liquid sludge.

There has been some successful use of land application of sludge for fertilization, particularly in silviculture operations. Forests and tree nurseries are far enough from population centers to minimize aesthetic objections, and the variable nature of sludge is not so problematical in silviculture as in other agricultural applications. Sludge may also be treated and packaged as fertilizer and plant food. The city of Milwaukee has pioneered the drying, disinfection, and deodorizing of sludge, which is packaged and marketed as the fertilizer Milorganite.

Transporting liquid sludge is often expensive, and volume reduction by dewatering is necessary. The solid sludge may then be deposited on land and disked in. A higher rate (tons/acre/year) may be achieved by trenching where 1-m^2 (3-ft^2) trenches are dug with a backhoe, and the sludge is deposited and covered. The sludge seems to assimilate rapidly, with undue leaching of nitrates or toxins.

In the last few years a method of chemically bonding the sludge solids so that the mixture "sets" in a few days has found use in industries that have especially critical sludge problems. Although *chemical fixation* is expensive, it is often the only alternative for besieged industrial plants. The leaching from the solid seems to be minimal.

The toxicity may be interpreted in several ways: toxicity to vegetation, toxicity to the animals (including people) who eat the vegetation, and poisoning of groundwater supplies. Most domestic sludges do not contain sufficient toxins such as heavy metals to cause harm to vegetation. The total body burden of heavy metals is of some concern, however. It is possible to precipitate out the metals during sludge treatment, but the most effective means of controlling such toxicity seems to be to prevent these metals from entering the sewerage system. Strong enforced sewerage ordinances are necessary and may be cost-effective. The regulatory aspects of sludge disposal are discussed further in Chapter 11.

CONCLUSION

Sludge disposal represents a major headache for many municipalities. Handling and treatment of sludge, particularly when it is concentrated, pose hitherto unrecognized hazards to the workers who perform these processes. Relief does not seem to be in sight. After all, sludge represents the true residues of our civilization, and its composition reflects our style of living, our technological development, and our ethical concerns. "Pouring things down the drain" that might cause damage or health problems in the ultimate disposal of the sludge is too often done without thought, or with malice. We return to the question of moral responsibilities in the chapter on environmental ethics (Chapter 26).

PROBLEMS

9.1   Using reasonable values, estimate the sludge production in a 10-mgd wastewater treatment plant. Express the sludges both as volumes per time and as dry solids per time.

9.2   A 1-L cylinder is used to measure the settleability of 0.5 percent SS sludge. After 30 minutes, the settled sludge solids occupy 600 mL. Calculate the SVI.

9.3   What measures of "stability" would you need if a sludge from a wastewater treatment plant was to be:
a. placed on the White House lawn.
b. dumped into a trout stream.
c. sprayed on a playground.
d. sprayed on a vegetable garden.

9.4   A sludge is thickened from 2,000 mg/L to 17,000 mg/L. What is the reduction in volume, in percent?

9.5   A laboratory batch thickening test produced the following results:

Solids Concentration (%)	Settling Velocity (cm/min)
0.6	0.83
1.0	0.25
1.4	0.11
1.8	0.067
2.2	0.041

Calculate the required thickener area if the desired underflow concentration (C_u) is to be 3 percent solids and the feed to the thickener is 0.5 m^3/min at a solids concentration of 2,000 mg/L.

9.6   A Büchner funnel test for specific resistance to filtration, conducted at a vacuum pressure of 6×10^4 N/m^2, with an area of 400 cm^2, and a solids concentration of 60,000 mg/L, yielded the following data:

Time (min)	Filtrate Volume (ml)
1	11
2	20
3	29
4	37
5	43

Calculate the specific resistance to filtration. Use $\mu = 0.011$ N · sec/m^2. Is this sludge amenable to dewatering by vacuum filtration?

9.7 Two geometrically similar centrifuges have the following machine characteristics:

	Machine 1	Machine 2
Bowl diameter, cm	25	35
Pool depth $(r_2 - r_1)$, cm	3	4
Bowl length, cm	30	80

If the flow rates are 0.6 and $4.0\,m^3/hr$ and the first machine operates at 1,000 rpm, at what speed should machine 2 be operated to achieve similar performance, if only the liquid loading is important?

LIST OF SYMBOLS

A = area of thickener, m^2
A_L = limiting area in a thickener, m^2
b = slope of filtrate volume versus time curve
C_i = solids concentration at any level i
C_u = underflow solids concentration, mg/L
C_o = influent solids concentration, mg/L
CST = capillary suction time
D = diameter, m
du/dy = rate of shear, or slope of the velocity (u)-depth (y) profile
G = flux in a thickener, $kg/m^2 \cdot sec$
G_L = limiting flux in a thickener, $kg/m^2 \cdot sec$
h = fraction of BOD not removed in the primary clarifier
i = fraction of BOD not removed in the biological treatment step
j = fraction of solids not destroyed in digestion
k = fraction of influent SS removed in the primary clarifier
L = length of the bowl cylinder, m
N = number of leads on a scroll
P = vacuum pressure, N/m^2
Q_o = influent flow rate, m^3/sec
r = radius, m
r = specific resistance to filtration, m/kg
S = scroll pitch, m
S_o = influent BOD, kg/hr
SVI = sludge volume index
v = interface velocity at solids concentration C_i
w = cake deposited per volume of filtrate, kg/m^3
X_e = effluent SS, mg/L
x_o = influent SS, kg/hr
Y = yield, $\Delta X/\Delta S$
z = depth of sludge in a bowl, m

β = beta factor

ΔS = net BOD utilized in secondary treatment, kg/hr

ΔX = net solids produced by the biological step, kg/hr

$\Delta\omega$ = difference between bowl and conveyor rotational velocity, rad/sec

η = plastic viscosity, $N \cdot sec/m^2$

μ = viscosity, $N \cdot sec/m^2$ (poise)

Σ = sigma factor

τ = shear stress, N/m^2

τ_y = yield stress, N/m^2

ω = rotational velocity, rad/sec

$\forall$ = filtrate volume, m^3

REFERENCES

1. Kormanik, R.A. "Estimating Solids Production for Sludge Handling," *Water and Sewage Works* 118:5(1972).
2. Vesilind P.A. "The Capillary Suction Time as a Fundamental Measure of Sludge Filterability." Unpublished. (Duke University, Durham, NC).
3. Ambler, C.M. "The Evaluation of Centrifuge Performance," *Chemical Engineering Progress* 48:3(1952).
4. Vesilind, P.A. *Treatment and Disposal of Wastewater Sludges* (Ann Arbor, MI: Ann Arbor Science Publishers, 1979).

Chapter 10
Nonpoint Source Water Pollution

As rain falls and strikes the ground, a complex runoff process begins, and nonpoint source water pollution is the unavoidable result. Even before people entered the picture, the rains came, raindrops picked up soil particles, muddy streams formed, and major water courses became clogged with sediment. Witness the formation of the Mississippi River delta, which has been forming for tens of thousands of years. We can safely surmise that, even before humanity, rivers were "polluted" by this natural series of events; sediment clogged fish gills and fish probably even died. This natural runoff is classified as "background" nonpoint source runoff and not generally labeled as "pollution."

Now view the world as it has been since the dawn of humankind—a busy place where human activities continue to influence our environment. For years, these activities have included farming, harvesting trees, constructing buildings and roadways, mining, and disposing of liquid and solid wastes. Each activity has led to disruptions in the surface of the earth's soil or has involved the application of chemicals to the soil. This increased transport of soil particles, i.e., increased sediment loads to watercourses and the application of chemicals to the soil, is generally labeled as pollution.

This "pollution" is the focus of this chapter, where we address:

- the runoff process
- loading functions for sediment, a critical pollutant
- control technologies applicable to nonpoint source pollution

Table 10–1 gives a list of nonpoint source categories, including sources ranging from agricultural practices to air pollution fallout to "natural" background. The focus of this chapter is the five major activities of concern: agriculture, urban stormwater, construction, silviculture, and urban stormwater runoff. Table 10–1 also summarizes the relative significance of the concentrations of pollutants that are attributable to the list of activities.

Table 10-1. Relative Significance of Concentrations of Pollutants[a]

NPS Category	Suspended Solids/Sediment	BOD	Nutrients	Toxic Metals	Pesticides	Pathogens	Salinity/TDS[c]	Acids	Heat
Urban storm runoff[b]	M	L-M	L	H	L	H	M	N	N
Construction[b]	H	N	L	N-L	N	N	N	N	N
Highway de-icing	N	N	N	N	N	N	H	N	N
In-stream hydrologic modifications	H	N	N	N-H	N	N	N-H	H	N
Non-coal mining[b]	H	N	N	M-H	N	N	M-H	H	N
Agriculture[b]									
Nonirrigated crop production	H	M	H	N-L	H	N-L	N	N	N
Irrigated crop production	L	L-M	H	N-L	L	N	H	N	N
Pasture and range	L-M	L-M	H	N	M-H	N-L	N-L	N	N
Animal production	M	H	M	N-L	N-L	L-H	N-L	N	N
Forestry[b]									
Growing	N	N	L	N	L	N-L	N	N	N
Harvesting	M-H	L-M	L-M	L-H	N	N	N	N	M
Residuals management[b]	N-L	L-H	L-M	L-H	L	L-H	N-H	N-H	N
On-site sewage disposal[b]	L	M	H	L-M	L	H	N	N	N
In-stream sludge accumulation/resuspension	H	H	M-H	L-H	M	L	N	N	N
Direct precipitation	N	N	N-M	L	L	N-L	N	N-M	N
Air pollution fallout	M	L	L-M	L-H	L-M	N-L	N	L-M	N
"Natural" background	L-H	L-H	M	N-M	N	N-L	N-H	N-H	N-M

[a] N = Negligible; L = Low; M = Moderate; H = High.
[b] Considered in this chapter as a major contributor of nonpoint source pollution.
[c] TDS = total dissolved solids.

THE RUNOFF PROCESS

The complex runoff process includes both the detachment and transport of soil particles and chemical pollutants. For the remainder of this chapter, we simply umbrella all human-induced soil erosion and chemical applications under the term "nonpoint source pollutant." Chemicals may be bound to soil particles or be soluble in rainwater; in either case, water movement is the prime mode of transport for solid and chemical pollutants. The *characteristics of the rain* indicate the ability of the rainwater to splash and detach the pollutants. This rain energy is defined by droplet size, velocity of fall, and the intensity characteristics of the particular storm.

Soil characteristics impact both the detachment and transport processes. Pollutant detachment is a function of an ill-defined motion of soil stability. Size, shape, composition, and strength of soil aggregates and soil clods all act to determine how readily the pollutants are detached from the soil to begin their movement to streams and lakes within a region. Pollutant transport is influenced by the permeability of the soil to the water. Soil permeability, the ease by which water passes through the soil, helps determine the infiltration capabilities and drainage characteristics of the surface receiving the rainfall. Pollutant transport is also a function of soil porosity, which affects storage and movement of water, and soil surface roughness, which tends to create a potential for temporary and long-term detention of the pollutants. (The porosity of selected materials is listed in Table 5–1.)

Slope factors also help define the transport component of the nonpoint source problem. The slope gradient as well as slope length influence the flow and velocity of runoff, which in turn influence the quantity of pollutants that are moved from the soil to the water course.

Land cover conditions also impact the detachment and transport of pollutants. Vegetative cover helps to:

- provide protection from the impact of raindrops, thus reducing detachment
- make the soil aggregates less susceptible to detachment by protecting soil from evaporation and thus keeping the soil moist
- furnish roots, stems, and dead leaves that help slow overland flow and hold pollutant particles in place

Table 10–2 shows the different types of erosion that may contribute to nonpoint source pollution.

Only a portion of the pollution detached and transported from upland regions in a watershed is actually carried all the way to a stream or a lake. In many cases, significant portions of the materials are deposited at the base of slopes or on flood plains. The portion of the pollution detached, transported, and actually delivered from its source to the receiving waterway is defined as the *delivery ratio*.

Table 10–2.   Types of Erosion that May Contribute to Nonpoint Source Pollution

Type of Erosion	Description
Raindrop	From direct impact of falling raindrops
	Detached soil and chemicals can be transported downhill
Sheet	From raindrop splash and runoff in wide, thin layers of surface water
	Wall of water transports pollutants downhill
Gully/rill	From concentrated rivulets of water cutting 5-cm to 10-cm rills in soil
	Rills grow into gullies
Stream/channel	From confluence of rills and gullies
	Increased volumes and velocities cause erosion along streams and riverbeds

Numerous factors influence the pollutant delivery ratio.[1,2] When chemical pollutants are involved, the whole spectrum of factors that determine reaction rates act to limit the delivery ratio: temperature, times of transport, presence of other chemicals, and presence of sunlight to name just a few. Whenever sediment or chemical pollutants become a problem, the list of physical factors becomes quite long, and includes:

1. *Magnitude of sediment sources.* Whenever the quantity of sediment available for transport is greater than the capability of the runoff transport system, disposition will occur and the delivery ratio will be decreased.

2. *Proximity of pollutant sources to receiving waterways.* Pollutants entrapped in runoff often move only short distances and, owing to factors such as surface roughness and slope, may be deposited far from the lake or stream. Areas close to a receiving waterway, or areas where channel-type erosion takes place, may be characterized with a relatively high delivery ratio.

3. *Velocity and volume of water.* The characteristics of the pollutant transport system, particularly the velocity and volume of water from a given storm, impact the delivery ratio. A small storm may not supply enough water to carry a load of pollution to a lake or stream, resulting in a zero or very low delivery ratio. A large, lengthy rainfall may have the opposite effect and transport a very large portion of the pollution that is detached from its source to the receiving waterway.

By understanding rainfall characteristics, soil properties, slope factors, and vegetative covers, the loads of different nonpoint source pollutants to lakes and rivers can be predicted and possibly controlled.

THE ART AND SCIENCE OF DEVELOPING
AND APPLYING LOADING FUNCTIONS FOR
SEDIMENT DETACHMENT AND TRANSPORT

Pollutant loading is defined as the quantity of pollution that is detached and transported into surface watercourses. For engineering studies, mathematical models exist that predict these loadings as a function of soil characteristics, slope, and vegetative cover, including models for estimating:[3]

- sediment from:
  sheet and rill erosion
  urban runoff
  feedlots
- nitrogen
- phosphorus
- organic matter
- pesticides
- insoluble pesticides
- salinity
- acid mine drainage
- heavy metals and radiation

In theory, a distinction is made between solutional transport and sediment transport, because one is essentially chemical and the other is mechanical. A second distinction that is also useful to note is the differences between processes that move large quantities of soil (mass movement) and those processes that move soil as individual dispersed particles in a fluid (fluid transport). Under mass movement conditions, transport of pollution is largely related to a component of the sediment's own weight. Under fluid transport, the motion of the water is the major controlling factor. The fluid mechanics that ties this theory to the engineering world is outside the scope of this presentation; the reader is encouraged to pursue additional reading on the subject.

In practice, the modeling of nonpoint source pollution becomes a mixture of art and science. Complicated mathematical expressions often boil down to a guess here or a gut feeling there by an engineer trying to map stormwater runoff.

In this section, we present sample calculations for using one empirical method of predicting sediment loads from sheet and rill runoff on agricultural land. Other mathematical models exist for sediment as well as other pollutants, and this representation has some limitations. However, throughout the discussion, we point to these limitations and draw on this limited methodology as one example for estimating sediment loadings.

This method for estimating the sediment from sheet and rill erosion is a loading function based on the often-criticized *universal soil loss equation* (USLE):

$$Y_{(S)_E} = \sum_{i=1}^{n} A_i (R \cdot K \cdot L \cdot S \cdot C \cdot P \cdot S_d)_i$$

where $Y_{(S)_E}$ = sediment loading from surface erosion, ton/yr

n = number of subareas identified for the region, with a subarea defined for each portion of the watershed having similar characteristics in terms of slope, vegetative cover, permeability

A_i = acreage of subarea, acre

R = rainfall factor, expressing the erosion potential of average annual rainfall in the locality. R is the summation of the individual storm products of the kinetic energy of rainfall (hundreds of ft-tons/acre) and the maximum 30-min rainfall intensity (in./hr) for all significant storms.

K = soil erodibility factor, ton/acre/R unit

L = slope-length factor (dimensionless ratio)

S = slope-steepness factor (dimensionless ratio)

C = cover factor (dimensionless ratio)

P = erosion control practice factor (dimensionless ratio) obtained from

S_d = sediment delivery ratio (dimensionless ratio) obtained from USDA Soil Conservation Service field office in each region of the nation

The USLE was developed primarily for agriculture by the U.S. Department of Agriculture (USDA), and the Soil Conservation Service. It is used in every state in the nation by federal, state, regional, and local agencies.

The terms on the right-hand side of the equation are best defined for pasture and cropland and are not necessarily well defined for silviculture, construction, mining, and other nonagricultural land uses. Although the model may not predict absolute values for sediment yields, it is useful for projecting relative changes in sediment yields if different nonpoint source control technologies are designed and constructed.

The *rainfall factor* expresses the erosion potential of precipitation in a given locality. This index is a measure of the summation of the individual storm products of the kinetic energy of rainfall E and the maximum 30-minute rainfall intensity I, for all significant storms occurring within the time period under consideration. The product $R = E \times I$, which has been calculated at the macro level for all regions of the nation, reflects the combined potential of rainfall impact and runoff turbulence to detach and transport soil particles (see Figure 10–1).

The *soil erodibility factor* is a quantitative measure of the rate at which a soil will erode. K is expressed as the soil loss in tons per acre unit of R, standardized for a plot of land with 9 percent slope, 72.6 feet long, under continuous cultivation. This factor has been developed for most soils in the

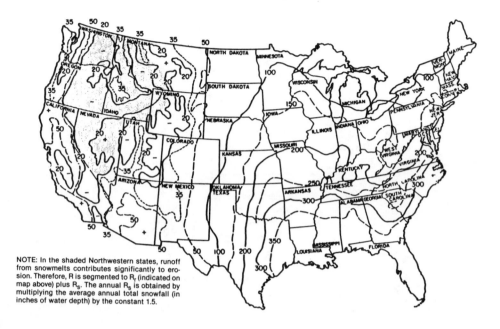

NOTE: In the shaded Northwestern states, runoff from snowmelts contributes significantly to erosion. Therefore, R is segmented to R$_r$ (indicated on map above) plus R$_s$. The annual R$_s$ is obtained by multiplying the average annual total snowfall (in inches of water depth) by the constant 1.5.

Figure 10–1.   Average annual values of rainfall factor, R.

nation, and the data are available from Soil Conservation Services offices. Figure 10–2 offers one method of determination.

The *topographic factor*, L·S, combines both the slope-length and slope-steepness factors and reflects the fact that soil loss is affected by both length of flow and steepness of slope. These two factors together account for the capability of runoff to detach and transport soil material. Slope-length is defined as the distance from the origin of overland flow to either the point where the slope decreases to the extent that disposition begins, or the point where runoff enters a well-defined channel that is part of the drainage network, whichever distance is less. Slope-length may be determined reasonably accurately by site inspection, by measurements from aerial photographs, or as a last resort, by topographic maps. Slope-steepness data may be obtained from similar sources of information or the Soil Conservation Service. Figure 10–3 presents estimates for this topographic factor.

The *cover factor*, C, presents the ratio of soil quantity eroded from land that has a vegetative cover to the soil quantity eroded from clean-tilled soil under identical slope and rainfall conditions. Local Soil Conservation Service offices keep records of recommended values of C for given localities, or they may be estimated from Table 10–3.

The *sediment recovery ratio*, S$_d$, is defined as the fraction of total erosion that is delivered to a stream. The historical method of determining an average

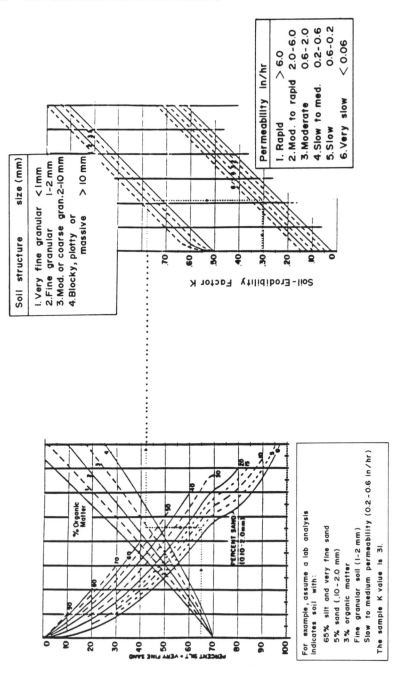

Figure 10–2. Soil erodibility nomograph for calculating values of K.

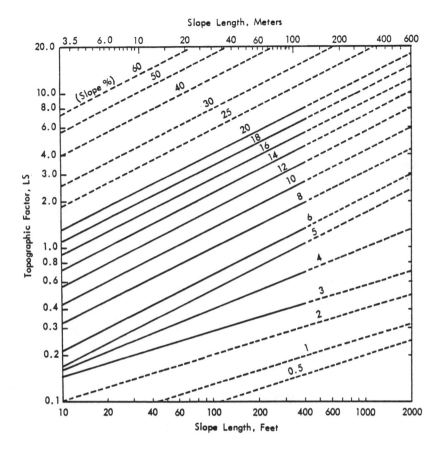

Figure 10–3. Slope effect for areas east of the Mississippi River.

NOTE: The dashed lines are estimates for slope dimensions beyond the range of lengths and steepness for which data are available.

delivery ratio is by comparing the magnitude of sediment yield at a given point in a watershed with the total amount of erosion. Estimates of delivery ratios are also available from the Soil Conservation Service. If S_d is omitted from the USLE, the projected sediment loading is the tonnage of soil that is eroded from a given area, but not necessarily the tonnage of soil that reaches a waterway. Some fraction of what is eroded is deposited before it enters the lake or stream; this variable is estimated for each region of the nation by the Soil Conservation Service.

This soil loss equation has many limitations. For one, it predicts soil loss from sheet and rill erosion, but does not predict losses from gullies, streambanks, landslides, or wind. This model was developed and calibrated primarily for croplands and has been applied mostly in the humid regions east of the Rocky Mountains if not mostly east of the Mississippi River. The model generally is not applicable for western croplands, and its utility in predicting

Table 10–3. Relative Protection of Ground Cover Against Erosion: Values for C Factor*

Land-Use Groups	Examples	Range of C Values
Permanent vegetation	Protected woodland Prairie Permanent pasture Sodded orchard Permanent meadow	0.0001–0.45
Established meadows	Alfalfa Clover Fescue	0.0004–0.3
Small grains	Rye Wheat Barley Oats	0.07–0.5
Large-seeded legumes	Soybeans Cowpeas Peanuts Field peas	0.1–0.65
Row crops	Cotton Potatoes Tobacco Vegetables Corn Sorghum	0.1–0.70
Fallow	Summer fallow Period between plowing and growth of crop	1.0

* Source: Reference 3, page 59.

sediment loading from silvicultural activities, construction sites, and mining locations is questionable.

Nevertheless, this model is useful because it can predict erosion rates by storm event, season, and annual averages, and a nationwide data collection effort has resulted in an adequate approximation of the independent factors in the model.

Example 10.1

An 830-acre watershed is located 5 miles south of Indianapolis in central Indiana. Given the information presented below, compute the projected sediment from sheet and rill erosion in terms of average daily loading.

Cropland: 180 acres

- Corn
- Conventional tillage, average annual yield of 40 to 45 bushels/acre
- Cornstalks remain in fields after harvest
- Contour strip-cropped planting method
- Soil is Fayette silt loam
- Slope is 6%
- Slope length is 250 feet

Pasture: 220 acres

- No canopy of trees or brush
- Cover at surface is grass and grasslike plants
- 80% ground cover
- Soil is Fayette silt loam
- Slope is 6%
- Slope length is 200 feet

Woodland: 430 acres

- 50% of area covered with tree canopy
- 80% of area covered with leaf and limb litter
- Undergrowth is managed and therefore minimal
- Slope is 12%
- Slope length is 150 feet

Calculations for loading from cropland:

$$Y(s)_{annual\ cropland} = R \cdot K \cdot L \cdot S \cdot C \cdot P \cdot S_d$$

$R = 200$ from Figure 10–1

$K = 0.37$ from Figure 10–2

$L \cdot S = 1.0\%$ from Figure 10–3

$C = 0.49$ from Table 10–3 and USDA Soil Conservation Service field office in Indianapolis

$P = 0.25$ from Table 10–4

$S_d = 0.60$ from USDA Soil Conservation field office in Indianapolis

$$Y(s)_{annual\ cropland} = 200 \cdot 0.37 \cdot 1.08 \cdot 0.49 \cdot 0.25 \cdot 0.60$$
$$= 5.87\ ton/acre/yr$$

$$Y(s)_{avg.\ daily\ cropland} = 5.87\ ton/acre/yr + 365\ days$$
$$= 0.016\ ton/acre/day$$
$$= 32\ lb/acre/day$$

Table 10–4.   P Values for Erosion Control Practices on Cropland*

| | | Type of Control Practice | | | |
Slope	Up- and Downhill	Cross-Slope Farming Without Strips	Contour Farming	Cross-Slope Farming with Strips	Contour Strip-Cropping
2.0–7	1.0	0.78	0.50	0.37	0.25
7.1–12	1.0	0.80	0.60	0.45	0.30
12.1–18	1.0	0.90	0.80	0.60	0.40
18.1–24	1.0	0.95	0.90	0.67	0.45

*Source: Reference 3, page 64.

Calculations for loading from pasture:

$$Y(s)_{annual\ pasture} = R \cdot K \cdot L \cdot S \cdot C \cdot P \cdot S_d$$
$$R = 200$$
$$K = 0.37$$
$$L \cdot S = 0.95$$
$$C = 0.013$$
$$P = 1.00$$
$$= 0.013$$

$$Y(s)_{annual\ pasture} = 200 \cdot 0.37 \cdot 0.95 \cdot 0.013 \cdot 1.00 \cdot 0.6$$
$$= 0.548 \text{ ton/acre/yr}$$
$$= 1,100 \text{ lb/acre/yr}$$

$$Y(s)_{avg.\ daily\ pasture} = 0.548 \text{ ton/acre/yr} \div 365 \text{ days}$$
$$= 0.0015 \text{ ton/acre/day}$$
$$= 3 \text{ lb/acre/day}$$

Calculations for loading from woodland:

$$Y(s)_{annual\ woodland} = R \cdot K \cdot L \cdot S \cdot C \cdot P \cdot S_d$$
$$R = 200$$
$$K = 0.32$$
$$L \cdot S = 2.75$$
$$C = 0.003$$
$$P = 1.00$$
$$S_d = 0.60$$

$$Y(s)_{annual\ woodland} = 200 \cdot 0.32 \cdot 2.75 \cdot 0.003 \cdot 1.00 \cdot 0.60$$
$$= 0.3168 \text{ ton/acre/yr}$$

$$Y(s)_{avg.\ daily\ woodland} = 0.3168 \text{ ton/acre/yr} + 365 \text{ days}$$
$$= 0.0009 \text{ ton/acre/day}$$
$$= 1.8 \text{ lb/acre/day}$$

Calculations for cross loading from watershed:

Cropland: 180 acres · 0.016 ton/acre/day = 2.88 ton/day
Pasture: 220 acres · 0.0015 ton/acre/day = 0.33 ton/day
Woodland: 430 acres · 0.0009 ton/acre/day = 0.39 ton/day

Total $Y(s)_{avg.\ total} = 3.60$ ton/day

Finally, this loading function is designed to predict long-term average loadings. Sediment loadings for a specific year may vary drastically from this long-term average because of differences in the number, size, and timing of erosive rainstorms.

Because of the large uncertainties in this model, it is advisable that all predictions be utilized only as reasonable estimates and not labeled as absolute values. Sensitivity analysis, the varying of one independent variable at a time to observe the change in predicted loadings, is advisable in all cases. The model is also useful in areas outside agriculture for projecting relative impacts of such nonpoint source control measures as the addition of vegetative covers and slope modifications. Data have been collected by the Soil Conservation Service since the early 1930s, and models have been developed and revised to the present form of the USLE. This model is used extensively by engineers in the Soil Conservation Service, and understanding its strengths and weaknesses is critical to environmental engineers.

CONTROL TECHNOLOGIES APPLICABLE TO NONPOINT SOURCE POLLUTION

The importance of controlling nonpoint source pollution and the obstacles to achieving that goal are becoming more and more evident. Control over municipal and industrial point sources of pollution historically has received considerable federal and corporate attention through the nationwide construction grants and permit programs. Yet, public and private investment to reduce point source pollution significantly may be ill spent in cases in which water quality is governed instead by nonpoint source discharges.

Agriculture

Many control measures exist for reducing nonpoint source pollution from agricultural areas. These options range from the management of surface cover and tillage to mechanical conservation measures. Alternative systems have been developed to reduce the erosion potential from tilled (inverted or plowed) land.

Terracing, which breaks the slope of a field into shorter segments, is often applied in fields where contouring and other tillage systems do not offer adequate soil stabilization. Terraces consist of a combination of ridges and channels constructed across the slope that collect surface water from above. Water is either held until it is absorbed or diverted in a controlled manner.

Diversions are large, individually designed terraces constructed across the slope of a field to intercept and divert runoff to a stable outlet. Usually constructed above croplands or above such critical erosion areas as gullies, they act to reduce the volume of runoff entering the problem area.

Other options available to control agricultural nonpoint source pollution include grassed waterways (the use of year-round grasses in areas of surface water movement), and cover plants (grasses, trees, and shrubs planted in critical areas to control severe erosion problems near stream banks).

Silviculture

Silviculture is defined as that part of forest management that deals with the process of utilizing forest crops. Surface water pollution results mainly from three silvicultural activities: the building and use of transportation systems, the harvesting of timber, and intermediate practices such as thinning and spraying for the control of fire, insects, and disease.

From a water quality standpoint, the most critical decisions are made during the planning phase of the crop utilization process. Transportation networks, the most significant nonpoint source of water pollution from silviculture in the Northwest and other areas of the United States, may be designed to minimize pollution runoff. Design considerations include optimizing the number of roadways and their layout and minimizing the number of times they cross natural water courses. The three surfaces associated with a given segment of roadway offer distinct control options: the cut-slope, the roadway surface itself, and the fill-slope. Geotechnical investigations should precede the construction of any road or staging area. In areas with a slope of greater than 60 percent, it is generally prescribed that roads be built only as a last resort; such log removal techniques as skylining, helicoptering, and ballooning should be investigated. The techniques by which logs are cut and transported influence the amount of nonpoint source pollution.

A component of the runoff problem may be controlled if transportation and harvesting techniques are planned to avoid or minimize the impact on environmentally sensitive flood plains, and buffer strips along waterways should be protected. In addition, chemical applications should follow best management practices. Erosion from staging areas may be controlled by diverting runoff around the site. Bars and ditches may be 70 percent effective in reducing sediment loading to receiving waterways.[4]

Construction

The erosion of soil may cause construction problems onsite as well as water quality problems offsite. Additionally, the loss of soil is often regarded as the loss of a valuable natural resource. Home buyers expect a landscaped yard, and lost topsoil is often costly for the contractor to replace. The builder of homes, highways, and other construction views soil erosion as a process that must be controlled to maximize economic gain.

The planning phase of construction activities considers controlled clearing of the proposed construction site, as the area to be disturbed during the construction and site restoration phases is held to a minimum. Environmentally sensitive areas must be designated, and if any clearing is required in such areas, it should be limited as much as possible. Such critical areas include: steep slopes, unaggregated soils (sands, etc.), natural sediment ponds, natural waterways (including intermittent streams), and flood plains.

The planning phase should also consider a comprehensive erosion control system. The components of such a system include planned access as well as techniques for use in the operational and site restoration phases of the construction activity.

During construction, several pollution-abatement techniques appear to be effective. Velocity regulation methods attempt to reduce the rate at which water moves over the construction site. Velocity reduction minimizes particle uptake by the water and may lead to particle deposition in instances when the reduction is sufficient. The result is a decrease in erosion. There are alternative methods of achieving velocity reduction, and all involve the application of some material to the exposed soil: filter inlets, jute mesh, seeding, fertilizing, and mulching can reduce sediment runoff. The best method, and its corresponding cost, must be determined on a site-by-site basis.

Stormwater deflection methods attempt to reduce the amount of water passing over the construction site by diverting it. However, the dike does not limit or control runoff generated by rain falling directly on the site, and its effectiveness is thus limited. Stormwater-channeling methods attempt to reduce the effect of water passing over the construction site by controlling its movement through the site. Chutes, flumes, and flexible downdrains are effective in certain areas, but their costs are quite high. Their outfall must be handled to ensure that a secondary source of pollution is not created.

Restoration of the site after construction is necessary if water pollution is to be controlled. Regrading costs are necessary expenditures at most construction sites if an effective revegetation process is to be undertaken. Regrading is only effective if it is followed by seeding, fertilizing, and mulching. Such revegetation costs depend on the type of mulch used, and the type of mulch applicable on a given site is highly dependent on the slope of the site. More expensive practices generally are required on steeper slopes. Wood fiber mulch applied by a hydroseeder is very effective on relatively flat landscapes. The more expensive and more permanent excelsior mats and jute netting are also available.

Mining

Mining is discussed in this section under two major headings: surface mining and subsurface mining. The major pollutants from both types of operations are sediments, toxic substances, and acids.

There are several forms of surface mining, including strip, open pit, and dredging. Strip mining requires removing a large amount of overlying material (overburden) to expose the desired ore. The distinction between strip mining and open pit mining is that open pit is used in areas where there is relatively little overburden. In such areas, most of the material being removed is the desired mineral, whereas most of the material being removed during strip mining is overburden and waste. Most strip mining is done to obtain coal, whereas open pit mining is performed to mine a range of minerals. Dredging is used to recover minerals from underwater mines, with gravel accounting for the majority of dredging production. Hydraulic mining is done by directing a high-velocity stream of water at a disperse mineral deposit. It is used almost exclusively for gold.

These forms of mining result in varying amounts of nonpoint source water pollution. Siltation is a major problem, and chemical pollution may occur when surface mining results in accelerating pollution-forming chemical reactions. Water pollution control is generally achieved by changing the conditions responsible for the pollution. Combinations of several techniques are usually required to achieve adequate water pollution abatement. As an example, regrading should be accompanied by revegetation and possibly by water diversion.

A given area of surface mining may simultaneously exhibit characteristics of three phases of operation: planning of a portion of the dig area may be taking place, mining operations may be taking place in a second portion of the dig area, and restoration may be taking place in a third portion of the dig area.

The planning phase of surface mining should consider a technique referred to as controlled mineral extraction. This procedure is based on the fact that pollution-forming materials are not usually distributed evenly throughout a mineral seam. Water quality data, hydrologic considerations, and core bore sampling often indicate where high pollution potentials exist in any prospective mining area. Planning mineral extraction around such problem spots may result in the abatement of nonpoint source water pollution from surface mines. The cost of sampling and the opportunity cost associated with not mining a critical area must be developed on an individual mine basis.

The planning phase must also consider a comprehensive erosion control system. Elements in such a system include access roads and transportation considerations. Roadways are a major source of pollution in active mine sites. Additionally, the erosion control system must consider water pollution from mine tailings.

During the operational phase of surface mining, several water pollution abatement strategies appear to be useful. The utility of each, however, is site specific, depending on each mine's geography and overburden characteristics. Overburden segregation has been used many times in the United States. When properly utilized with regrading and revegetation, it is believed to be one of the most effective methods for controlling water pollution from surface mines. The cost of overburden segregation is site specific. In the contemporary reclamation method, only the overburden and waste from the first dig is placed on land adjacent to the mine. The overburden and waste from successive digs is placed in the mined portion of the area. The technique appears to be a good method to control water pollution. It has the distinct advantage to the mineral industry in that most of the overburden is handled only once, and a cost savings is available. Further, regrading and revegetation costs are often reduced. Water diversion is a technique that focuses on collecting surface water before it enters the mine area and diverting it around the mine site. The technique is effective in reducing nonpoint source water pollution, and it may be applied to any mine surface or mine waste pile. The cost of the technique may prove to be excessive for large mining areas.

During the site restoration phase of a surface mining operation there are basically two types of water pollution control techniques: regrading and revegetation. Regrading is undertaken both to prevent significant erosion and to restore the site to a usable form. In general, regrading will significantly reduce surface water pollution from surface mining. To be totally effective, regrading must be accompanied by revegetation. Revegetation includes not only seeding and planting an abandoned mine site, but also preparing the soil, burying harmful substances, and maintaining the vegetation cover through several growing seasons to ensure that the vegetation is permanent. If this method is properly undertaken and coupled with regrading, it may eliminate essentially all pollution resulting from a surface mining activity. However, many questions remain unanswered concerning the economics of both regrading and revegetation.

Subsurface mines are identified as nonpoint sources of water pollution because the water that runs out of mine openings is often extremely acidic (with a pH as low as 2) and high in metals that may be toxic to aquatic and human life. The source of the water in the mines is usually groundwater that seeps through the porous structure, boreholes, and fractures in the strata adjacent to the mine. The pollutants carried in the water are a result of ores that have been oxidized because of exposure to air. Therefore, pollution controls for subsurface mines are aimed at restricting waterflows from the mines or maintaining anaerobic atmosphere in the mine or both.

Several actions may be implemented during the planning phase of a project to reduce the potential for pollution. The actual location of the mine may be selected, and areas with high pollution potential within the site may be avoided. Next, an overall erosion control program may be designed, accounting for surface disturbance activities such as milling and mine tailing disposal. The mine itself may be designed to minimize the production of pollutants through planning contemporaneous backfilling and roof fraction control.

Controls applicable during the actual operation of the mine include many different technologies. The most commonly applied method is the plugging of boreholes and the grouting of fracture zones. Other methods call for the diversion of groundwater flows, the restriction of free oxygen into the mine, the reduction of rainwater infiltration in the watershed above, and other diverse ideas. The costs of these activities are not estimable for a typical mine and will vary significantly from site to site. Along with these controls, the structural controls that require preplanning to be most effective, such as backfilling and the use of roof supports, are implemented during this phase.

After the production phase is completed, site restoration activities must be initiated. For subsurface mines this involves sealing the openings to the mines and at times inducing cave-ins to fill the void spaces. The cost of these activities is dependent on the number of openings to the mine and the strata around the deposit.

Urban Stormwater Runoff

The methods for controlling urban stormwater runoff range from nonstructural urban housekeeping practices, such as litter control regulations, to structural collection and treatment systems, such as settling tanks and possibly even secondary treatment. No single control can be used in all situations or locations. Factors affecting the choice of controls for a given site include:

- the type of sewerage system (separate or combined)
- the status of development in the area (planned, developing, or established urban area)
- the land use (residential, commercial, or industrial)

Controls are grouped into three categories: planning controls, pollutant accumulation controls, and collection and treatment controls. These control categories correspond to those applicable to planned, developing, and developed urban areas.

Several planning controls may be undertaken to reduce the number of pollution sources. Street litter may be reduced by passage and enforcement of antilittering laws. Air pollution, which becomes a source of water pollution when it settles on urban surfaces (particularly streets and rooftops), may be reduced via effective air pollution abatement planning. Transportation residues such as oil, gas and grease from cars and particulates from deteriorating road surfaces may be reduced via transportation planning, selection of road surfaces that are less susceptible to deterioration, and automobile inspection programs. Preventative actions may be taken as part of land use planning strategies to reduce potential runoff pollution, such as avoiding development in environmentally sensitive areas or in areas where urban runoff is an existing problem. Floodplain zoning, one type of land use regulation, often creates a buffer strip that is effective in reducing urban runoff pollution by filtering solids from overland flows and by stabilizing the soils of the floodplain.

Several control techniques prevent pollutants from accumulating on urban surfaces such as streets and parking lots. By preventing such buildup, the total pollutant loadings and the concentration of pollutants in the first flush of an area are reduced. The high concentration of pollutants in this first flush is the cause of many negative water quality impacts associated with urban runoff. The most common methods of street cleaning include street sweeping, street vacuuming, and street flushing. Street sweeping is the oldest technique and is used in most urban areas. It is also the least expensive of the three. Street sweeping will reduce soil loadings in the runoff, but fails to pick up the finer particulates, which often are the more significant source of pollution (biodegradables, toxic substances, nutrients, etc.). Street vacuuming is more efficient in collecting the small particulates, but is more expensive.

Catch basin cleaning refers to the periodic removal of refuse and other solids from catch basins. Significant reductions in biodegradables, nutrients, and other pollutants may result from regular cleaning. The basins may be cleaned by hand or by vacuum.

Urban stormwater runoff pollution may also be controlled after it enters the stormwater drainage system. Detention systems reduce runoff pollutant loadings by retarding the rate of runoff and by encouraging the settling of suspended solids. These systems range from low-technology controls such as rooftop storage to intermediate-technology controls such as small detention tanks interspersed in the collection network. In general, the size and number of units are directly proportional to the effectiveness of the system. Detention basins act as settling tanks and can be expected to remove 30 percent of the biodegradables and 50 percent of the suspended solids in the stormwater.

In storage and treatment systems, the first flush of an area is retained in the collection network, in a storage unit, or in a flow equalization basin. The stormwater is then treated at a nearby wastewater treatment facility when the sanitary flow volume and the design capacity of the facility allow. Several innovative storage methods exist, ranging from storing the stormwater in the drainage network, to routing the flow by using a computerized network of dams and regulators (in Seattle), to the digging of a subterranean storage tunnel (in Chicago). The effectiveness of these systems depends on the quantity of pollutants captured and the sizing of the storage units and the extent of treatment received at the local wastewater facility. The storage capacity is the most critical factor. A debate exists on what size storm should be used as the design storm. The larger the design storm, the more costly the system and (generally) the more effective the system. The discussion focuses on whether treatment should be planned for a 1-month, a 6-month, or even a less common storm.

CONCLUSION

Nonpoint sources contribute major pollutant loadings to the waterways throughout the nation. Control techniques are readily available but vary considerably in both cost and effectiveness and are typically implemented through a generally confused institutional framework. This institutional setting sometimes poses an obstacle to nonpoint source abatement.

In addition, costs of control for abatement of nonpoint source pollution vary widely and are site specific. Planning controls (preventative measures), however, are relatively inexpensive when compared with operational and site restoration controls (remedial measures) and thus may be the most efficient way to control nonpoint pollution.

PROBLEMS

10.1   Use Table 10–1 and rank the concentrations of nonpoint source pollutants coming from construction sites. How would you rank these same pollutants in terms of harm to the environment? Compare and discuss both rankings.

10.2   Contact the Soil Conservation Service agent in your area and calculate the soil loss from 500 acres of cropland using the USLE. Assume:

- conventional tillage
- cornstalks remain in field after harvest
- farmer uses contour stripped planting
- soil is Fayette silt loam
- slope $\approx 7\%$
- slope length ≈ 200 feet

10.3   Onsite wastewater treatment and disposal systems are also criticized for creating nonpoint source water pollution. Describe the pollutants that are factors in backyards utilizing such systems and identify alternative technologies to control the problems.

10.4   Discuss the wastewater treatment technologies in Chapter 8 that are particularly applicable to help control pollution from urban stormwater runoff.

10.5   Calculate the cost for controlling runoff from the construction of a 5-mile highway link across rolling countryside near your home town. Assume hay bales are sufficient along the construction site if they are coupled with burlap barriers located at key locations. No major collection and treatment works are required. Document your assumptions, including labor and materials charges, as well as time commitments.

LIST OF SYMBOLS

A_i = acreage of subarea i, acres
C = cover factor (dimensionless ratio)
K = soil erodibility factor, ton/acre/R unit
L = slope-length factor (dimensionless ratio)
n = number of subareas identified in a region
P = erosion control practice factor (dimensionless ratio)
R = rainfall factor
S = slope-steepness factor (dimensionless ratio)
S_d = sediment delivery ratio (dimensionless ratio)
USDA = U.S. Department of Agriculture
USLE = Universal Soil Loss Equation
$Y_{(s)E_e}$ = sediment yield from surface erosion, ton/yr

REFERENCES

1. USDA Soil Conservation Service. *National Engineering Handbook*, Section 3, "Sedimentation" (Washington, DC: U.S. Government Printing Office, 1978).
2. USDA Agricultural Research Service. "Present and Prospective Technology for Predicting Sediment Yield and Sources," Proceedings of the Sediment Yield Workshop (Oxford, MS: U.S. Department of Agriculture Sedimentation Laboratory, 1972).
3. U.S. EPA Office of Research and Development. *Loading Functions for Assessment of Water Pollution from Nonpoint Sources* (Springfield, VA: National Technical Information Service, 1976).
4. U.S. EPA Environmental Research Laboratory. *Silvicultural Activities and Non-Point Pollution Abatement: A Cost-Effectiveness Analysis Procedure* (Athens, GA: U.S. Environmental Protection Agency), 63–74.

Chapter 11
Water Pollution Law

A complex system of laws requires industries and towns to treat their waste-water flows before discharge to receiving waterways. In this system, *common law* and *statutory law* are intertwined to form the regulatory basis for pollution control.

The American legal tradition is based on common law, a body of law vastly different from statutory law as written by Congress and state governments across the nation. This common law is the aggregate body of decisions made in courtrooms as judges decide individual cases. An individual or group of individuals damaged by water pollution or any other wrong (the plaintiff) historically could seek relief in the courtroom in the form of an injunction to stop the polluter (the defendant) or in the form of payment for damages. Because of the possibility that the polluter would have to pay damages, the pollution of waterways can be reduced.

Court rulings in these cases were, and are, based on *precedents*. The underlying theory of precedents is that if a similar case or cases were brought before any court in the past, then the present-day judge would violate the rules of fair play if the present-day case were not decided in the same manner and for the same party that the precedent cases dictated. Similar cases, defined to be so by the judge, theoretically have similar endings. If no precedent exists, then the plaintiff essentially rolls the dice in hopes of convincing the court to make a favorable precedent-setting decision.

Statutory law, on the other hand, is a set of rules mandated by a representative governing body, be it the Congress of the United States or the state legislatures. Such legislation supplements or changes the effect of existing common law in areas in which Congress or state legislatures perceive shortcomings. For example, environmental quality in general and public health in particular were continually harmed under the common laws as they related to dirty water. Common law courts were taking years to reflect changed societal conditions because the courts were bound to precedents set during times in the nation's history when clean water was plentiful and essentially free. Finally,

Congress decided to take the initiative with a series of laws aimed at abating water pollution and cleaning the surface waters of the nation.

In this chapter, we discuss the evolution of water law from the common law courtrooms through the legislative chambers of Congress to the administrative offices of the U.S. Environmental Protection Agency (EPA) and state agencies.

COMMON LAW

To date, common law has concerned itself with the disposition of surface water, rather than with groundwater. There are two major theories of common law as it applies to water. One theory, labeled the *riparian doctrine*, says that conflicts between plaintiffs and defendants must be decided by the ownership of the land underlying or adjoining a body of surface water. The second theory, known as the *prior appropriations doctrine*, takes a different focus and simply states that water use is rationed on a first-come, first-serve basis, regardless of land ownership. Note that in both of these doctrines, the focus is on water quantity, on deciding how to apportion a finite body of clean surface water. Common law is generally unclear about water quality consideration.

The principle underlying the riparian doctrine is that water is owned by the owner of the land underlying or adjoining the stream and that the owner generally is entitled to use the water as long as the quantity is not depleted nor the quality degraded. Riparian land is land bordering a surface waterway.

The doctrine has a somewhat confused history. It was originally introduced to the New World by the French and adopted by several colonies. The English court system eventually adopted the theory as common law in several court cases, and it thus eventually became the official law of the New World. In colonial courtrooms, the riparian doctrine was a workable concept. The land owner was entitled to use the water for domestic purposes, such as washing and watering stock, but common law held that it could not be sold to nonriparian parties, simply because the water in the stream or river would be diminished in quantity and the downstream user would then not have access to the total flow. Even at the present time in sparsely populated farming areas where water is plentiful, this system is still applicable.

In urban and more densely populated areas, courts generally found that the riparian doctrine could not be applied in its pure form. Accordingly, several variations or ground rules were developed in the courts: the *principle of reasonable use* and the *concept of prescriptive rights*.

The *principle of reasonable use* holds that a riparian owner is entitled to make reasonable use of the water, taking into account the needs of other riparians. Reasonable use is defined on a case-by-case basis by the courts. Obviously, this opens tremendous loopholes, which have been used in numerous litigations. Possibly the most famous example is the case of *New York City vs.*

the States of Pennsylvania and New Jersey. In the 1920s, New York City began to pipe drinking water from the upper reaches of the Delaware River, and as the city grew, the demand increased until the people downstream from these impoundments found that their rivers had disappeared. Many resort owners simply went out of business. After prolonged court battles it was finally determined that since the city did own the land around the impounded streams, and the use of this water was "reasonable," the city could continue to use the water. There was some monetary compensation for the downstream riparian owners, but in retrospect the failure of common law is quite evident.

The *concept of prescriptive rights* evolved to the point at which, if a riparian owner doesn't use the water and an upstream user "openly and notoriously" abuses the water quantity or quality, then the upstream user is entitled to continue this practice. This concept holds that, through lack of use, the downstream riparian has forfeited the water rights.

This concept was established in a famous case in 1886: *Pennsylvania Coal Co. vs. Sanderson.* Anthracite coal mines north of Scranton, Pennsylvania, at the headwaters of the Lackawanna River, were polluting the river and eventually made it unfit for aquatic life and human consumption. Mrs. Sanderson, a riparian landowner, built a house near the river before the polluted water quality conditions became noticeable. Her intent was to live there indefinitely, but the water quality soon deteriorated, eliminating her opportunities to benefit from the resource. She took the mining company into a courtroom—and lost. The court basically held that the use of the river as a sewer was "reasonable," since the company had been in operation before Mrs. Sanderson had built her house, and that since water pollution was a necessary result of coal mining, the coal company could continue its open and notorious practice. This illustrates another example of where common water law started to break down in its ability to serve the people.

Although historically important, the riparian doctrine is declining in use. It is, after all, applicable to sparsely populated areas with no severe water supply and water quality problems. Most of its applicability is limited to areas east of the Mississippi River where there is sufficient rainfall to enable the system to work.

The other important water law concept is the appropriations doctrine, which states that water users "first in time" are necessarily "first in line." In other words, if a user put surface water to some "beneficial use" before the next person, the first user is guaranteed that quantity of water for as long as the use demands. Land ownership and user location, upstream or downstream, are irrelevant.

The doctrine began in the mid-1800s as gold miners and ranchers in the arid western United States sought to stake claims to water in the same manner that they could stake mining claims. A farmer or rancher who irrigated had a particular concern that development of an upstream farm or ranch would reduce or eliminate stream flow to which he had a prior claim. The Reclamation

Act of 1902, which provided cheap irrigation water to develop the West, made this need particularly critical.

Since the flow of most streams is highly variable, it is possible to own a water right and a dry stream bed simultaneously. This conflict is resolved under the appropriation doctrine by prior claim. For example, if a user has first claim of 1 million gallons per day (mgd), a second claimant has 3 mgd, and the third has 2 mgd, as long as the river flows at 6 mgd everyone is happy. If the flow drops to 4 mgd, the third claimant is completely out of business. If the third claimant happens to be upstream from 1 and 2, the 4 mgd of water must be permitted to flow past that claimant's water intake without the removal of even a single drop.

The Colorado River and its tributaries are completely appropriated. To set aside water for oil shale and uranium mining in the Colorado Basin, the developing corporations had to purchase water rights. The cities located along the Front Range of the Rocky Mountains—Denver, Laramie, Colorado Springs—have purchased water rights west of the Rockies and divert water through tunnels under the mountains. Albuquerque, Phoenix, and Tucson have also purchased water rights for urban development and divert water from the Rio Grande drainage and the lower Colorado drainage, respectively. The looming water shortage in the Colorado Basin has led to suggestions that water be diverted from the Columbia River to the Colorado, or even from the Yukon–Charlie system in northern Canada to the Colorado. In 1968, Congress enacted a national water policy that prohibits these massive diversions; this policy is still in force. Clearly, a doctrine of water conservation is needed as a supplement to the prior appropriations doctrine.

As water supply decreases, the monetary value of water rights, which can be bought and sold, increases, leading to overappropriation and the mining of groundwater. Overappropriation and over-use of irrigation water result in concentration of pollutants in that water. The recycling of irrigation water in the Colorado River has resulted in an increase in dissolved solids and salinity so great that, at the Mexican border, Colorado River water is no longer fit for use in irrigation.

As is the case with the riparian doctrine, the appropriations doctrine says very little about water quality. Under the modified appropriations doctrine, the upstream user who is senior in time generally may pollute. If the downstream user is senior, the court has in the past directed payments for losses. If the cost of cleanup is greater than the downstream benefits, however, courts have often found it "reasonable" to allow the pollution to continue. Under the appropriations doctrine, a downstream owner who is not actually using the water has no claim whatever.

The common law theories of public and private nuisance have been found to have applicability in certain cases. Nuisance, however, has found more application in air pollution control and is discussed in some detail in Chapter 22.

STATUTORY LAW

Citing the shortcoming in common law and continued water pollution problems, Congress and state governments have passed a series of laws designed to clean the surface waters across the nation. Although most states had some laws regulating water quality, it wasn't until 1965 that a concerted push was made to curb water pollution. In that year the U.S. Congress passed the Water Quality Act, which among other provisions required each state to submit a list of water quality standards and to classify all streams by these standards.

Most states adopted a system similar to the *ambient water quality stream classifications* shown in Table 11–1. Streams were classified by their anticipated maximum beneficial use. This allowed some states to classify certain streams as low-quality waterways and others as virgin trout streams. The method of stream classification theoretically forces the states to limit industrial and municipal discharges and prevents a stream from decaying further. As progress in pollution control is made, the classifications of various streams may be improved. Lowering a stream classification is generally not allowed by state and federal regulatory agencies.

To attain the desired water quality, restrictions on wastewater discharges are necessary. Such restrictions, known as *effluent standards*, have been used by various levels of government for many years. For example, an effluent standard for all pulp and paper mills may require that the discharge not exceed 50 mg/L BOD. The total loading (in pounds of BOD per unit time) of the pollution or its effect on a specific stream is thus not considered. Using perfectly "reasonable" effluent standards, it is still possible that the effluent from a large mill, although it meets the effluent standards, completely destroys a stream. On the other hand, a small mill on a large river, which may in fact be able to discharge even untreated effluent without producing any appreciable detrimental effect on the water quality, must meet the same effluent standards.

This dilemma may be resolved by developing a system in which minimum effluent standards are first set for all discharges, and then these standards are modified based on the actual effect the discharge would have on the receiving watercourse. For example, the large mill above may have a BOD standard of 50 mg/L, but because of the severe detrimental impact it has on the receiving water quality, this may be reduced to 5 mg/L BOD. This concept requires that each discharge be considered on an individual basis, a process spelled out in the 1972 Federal Water Pollution Control Act, which also established a nationwide policy of zero discharge by 1985. In plain terms, Congress mandated the EPA to ensure that all waste be removed before discharge to a receiving waterway. The control mechanism to achieve a reduction in pollution was the EPA's prohibition of any discharge of pollutants into any public waterway unless authorized by a permit. The permit system, known as the National Pollutant Discharge Elimination System (NPDES), is administered by the EPA, with direct permitting power transferred to states able to convince the EPA that their administering

Table 11-1. Washington State Stream Classification System

Class	Best Use	Fresh Waters				Marine Waters			
		Minimum DO (mg/L)	Maximum Temperature (°C)	pH Range	Coliforms per 100 mL	Minimum DO (mg/L)	Maximum Temperature (°C)	pH Range	Coliforms per 100 mL
AA	All, fisheries	9.5	16	6.5–8.5	50	7.0	13	7.0–8.5	14
A	All, fisheries	8.0	18	6.5–8.5	100	6.0	16	7.0–8.5	14
B	No fish spawning	6.5	21	6.5–8.6	200	5.0	19	7.0–8.5	100
C	Fish passage, boating		24	6.5–9.0		4.0	22	6.5–9.0	200
Lake	All	Natural conditions			50				

agency has the authority and expertise to conduct the program. In Wisconsin, for example, the state government has been granted the authority to administer the Wisconsin Pollutant Discharge Elimination System, or WPDES.

In situations in which an industry wishes to discharge into a municipal sewage treatment system, the industry must agree to contractual arrangements developed with the local governments to ensure compliance with federal industrial pretreatment requirements. For selected industries, pretreatment guidelines are being developed that require facilities to treat their wastewater flows before discharge to municipal sewer systems. These pretreatment rules are discussed in detail later in this chapter.

The 1977 Amendments to the Clean Water Act have, in recognition of limits of technology and management of wastewater treatment systems to achieve a zero discharge, proposed that eventually all discharges be treated by using "best conventional pollutant control technology," even though this would not be 100 percent removal. In addition, the EPA is responsible now for setting effluent limits to a list of about 100 toxic pollutants, which must be controlled by using "best available control technology."

An industrial facility has two choices in the disposal of wastewater:

- discharge to a watercourse—in which case an NPDES permit is required and the discharge will have to be continually monitored
- discharge to a public sewer

The latter method may be preferable, if the local publicly owned treatment works (POTW) have the capacity to handle the industrial discharge and if the discharge contains nothing that will poison the POTW secondary treatment system. Some industrial discharges may, however, cause severe treatment problems in the POTWs, and tighter restrictions on what industries may and may not discharge into public sewers have become necessary. This has evolved into what is now known as the *pretreatment* program.

Pretreatment Guidelines

The pretreatment regulations have been developed by the EPA.[1] Under these general regulations, any municipal facility or combination of facilities operated by the same authority with a total design flow greater than 5 mgd and receiving pollutants from industrial users is required to establish a pretreatment program. The EPA regional administrator may require that a municipal facility with a design flow of 5 mgd or less develop a pretreatment program if it is found that the nature or the volume of the industrial effluent disrupts the treatment process, causes violations of effluent limitations, or results in the contamination of municipal sludge.

In addition to these general pretreatment regulations, the EPA is developing specific regulations for the 34 major industries listed in Table 11–2. The regulations for each industry are designed to limit the concentration of certain pollutants that may be introduced into sewerage systems by the respective industries. The standards require limitations on the discharge of pollutants that are toxic to human beings as well as to aquatic organisms, such as cadmium, lead, chromium, copper, nickel, zinc, and cyanide.

Table 11–3 summarizes the effect of pretreatment on the sludge quality at the Northeast Water Pollution Control Plant in Philadelphia. Note that although the *identified* industrial contribution of Cd is only 25 percent, the effect of the pretreatment regulations reduced the cadmium concentration in the sludge by 90 percent.

Table 11–2. Pretreatment Industries

Timber processing	Laundries
Leather tanning	Soaps and detergents
Steam electric	Machinery
Petroleum refining	Copper working
Iron and steel	Aluminum
Paving and roofing	Plastics
Nonferrous	Batteries
Paint and ink	Coated coils
Printing	Enamel products
Coal mining	Photographic supplies
Ore mining	Foundries
Organics	Adhesives
Inorganics	Explosives
Plastics and synthetics	Gum and wood
Textiles	Pharmaceuticals
Pulp and paper	Pesticides
Rubber	Electroplating

Table 11–3. Reduction in Heavy Metal Concentrations in Sludge*

Metal	Industrial Contribution (%)	Percent Reduction in Sludge
Cd	25	90
Cr	47	89
Cu	24	64
Ni	23	73
Pb	12	88
Zn	22	76

* *Source*: Martin Goldberg, City of Philadelphia.

Drinking Water Standards

Drinking water standards are equally if not more important to public health than stream standards. These standards have a long history. In 1914, faced with the questionable quality of potable water in the towns along their routes, the railroad industry asked the U.S. Public Health Service (USPHS) to suggest standards that would describe a good drinking water. As a result of this problem, the first USPHS Drinking Water Standards were born. There was no law passed to require that all towns abide by these standards, but it was established that interstate transportation would not be allowed to stop at towns that could not provide water of adequate quality. Over the years most water supplies in the United States have not been closely regulated, and the high-quality water provided by municipal systems has been as much the result of the professional pride of the water industry personnel as any governmental restrictions.

Because of a growing concern with the quality of some of the urban water supplies and reports that not all waters are as pure and safe as people have always assumed, the federal government passed the Safe Drinking Water Act in 1974. This law authorizes the EPA to set minimum national drinking water standards. The EPA has published some of these standards, which are quite similar to the USPHS Water Standards. Some of these representative numbers are shown in Table 11–4. Potable water standards used to describe these contaminants may be divided into three categories: physical, bacteriological, and chemical.

Table 11–4.   Selected EPA Drinking Water Standards

Standard	Amount	Suggested (mg/L)	Maximum (mg/L)
Physical			
Turbidity	5 units		
Color	15 units		
Odor	3 (threshold odor)		
Bacteriological			
Coliforms	1 coliform/100 mL		
Chemical			
Arsenic		0.01	0.05
Chloride		250	
Copper		1	
Cyanide		0.01	0.2
Iron		0.3	
Phenols		0.001	
Sulfate		250	
Zinc		5	

Physical standards include color, turbidity, and odor, all of which are not dangerous in themselves but could, if present in excessive amounts, drive people to drink other, perhaps less safe, water.

Bacteriological standards are in terms of coliforms, the indicators of pollution by wastes from warm-blooded animals. Tests for pathogens are almost never attempted. The present EPA standard calls for a concentration of coliforms of less than 1 per 100 mL of water. This standard is a classical example of how the principle of expediency is used to set standards. Before modern water treatment plants were commonplace, the bacteriological standard stood at 10 coliforms/100 mL. In 1946, this was changed to the present level of 1/100 mL. In reality, with modern methods we can attain about 0.01 coliforms/ 100 mL. It is, however, not expedient to do this because the extra measure of public health attained by lowering the permissible coliform level would not be worth the price we would have to pay.

Chemical standards include a long list of chemical contaminants beginning with arsenic and ending with zinc. Two classifications exist, the first being a suggested limit, the latter a maximum allowable limit. Arsenic, for example, has a suggested limit of 0.01 mg/L. This concentration has, from experience, been shown to be a safe level even when ingested over an extended period. The maximum allowable arsenic level is 0.05 mg/L, which is still under the toxic threshold but close enough to create public health concern. On the other hand, some chemicals such as chlorides have no maximum allowable limits since at concentrations above the suggested limits the water becomes unfit to drink on the basis of taste or odor.

At present, the only legislation that directly protects groundwater quality is the Safe Drinking Water Act. Increasing pollution of groundwater from landfill leachate and inadequately stabilized waste sites is a matter for public concern. Products of the anaerobic degradation of plastics and other synthetic materials are found in groundwater in increasing concentration. Some provisions of the Resource Conservation and Recovery Act (RCRA), particularly the provision prohibiting landfill disposal of organic liquids and pyrophoric substances, also provide groundwater protection.

CONCLUSION

Over the years, the battles for clean water have moved from the courtroom through the congressional chambers to the administrative offices of the EPA and state departments of natural resources. Permitting systems have replaced inconsistent, one-case-at-a-time judicial proceedings as ambient water quality standards and effluent standards are sought. Tough decisions lie ahead as current water programs are administered, particularly the NPDES permits for polluters discharging to waterways and the pretreatment guidelines for polluters discharging to municipal sewer systems. Even tougher decisions must be faced in the future as regulations are developed for the control of toxic substances.

PROBLEMS

11.1 In your home town, describe the NPDES reporting requirements for the local wastewater treatment facility. What data are required, how often are summary forms completed, and what agency reviews the data on the forms?

11.2 Health departments often require that chlorine be added to water as it enters municipal distribution systems. Discuss the benefits and risks associated with these rules, and describe alternative ways to ensure potable water at the household tap.

11.3 Many industrial processes are water intensive. That is, to produce a product that will sell in the marketplace, many gallons of water must flow into the factory. Develop a sample listing of such industries and discuss the legal and administrative problems generally associated with securing this water for new factories. Compare and contrast these problems with respect to the generally wet eastern states and dry western states.

11.4 Federal regulations are designed to achieve "zero discharge" of pollutants from point sources located along surface waterways. Land application of liquid waste is an option often proposed in many sections of the nation. Discuss the advantages and disadvantages of such systems particularly in terms of heavy metal pollutants, and outline possible restrictions on land where such wastes have been applied.

11.5 Assume you work for the EPA and are assigned to propose a standard for the allowable levels of arsenic for household drinking water. What data would you collect, where would you go to get the data (literature or laboratory), and in what professions would you seek experts to help guide you?

LIST OF SYMBOLS

> BOD = biochemical oxygen demand
> EPA = U.S. Environmental Protection Agency
> NPDES = National Pollutant Discharge Elimination System
> POTW = publicly owned (wastewater) treatment works
> USPHS = U.S. Public Health Service
> WPDES = Wisconsin Pollutant Discharge Elimination System

REFERENCE

1. "General Pretreatment Regulations for Existing and New Sources of Pollution," *Federal Register* Part IV (June 28, 1978), pp. 27, 736–73.

Chapter 12
Solid Waste

In this chapter, solid wastes other than hazardous materials and nuclear wastes are considered. Such solid wastes are often called *municipal solid waste* (MSW) and consist of all the solid and semisolid materials discarded by a community. The fraction of MSW produced by a household is called *refuse*.

Refuse until fairly recently was mostly food waste, but new materials such as plastics, new packages for products such as beer cans, and new products such as garbage grinders have all changed the composition of municipal solid waste. Industry creates about 2,000 new products each year, all of which eventually find their way into municipal refuse and contribute to individual disposal problems.

The components of refuse are *garbage* and food wastes; *rubbish*, which includes glass, tin cans, and paper; and *ashes*, still a problem where coal is used for heating homes. Last, *trash* refers to such larger items as tree limbs, old appliances, etc., which are not normally deposited into garbage cans.

The relationship between solid wastes and human disease is intuitively obvious, but difficult to prove. For example, if a flea, sustained by a rat, which in turn is sustained by an open dump, transmits murine typhus to a human, the absolute proof of the pathway is to find *the* rat and *the* flea, an obviously impossible task. Nevertheless, we are certain that improper solid waste disposal is a true health hazard, for we know that at least 22 human diseases are associated with solid wastes.

The two most important vectors* of human disease in regard to solid wastes are rats and flies. The fly is a prolific breeder (70,000 flies can be produced in 1 cubic foot of garbage) and a carrier of many diseases, e.g., bacillary dysentery. Rats not only destroy property and infect by direct bite, but are also dangerous as carriers of insects that may also act as vectors. For example, the plagues of the Middle Ages were directly associated with the rat populations.

---

* Vectors are means by which disease organisms are transmitted. Water, air, and food may all be vectors.

Groundwater and drinking water contamination by leachate from solid waste disposal has been growing over the past two decades. Leachate is formed when rain water collects in landfills, pits, ponds, or lagoons and stays in contact with the material in the dump long enough to leach out and dissolve some of its chemical and biochemical constituents. Leachate may be a major groundwater and surface water contaminant, particularly where there is heavy rainfall and rapid percolation through the soil.

As serious as the public health aspects of solid wastes are, health is seldom of primary concern when municipalities decide on a method of disposal. The overriding criterion is still money! What is the cheapest way to "get rid of" this stuff?

In this chapter, the quantities and composition of this "stuff" are discussed first, followed by a brief introduction to disposal options and the specific problem of litter. In the following chapter, disposal is discussed further and Chapter 14 is devoted to the problems and promises of recovering energy and materials from refuse.

QUANTITIES AND CHARACTERISTICS OF MUNICIPAL SOLID WASTE

The quantities of MSW generated in a community may be estimated by one of three techniques:

- input analysis
- secondary data analysis
- output analysis

In the first case, the MSW is estimated based on the products people use. For example, if a community purchases 100,000 steel beer cans per week, it can be expected that MSW (including litter pickup) will include about 100,000 beer cans per week. Unfortunately, this technique is very difficult to use in any but very isolated small communities.

Secondary data may be used to estimate solid waste production by some empirical relationship. For example, one study[1] concluded that solid waste generation could be predicted as

$$W = 0.0179(S) - 0.00376(F) - 0.00322(D) + 0.0071(P) - 0.0002(I) + 44.7$$

where W = waste generated, ton (2,000 lb)
  S = number of stops made by the truck
  F = number of families served
  D = number of single-family dwellings
  P = population
  I = adjusted gross income per dwelling unit ($)

Such models are rarely sufficiently accurate, however, and solid waste generation is almost always measured directly by an output analysis.

In weighing the trucks arriving at a disposal site, it is important to keep in
mind that refuse generation varies with time—day of the week and week of the
year. The weather conditions also affect the weight of the refuse, since the
moisture content may vary from 15 to 30 percent. If the solid waste disposal site
has truck scales, refuse weights can be easily calculated. Without such scales,
however, wheel scales must be used to weigh incoming and exiting trucks. In
such surveys, seldom is it possible to weigh all trucks, and thus a sample must
be used. This introduces another error. Statistical methods for estimating the
confidence level in such a sampling survey must be used.

Characteristics

Among the characteristics of refuse that are important for developing engineered
solutions to refuse management are

- moisture
- particle size
- chemical makeup   5 .tb√s
- density
- composition

The moisture concentration of MSW may vary between 15 and 30 percent
water, with 20 percent being normal. The moisture is measured by drying a
sample at 77°C (170°F) for 24 hours and calculating it as

$$M = \frac{w - d}{w} \times 100$$

where M = moisture content, percent
       w = initial (wet) weight of sample
       d = final (dry) weight of sample

The particle size distribution is especially important in refuse processing for
resource recovery, and this is discussed further in Chapter 14.

The chemical composition of typical refuse is shown in Table 12–1. The use
of both the proximate and ultimate analysis in the combustion of MSW and its
various fractions is discussed further in Chapters 13 and 14.

The density of refuse varies depending on where it is. Table 12–2 shows
some typical densities of MSW.

Refuse composition is possibly the most important characteristic affecting
its disposal or the recovery of materials and energy from refuse. Composition

Table 12–1. Proximate and Ultimate Chemical Analysis of MSW*

Proximate analysis	
Moisture (%)	15–35
Volatile matter (%)	50–60
Fixed carbon (%)	3–9
Noncombustibles (%)	15–25
Higher heat value (Btu/lb)	3,000–6,000
Ultimate analysis	
Moisture (%)	15–35
Carbon (%)	15–30
Hydrogen (%)	2–5
Oxygen (%)	12–24
Nitrogen (%)	0.2–1.0
Sulfur (%)	0.02–0.1

* *Source*: U.S. Department of Health, Education, and Welfare. "Incinerator Guidelines" (1969).

Table 12–2. Refuse Densities

	kg/m^3	(lb/yd^3)
Loose refuse	60–120	(100–200)
Refuse from a collection vehicle, after dumping	200–240	(350–400)
Refuse in a collection vehicle	300–400	(500–700)
Refuse in a landfill	300–540	(500–900)
Baled refuse	470–700	(800–1,200)

may vary significantly from one community to the next and with time in any given community.

Refuse composition is expressed in terms of either "as generated" or "as disposed" since during the disposal process moisture transfer takes place, thus changing the weights of the various fractions. Table 12–3 shows typical breakdowns of average refuse components for the United States. It should be re-emphasized that such numbers are useful only as guidelines and that each community has unique characteristics that influence its solid waste production and composition.

Table 12–3. Average Composition of MSW in the United States

Category	As Generated		As Disposed	
	millions of tons	%	millions of tons	%
Paper	37.2	29.2	44.9	35.3
Glass	13.3	10.4	13.5	10.6
Metal	10.1	7.9	10.1	7.9
Ferrous	8.8	6.9	8.8	6.9
Aluminum	0.9	0.7	0.9	0.7
Other nonferrous	0.4	0.3	0.4	0.3
Plastics	6.4	5.0	6.4	5.0
Rubber and leather	2.6	2.0	3.4	2.7
Textiles	2.1	1.6	2.2	1.7
Wood	4.9	3.8	4.9	3.8
Food waste	22.8	17.9	19.1	15.0
Yard waste	26.0	20.4	20.0	15.7
Miscellaneous	1.9	1.5	2.8	2.1
Total	127.3		127.3	

COLLECTION

In the United States and most other countries, solid waste is collected by trucks. In some instances, these are open-bed trucks that carry trash or bagged refuse. The usual vehicle, however, is the packer, a truck that uses hydraulic rams to compact the refuse to reduce its volume and thus is able to carry larger loads (Figure 12–1). Commercial and industrial collections are facilitated by the use of containers, which are either emptied into the truck with a hydraulic mechanism, or the entire container is carried by the truck to the disposal site (Figure 12–2).

On the average, of the total cost of solid waste management, a full 80 percent is spent on collection. The common method of collection is by packer truck with three workers: one driver and two loaders. These workers fill the truck and then drive it to the disposal area. The entire operation is a study in inefficiency. The time spent by the loaders traveling to and from the dump, for example, is pure waste. Accordingly, many new devices and methods have been proposed to cut collection cost. Some that are already in use are discussed below.

Garbage grinders reduce the amount of garbage in refuse. If all homes had garbage grinders, the frequency of collection could be cut in half, since the twice-a-week collection in most communities is necessary only because of the

Figure 12–1.   Packer truck used for residential refuse collection.

rapid decomposition of the garbage component. Obviously, garbage grinders put an extra load on the wastewater treatment plant, but sewage is relatively dilute and ground garbage can easily be accommodated both in the sewers and in treatment plants.

Pneumatic pipes have been installed in some small communities, mostly in Sweden and Japan. The refuse is ground at the residence and sucked through underground lines. One system in the United States is in Walt Disney World in Florida. Collection stations scattered throughout the park receive the refuse, and pneumatic pipes deliver the waste to a central processing plant (Figure 12–3). There are no garbage trucks in the Magic Kingdom.

It is not at all unreasonable to expect that pneumatic pipes will be the collection method of the future.

Compactors in the kitchen, the first new appliances introduced to the household market in many years, have distinct possibilities of reducing collection costs, but only if everyone has one. It is obvious that if all residences in a community had garbage grinders and refuse compactors, the cost of solid waste disposal (and thus local taxes) could be reduced considerably. Stationary compactors for commercial establishments and apartment houses have already had significant influence on collection practices.

Transfer stations are applicable to almost all larger communities. A typical system (Figure 12–4) involves several stations scattered around a city to which ordinary collection trucks bring the refuse. The drive to the nearest station is fairly short for each truck, so the workers spend more time collecting and less traveling. At the transfer station, bulldozers cram this refuse into large cans, in which the material is taken to the ultimate disposal. In some towns the refuse is baled into convenient desk-sized blocks before disposal in a landfill.

Figure 12–2. Containerized collection system. [Courtesy of Dempster Systems.]

Green cans on wheels are now widely used for the transfer of refuse to the truck. Shown in Figure 12–5, the green cans are pushed curbside by the householders and emptied with a hydraulic lift. Not only does this system save money, but it has a dramatic effect on the incidence of injuries to solid waste collection personnel, who have by far the highest lost-time accident rate of any municipal or industrial workers.

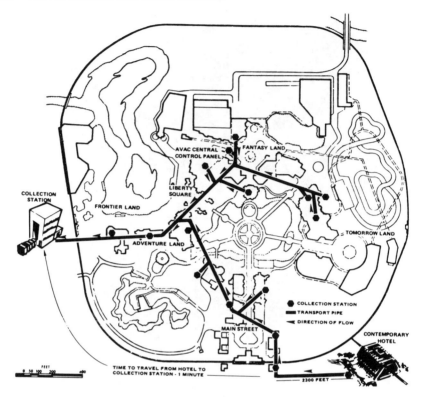

Figure 12–3. Solid waste collection system at Disney World. [Courtesy of AVAC Inc.]

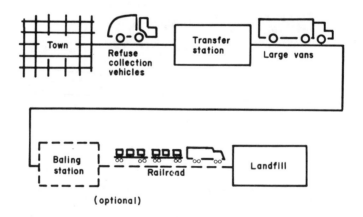

Figure 12–4. Transfer station method of solid waste collection.

Figure 12–5. The "green can" system of solid waste collection.

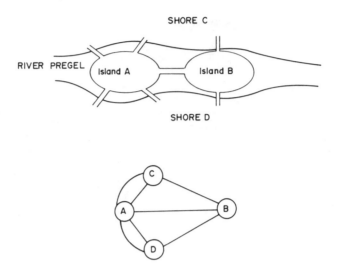

Figure 12–6.   The seven bridges of Königsberg; the routing problem presented to Euler.

Route optimization may result in significant savings to a city. Several computer programs are available for selecting the least-cost routes and collection frequencies. Such optimization techniques have resulted in increased effectiveness and lower cost of refuse collection.

The problem of route optimization was actually first addressed in 1736 by the famous Prussian mathematician, Leonard Euler. He was asked to design a parade route for Königsberg such that the parade would not cross any bridge over the River Pregel more than once (Figure 12–6). Not only did Euler show that this was not possible, but he generalized the problem by observing that *nodes* must have an even number of *links* to make such an *Euler's tour* possible. Clearly, the bridge problem had four nodes (two shores, C and D, and two islands, A and B) and seven links (the roads), with an odd number of links to each node.

The objective of truck routing is to create an Euler's tour and eliminate *deadheading, traveling* twice down the same street. Although sophisticated computer programs are available for routing trucks to achieve the most efficient tour, it is often just as simple to develop a route by common sense or *heuristic* means. Some common sense rules for routing trucks are as follows:[2,3]

1. Routes should not overlap, but should be compact and not fragmented.
2. The starting point should be as close to the truck garage as possible.
3. Heavily traveled streets should be avoided during rush hours.
4. One-way streets that cannot be traversed in one line should be looped from the upper end of the street.

5. Dead-end streets should be collected when on the right side of the street.
6. On hills, collection should proceed downhill so the truck can coast.
7. Clockwise turns around blocks should be used whenever possible.
8. Long, straight paths should be routed before looping clockwise.
9. For certain block patterns, standard paths, as shown in Figure 12–7, should be used.
10. U-turns may be avoided by never leaving one two-way street as the only access and exit to the node.

Figure 12–7 shows three examples of heuristic routing. The first two show a route where each side of a street is to be collected separately, and the third example shows a system where both sides of the street are collected at once.

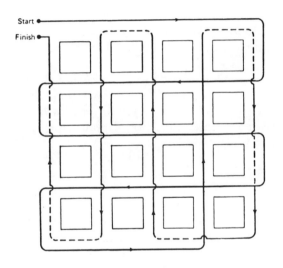

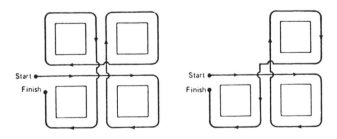

Figure 12–7. Heuristic routing examples.

DISPOSAL OPTIONS

The Romans invented city dumps, and MSW was not perceived as problematic from Roman times until the present century. As cities grew closer together, suburbs grew, and the use of disposable containers and packaging increased, the disposal of MSW became a serious issue. Many cities in the United States encouraged home incineration of trash ("backyard burning") to reduce MSW volume. After World War II, building codes mandated the installation of garbage grinders in new homes. Cities such as Miami, which simply had no dump sites, built MSW incinerators.

Growing concern about air and water pollution has resulted in widespread prohibition of backyard burning, even of leaves and grass clippings, and a de-emphasis of municipal incineration. Spontaneous dump fires, as well as the spread of disease, have resulted in the substitution of sanitary landfill for dumps. The sanitary landfill became the accepted method of disposal because it was considered environmentally sound and reasonably inexpensive. The Resource Conservation and Recovery Act (RCRA) prohibited open dumps after 1980 and limited the material that could be landfilled. Landfilling is discussed in the next chapter.

Unfortunately, landfilling is not always successful and is growing more and more expensive, and there was well-placed concern about throwing away materials that might prove useful. As a result, the idea of processing wastes to reclaim materials and energy was born. Options for resource recovery are discussed further in Chapter 14.

LITTER

Litter is not only unsightly, it is also unhealthy (as a breeding ground for rats and other rodents) and damaging to wildlife. Deer and fish, attracted to aluminum can pop-tops, ingest them and die in agony. Plastic sandwich bags are mistaken for jellyfish by tortoises, and death results. Litter is, in short, a most uncivilized by-product of our civilization, and considerable efforts have recently been directed toward curbing this insult.

Public awareness campaigns have been ongoing for many years. The most recent efforts toward volunteer participation have been funded by bottle manufacturers and bottlers, hoping to avert legal restrictions on their products. Unfortunately, people still litter, and alternative solutions to the problem are being sought.

One report has recommended that from the purely economic standpoint the best way to reduce litter is to hire more street cleaners and road crews to collect the litter. Common sense has happily prevailed, and this solution has not been suggested as the final solution to the litter problem.

The availability of trash cans seems to be an important factor influencing littering. A person will walk or ride only a short distance out of the way to deposit waste into a litter can. Increasing the availability and the frequency of maintenance of litter deposits can have a marked effect on litter. Incidentally, experiments have also determined that the type, construction, and color of the litter deposit can all influence the frequency of volunteer participation.

A much more drastic assault on the litter problem is restrictive legislation. The most notorious legislative act is the "Oregon Bottle Law," which prohibits the use of pop-top cans and discourages nonreturnable glass beverage bottles. The effect of the law on litter on Oregon has been either negligible or fantastic, depending on whose press releases one believes. It has been reported that the percentage of beverage containers collected as litter has dropped by 92 percent, and most of the 8 percent comes from other states, a statistic indicating success. Oregonians do not mind the mild inconveniences and seem to be quite pleased with the results. Similar legislation has been passed or is pending in other states.

The bottling people argue that it is people, not bottles, that cause litter, and they are right in principle. The trick is to solve the problems by changing the habits of the people without resorting to laws. Thus far all such efforts have been unsuccessful.

CONCLUSION

The solid waste problem has three facets: source, collection, and disposal. We have little chance of coping with this problem unless we begin to attack all three areas. The methods of collection and disposal are discussed in the next chapter. The source of solid waste is perhaps the most difficult of the three to tackle. New concepts in packaging, use of natural resources, and evaluation of planned obsolescence are necessary if progress is to be made. In this respect, however, we are swimming against the economic current. The economies in all political systems are oiled with money, and the worths of objects are in terms of cost in money. It is not unreasonable to speculate that this method may some day need to be changed. We must, in short, create a "new economy" with regard to productivity, obsolescence, and waste before we can begin to be at peace with our ecosystem and have any hopes of long-term survival.

PROBLEMS

12.1 Walk along a stretch of road and collect the litter into two bags, one for beverage containers only, and one for everything else. Calculate the:

a. number of items per mile
b. number of beverage containers per mile
c. weight of the litter per mile
d. weight of the beverage containers per mile
e. percent of beverage containers by count
f. percent of beverage containers by weight

If you were working for the bottle manufacturers, how would you report your data, as e or f? Why?

12.2 How would it be possible to tax the withdrawal of natural resources? What effect would this have on the economy?

12.3 What effect will the following have on the composition of municipal solid waste: (a) garbage grinders, (b) home compactors, (c) no more returnable beverage containers, and (d) a newspaper strike?

12.4 What effect would the Oregon Bottle Law have on your consumption practices? How would you change your life style?

12.5 Drive along a measured stretch of highway and count the pieces of litter visible from the car. (It's best to do this with a friend riding shotgun who does the counting.) Then walk along the same stretch and pick up the litter, counting the pieces and weighing the full bags. What percentage of litter (by weight and by piece) is visible from a car?

12.6 On a map of your campus (or any other convenient map) develop an efficient route for refuse collection, assuming that every blockface must be collected.

12.7 Using a study hall or social lounge as a laboratory, study the prevalence of litter by counting the items in the receptacles versus the items improperly disposed of. Each day (weekdays only), vary the conditions as follows:

Day 1: Normal conditions (baseline)
Day 2: Remove all receptacles except one
Day 3: Add additional receptacles (more than normal)

If possible, run several more experiments with different numbers of receptacles. Plot the percentage of properly disposed of material versus number of receptacles. Discuss the implications.

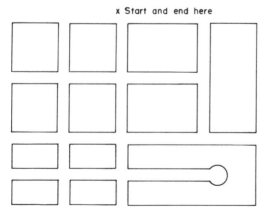

x Start and end here

Figure 12–8. Route, for Problem 12.8.

12.8   Using heuristic routing, develop an efficient route for the map shown in Figure 12–8, if (a) both sides of the street are to be collected together or (b) one side of a street is to collected at a time.

REFERENCES

1. Shell, R.L., and D.S. Shute. "A Study of the Problems Predicting Future Volumes of Wastes," *Solid Waste Management* (March 1972).
2. Liebman, J.C., J.W. Male, and M. Wathne. "Minimum Cost in Residential Refuse Vehicle Routes," *Journal of the Environmental Engineering Division of the American Society of Civil Engineers* 101(EE3):339(1975).
3. Shuster, K.A., and D.A. Schur. "Heuristic Routing for Solid Waste Collection Vehicles," U.S. Environmental Protection Agency OSWMP SW-113 (1974).

Chapter 13
Solid Waste Disposal

The disposal of solid wastes is defined as placement of the waste so it no longer impacts society. This is achieved either by assimilating the residue so it can no longer be identified in the environment (e.g., fly ash from an incinerator) or by hiding the wastes well enough so they cannot be readily found.

Solid waste may also be processed so that some of its components may be recovered, a procedure popularly known as recycling. Before disposal or recycling, however, the waste must be collected. All of these—collection, disposal, or recovery—form a part of the total solid waste management system. The collection operation is discussed in the previous chapter, and the next chapter is devoted to the recovery of energy and materials from refuse. This chapter covers the disposal of solid wastes.

Refuse may be disposed of either as is, or after suitable processing. This processing may be thermal or physical and is performed only for the purpose of converting refuse to a more readily disposable form, and not as a method of energy or materials recovery.

DISPOSAL OF UNPROCESSED REFUSE

The only two realistic options for disposal are in the oceans and on land. In the United States, the former is presently forbidden by federal law, and it is becoming similarly illegal in most other developed nations. Little else need thus be said of ocean disposal, except perhaps that its use was a less than glorious chapter in the annals of public health and environmental engineering.

Until the 1970s, the place for solid waste disposal on land was called a *dump* in the United States and a *tip* in Great Britain (as in "tipping"). The dump was by far the least expensive means of solid waste disposal and thus was the original method of choice for almost all inland communities. The operation of a dump was simple, and involved nothing more than making sure that the trucks emptied out at the proper spot. Volume was often reduced by setting the refuse on fire, which prolonged dump life.

The problems with rodents, odor, air pollution, and insects at the dump, however, became serious public health and aesthetic problems, and an alternative method of refuse disposal then became necessary. For larger communities, this alternative often was the incinerator. Smaller towns, however, could not afford such a capital investment and opted for land disposal.

Under the Federal Resource Conservation and Recovery Act of 1976 (RCRA), dumps were rendered obsolete in the United States, and have been replaced by sanitary landfills.

The term *sanitary landfill* was first used for the method of disposal employed in the burial of waste ammunition and other material after World War II. The concept of refuse burial had, however, been used by several communities in the Midwest and had proven highly successful.

The sanitary landfill differs markedly from open dumps in that the latter are simply places to dump wastes, but sanitary landfills are engineered operations, designed and operated according to acceptable standards.

Sanitary landfilling involves two principles: compaction of the refuse and cover. Typically, refuse is unloaded, either into a trench or at the surface, compacted with bulldozers, and covered with compacted soil. Daily cover of the refuse is the single feature that renders a landfill much less of a nuisance and health hazard than a dump. This cover is from 6 to 12 in. thick, depending on the soil composition, and a final cover at least 2 feet thick is used to close the landfill (Figure 13–1). After closure, a landfill continues to subside, so perma-

Figure 13–1. The sanitary landfill.

nent structures cannot be built onsite. Closed landfills do have potential uses as golf courses, playgrounds, tennis courts, winter recreation, or even a park or greenbelt.

Landfilling may be done by either the trench method or the area method. In the trench method, a trench is dug (the excavated material is used as cover) and is gradually filled with compacted refuse and cover until grade level is reached. Trench landfills are limited in size, but are less of an eyesore than area landfills.

In the area method, a site is excavated and refuse and cover are built up on the excavated site according to a predetermined plan. Generally, cover material is dug from one part of the site for use in another part. The refuse is built into compacted mounds, called cells, which are approximately trapezoidal in cross-section. A completed layer of cells, or lift, is covered again with cover material, and another lift is built on the lower lift. The completed landfill may extend well above grade. A typical arrangement of cells is shown in Figure 13–2. Cells are generally 6, 8, or 10 feet deep. Figure 13–3 shows the relationship between the refuse disposal rate and the amount of cover material needed for a landfill.

The selection of a landfill site is a sticky problem. The engineering aspects include: (1) drainage—rapid runoff will lessen mosquito problems, but proximity to streams or well supplies might result in water pollution; (2) wind—it is preferable that the landfill be downwind from the community; (3) distance from collection; (4) size—a small site with limited capacity is generally not acceptable since the trouble of finding a new site is considerable; and (5) ultimate use—can the area be utilized for public or private use after the operation is complete?

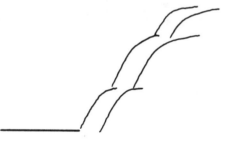

Figure 13–2. Arrangement of cells in an area-method landfill.

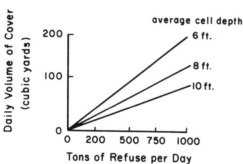

Figure 13–3. Daily volume of cover versus refuse disposal rate.

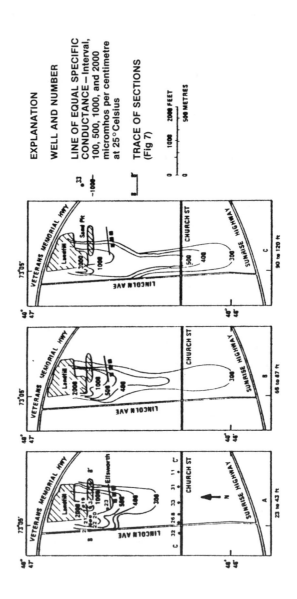

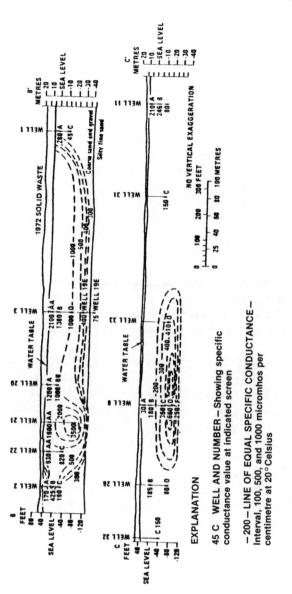

Figure 13–4.   Pollution of groundwater from a sanitary landfill.

Although daily cover helps to limit disease vectors, a working landfill still has a marked and widespread odor during the working day. Flocks of birds that feed at worked landfills are both a nuisance and a hazard to low-flying aircraft and nearby airports. In addition, there are the social and psychological problems of a landfill as a neighbor: no one wants one in the back yard.

Before RCRA, "sanitary landfills" were often indistinguishable from dumps, and the process gained a reputation as being a bad neighbor. In recent years, as more landfills have been operated properly, it has even been possible to enhance property values with a closed landfill site, since such a site must remain open space. Acceptable operation and eventual enhancement of the property are understandably difficult to explain to a community.

The landfill operation is actually a biological method of waste treatment. Municipal refuse deposited as a fill is anything but inert. In the absence of oxygen, anaerobic decomposition steadily degrades the organic material to more stable forms. But this process is very slow. After 25 years the decomposition may still be going strong.

The liquid produced during the decomposition process, as well as the water that has seeped through the groundcover and worked its way out of the refuse, is known as leachate. This liquid, although small in volume, is extremely high in pollutional capacity. Table 13–1 shows some typical values of leachate composition.

The effect of leachate on groundwater may be severe. In a number of cases, the leachate has polluted wells around a landfill to the point at which they ceased to be a source of potable water. One example of such pollution is shown in Figure 13–4, which depicts the problems encountered with a landfill on Long Island. The town of Islip's Sayville Landfill was started in a sand and gravel pit in 1933 and is still in operation. The waste at this disposal site, which extends from about 20 feet above grade to the water table about 30 feet below grade, covers 17 acres. Initially, the site was an open dump and received all types of waste. Presently, it receives mostly incinerator residue and some individually hauled residential wastes. The site is underlain by mostly coarse sand with streaks of gravelly sand.

Table 13–1.   Typical Sanitary Landfill Leachate Composition

Component	Typical Value
BOD_5	20,000 mg/L
COD	30,000 mg/L
Ammonia nitrogen	500 mg/L
Chloride	2,000 mg/L
Total iron	500 mg/L
Zinc	50 mg/L
Lead	2 mg/L
pH	6.0

The leachate plume at Islip's Sayville Landfill extends more than 5,000 feet downgradient of the site, 170 feet in depth, and up to 1,300 feet in width. About 0.22 square mile and one billion gallons of groundwater have been contaminated. Three residential wells near the disposal size and in the leachate plume were contaminated and had to be abandoned. A laundry, sink fixtures, pipes, and a water heater were among the items damaged as a result of the wells' contamination.

The amount of leachate produced by a landfill is difficult to predict. The only available method is to use a water balance—that is, the water entering a landfill has to equal the water flowing out of a landfill (i.e., the leachate).

The total water entering the top soil layer is

$$C = P(1 - R) - S - E$$

where C = total percolation into the top soil layer, mm
P = precipitation, mm
R = runoff coefficient
S = storage, mm
E = evapotranspiration, mm

The percolation for three typical landfills is shown in Table 13–2.

Using these figures, it is possible to predict when landfills produce leachate. Obviously, Los Angeles' landfills may "never" produce leachate; it will take 15 years to leach through a 7.5-m (25-foot)-deep landfill in Orlando; and it will take only 11 years for a 20-m (65-foot)-deep landfill in Cincinnati to produce leachate. Leachate production depends, to a certain extent, on rainfall patterns as well as on the amount of precipitation. The figures given for Cincinnati and Orlando are typical of a "summer thunderstorm" climate, which exists in most of the United States. Exceptions are those parts of Washington, Oregon, and California that are west of the Pacific Coast Range. These experience a maritime climate, in which rainfall is spread more evenly throughout the year, and in which leachate production is enhanced.

Table 13–2. Percolation in Three Landfills*

Location	Precipitation, P (mm)	Runoff Coefficient, R	Evapotrans-piration, E (mm)	Percolation, C (mm)
Cincinnati, OH	1,025	0.15	568	213
Orlando, FL	1,342	0.07	1,172	70
Los Angeles, CA	378	0.12	334	0

* Tenn, D.G., K.J. Haney, and T.V. Degeare. *Use of the Water Balance Method for Predicting Leachate Generation from Solid Waste Disposal Sites* (U.S. Environmental Protection Agency, OSWMP, SW-168, 1975).

A second by-product of a landfill is gas. Since landfills are anaerobic biological reactions, they produce mostly methane and carbon dioxide.

Landfills go through four distinct stages. As illustrated in Figure 13–5, the first stage is aerobic and may last from a few days to several months, during which time aerobic organisms are active and affect the decomposition. As the organisms use up all available oxygen, however, the land fill enters the second stage, at which anaerobic decomposition begins, but at which methane-forming organisms have not yet taken hold, and the acid formers cause a buildup of CO_2. This stage may also vary with environmental conditions. The third stage is the anaerobic methane production buildup stage, during which the percentage of CH_4 progressively increases along with an increase in landfill temperature to about 55°C (130°F). The last steady-state condition occurs when the fractions of CO_2 and CH_4 are about equal and microbial activity has stabilized.

The amount of methane produced from a landfill may be estimated using the following empirical relationship:[1]

$$CH_aO_bN_c + \tfrac{1}{4}(4 - a - 2b + 3c)H_2O \rightarrow \tfrac{1}{8}(4 - a + 2b + 3c)CO_2$$
$$+ \tfrac{1}{8}(4 + a - 2b - 3c)CH_4 + CNH_3$$

This equation is useful only if the chemical composition of the waste is known.

The rate of gas production from sanitary landfills may be controlled by varying the particle size of the refuse (shredding the refuse before placing it in the landfill) and by changing the moisture content. Gas production may be minimized with the combination of low moisture, large particle size, and high density. Unwanted migration of the gas may be prevented by escape vents

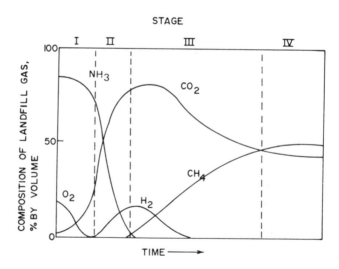

Figure 13–5. States in the decomposition of organic matter in landfills.

installed in the landfills. These vents, called "tiki torches," are kept lit, and the gas is burned off as it is formed.

Improper venting may lead to dangerous accumulation of methane. In 1986, a dozen homes near the Midway Landfill in Seattle were evacuated because potentially explosive quantities of methane had leaked through underground fissures into the basements. Venting of the accumulated gas, so that these homes may be reoccupied, is expected to take about 2 years.

Since landfills produce considerable quantities of methane, interest has recently been shown in the capture and use of landfill gases. The gas must, however, be cleaned (CO_2 and other contaminants removed) before it can be used as pipeline gas. Such cleaning is both expensive and troublesome. A more reasonable use of landfill gas is to burn it as is, perhaps in industrial applications such as brick kilns.

The biological aspects of landfills as well as the structural properties of compacted refuse limit the ultimate uses of landfills. Uneven settling is often a problem, and it is generally suggested that nothing be constructed on a landfill for at least 2 years after completion. With poor initial compaction, it is not unreasonable to expect 50 percent settling within the first 5 years. The owners of the motel shown in Figure 13–6 learned this fact the hard way.

Landfills should never be disturbed. Not only will this cause additional structural problems, but trapped gases may be a hazard. Buildings constructed on landfill sites should have spread footings (large concrete slabs) as foundations, although some have been constructed on pilings that extend through the fill and onto rock or other adequate strong material.

Figure 13–6. A motel that was built on a landfill and experienced differential settling.

The cost of operating a landfill varies from about $6 to $30 per ton of refuse and usually represents the least-cost method of acceptable solid waste disposal.

VOLUME REDUCTION BEFORE DISPOSAL

Refuse is a bulky material that does not compact easily, and thus the volume requirements in landfills are significant. Where land is expensive, the costs of landfilling may be high. Some of our larger cities, for example, pay as much as $30/ton of refuse for their landfill. Accordingly, various methods of reducing the volume of refuse to be disposed of have been found to be effective.

Under the right circumstances, incineration may be an effective treatment of municipal solid waste. Incineration can reduce the volume of waste by a factor of 10 or 20, and the incinerator ash is usually more stable than the municipal solid waste itself. Disposal of the ash may be problematic, since heavy metals and some toxic materials will be concentrated in the ash. Incinerators have high capital costs, usually exceeding $10,000/ton of 24-hour capacity.* Similarly, operating expenses may be high. Air pollution control has increased the cost of incineration to anywhere from $30 to $100 per ton of refuse. Understandably, the United States is littered with incinerators that are too expensive to operate.

A schematic of a typical large incinerator is shown in Figure 13–7. The grapple bucket lifts the refuse from a storage pit and drops it into the charging chute. The stoker (in this case a traveling grate) moves the refuse to the furnace area. Combustion occurs both on the stoker and in the furnace. Air is fed under and over the burning refuse. The walls of the furnace are cooled by pipes filled with water. The flue gases exit through an electrostatic precipitator for controlling particulates and then up the stack.

Smaller incinerators, known popularly as modular incinerators, have been widely used. These units (Figure 13–8) have two chambers, the first chamber being a "smoldering" pit, where the refuse is combusted with limited air. The fly ash carried off is combusted in an afterburner, using oil or natural gas as supplemental fuel. This process produces a clear stack gas and can reduce the volume of refuse by 90 percent. It is, however, expensive and dependent on the availability of oil and natural gas for its operation. Modular incineration in larger facilities is composed of a series of smaller units, each operating independently.

---

* Incinerators are sized on the basis of 24 hours of operation. For example, if an incinerator is fed 20 tons of refuse per day during a normal 8-hour shift, the incinerator capacity is rated as 60 tons.

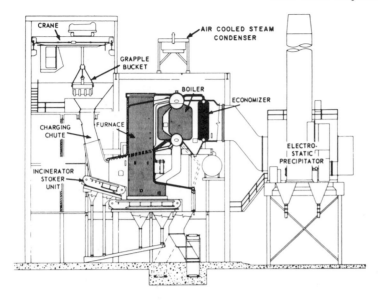

Figure 13–7.   Schematic of a typical solid waste incinerator.

2. *Shredding* solid wastes and then spreading the material on fields has been successful in a number of places. The organics are too ground up to interest rats, and spreading dries the refuse, thus avoiding odor and fly problems.

The shredded material does not have to be covered with dirt—a significant advantage over the landfill. Landfill operations are at best difficult during wet or freezing weather, and the capacities of many landfill sites are limited not by available volume but by the dirt available for covering the refuse.

3. *Pyrolysis* is combustion in the absence of oxygen. One stated advantage of pyrolysis is that the residues have economic value. However, these products— combustible gas, tar, and charcoal—have thus far not found acceptance as a raw material. Most of the gas is in fact used in making the process go. The tar is full of water and must be refined. The charcoal is full of glass and metal, which must be separated before the charcoal can be useful. This separation, however, makes the charcoal too expensive when compared with charcoal derived from wood.

Pyrolysis reduces the volume considerably, produces a stable end product, and has fewer air pollution problems (small volume of gas to treat). On a large scale, such as for some of our larger cities, pyrolysis as a method of volume reduction has significant advantages over incineration. It also may be combined with sludge disposal, thus solving two major solid waste problems for a community. Such systems, however, are still to be proven in full-scale operation.

Figure 13–8.   Modular incinerators used for solid waste processing. [Courtesy of Consumat.]

CONCLUSION

We started this chapter by defining the objective of solid waste disposal as the placement of solid waste so that it no longer impacts society. This was at one time fairly easy to achieve. In fact, dumping solid waste over the city walls was quite adequate as a method of disposal. In our modern civilization, however, this is no longer possible, and it is becoming increasingly difficult to get rid of the stuff so it no longer impacts society.

One potential solution would be to simply redefine solid waste as a resource and use it for the production of goods for people. This idea is explored in the next chapter.

PROBLEMS

13.1 Suppose the municipal garbage collectors in a town of 10,000 go on strike, and as a gesture to the community your college or university decides to

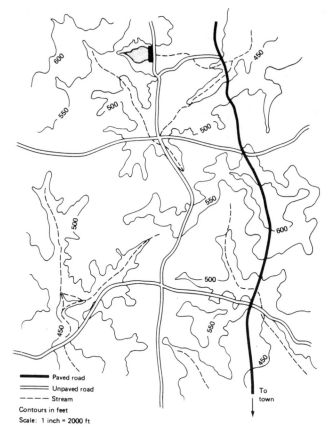

Figure 13–9. Map for siting a landfill.

accept all the city refuse temporarily and pile it on the football field. If all the people did indeed dump the refuse into the stadium, how many days must the garbage men be on strike before the stadium is filled to 1 yard deep? (Note: Assume density of refuse as 300 lb/yd^3, and assume the dimensions of the stadium as 120 yards long and 100 yards wide).

13.2 If a town has a population of 100,000, what is the daily production of wastepaper?

13.3 Describe how you would run a composting operation for leaves in your community. Be sure to include the technical, economical, and psychological considerations.

13.4 What would be some environmental impacts and effects of depositing dewatered (but sloppy wet) sludge from a wastewater treatment plant into a sanitary landfill?

13.5 If all of the readily decomposable organics refuse in a landfill were cellulose, $C_6H_{10}O_5$, how much gas (CO_2 and CH_4) would be produced (m^3/kg of refuse)?

13.6 Choose a place for a 25-acre landfill on the map shown in Figure 13–9. What other information would you need? Justify your selection of the site.

LIST OF SYMBOLS

A = area, m^2
C = total percolation of rain into the soil, mm
E = evaporation, mm
h = depth of landfill, m
P = precipitation, mm
Q = flow rate of gas, m^3/sec
R = radius of influence of a gas withdrawal well, m
R = runoff coefficient
S = storage, mm

REFERENCE

1. Chian, E.S.K., F.B. DeWalle, and E. Hammerberg. "Effect of Moisture Regime and Other Factors on Municipal Solid Waste Stabilization," in *Management of Gas and Leachate*, S.K. Banerji, ed. (U.S. Environmental Protection Agency 600/9-77-026 1977).

Chapter 14
Resource Recovery

It is becoming increasingly difficult to find new sources of energy and materials to feed our industrial society. Concurrently, we are finding it more and more difficult to locate solid waste disposal sites. Mainly because of transportation requirements, the cost of disposal is escalating exponentially. These two factors—energy and material shortages, and fewer and more expensive disposal options—have fueled a new technology called *resource recovery*.

Except for the process of mass burning of raw refuse, discussed in the previous chapter, the recovery of resources from refuse is primarily a quest for purity. Pure materials may be obtained from mixed municipal solid waste in one of two ways:

1. Separation of the materials is performed by the user, the person who decides to discard the various "consumer products."*
2. Separation is performed after the mixed refuse is collected at a central processing facility.

SOURCE SEPARATION

Materials separation by the public, commonly known as *source separation*, has not been very successful in the United States. There are only two incentives that could be used to convince the public to undertake source separation. The first is regulatory, in which the governmental agency *dictates* that only separated material will be picked up. Unfortunately, this is not a successful approach in a democracy, since public officials advocating unpopular regulations can be removed from office. Source separation in totalitarian regimes, on the other hand, is easy to implement, but this is hardly a compelling argument to abolish democracy.

---

* The word consumer is clearly a misnomer. If we all were true consumers, and not users, there would be no solid waste.

The second means of achieving cooperation in source separation programs is to appeal to the sense of community spirit and the ethics of environmental concern. Indeed, studies conducted on the feasibility of community source separation programs, in which householders were asked whether they would participate in such programs, attained a 90 to 95 percent positive response. Unfortunately, the active response, or the participation in a source separation project, has seldom exceeded 5 percent of the households. There is a wide gap between what people say they will do (especially if they perceive that the question contains a value component) and how they will actually perform. A further complication is that in some inner cities it is difficult to convince people to put refuse in trash cans, much less convince them to separate the refuse into components. Although source separation appears to be the least expensive and least energy-intensive method of resource recovery, it has not been successfully implemented on a large scale in the United States, and effort therefore has been directed toward separation of mixed and collected refuse.

SOLID WASTE SEPARATION PROCESSES

Most processes for separation of the various materials in refuse rely on a characteristic or property of the specific material as a code, and this code is used to separate the material from the rest of the mixed refuse. Before such separation can be achieved, however, the material must be in separate and discrete pieces, a condition clearly not met by most components of mixed refuse. A common "tin can," for example, contains steel in its body, zinc on the seam, a paper wrapper on the outside, and perhaps an aluminum top. Other common items in refuse provide equally or even more challenging problems in separation.

One means of assisting in the separation process is to decrease the particle size of refuse, thus increasing the number of particles and achieving a greater number of "clean" particles. This size reduction step, although not strictly materials separation, is commonly a first step in a solid waste processing facility.

Size Reduction

Commonly called *shredding*, the process of size reduction usually consists of a brute force breakage of particles by swinging hammers in an enclosure. Two types of shredder are used in solid waste processing: the vertical and horizontal hammermills, as shown in Figure 14–1. In the former, the refuse enters the top and must work its way past the rapidly swinging hammers, clearing the space between the hammer tips and the enclosure. Particle size is controlled by adjusting this clearance. In the horizontal hammermill, the hammers swing over a grate that may be changed depending on the size of product required.

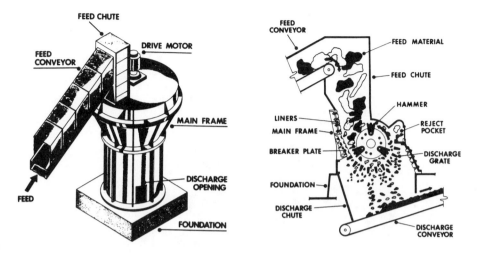

Figure 14–1. Vertical and horizontal hammermills.

Figure 14–2. A shredding facility, showing the conveyor belt leading to a vertical shredder.

Figure 14–3. Inside a vertical hammermill. The hammers have been worn down by the shredding process. [Photo courtesy of W.A. Worrell.]

Figure 14–2 shows a solid waste processing facility with a conveyor belt leading up to a vertical shredder. The control room is above and to the left of the conveyor. The hammers inside the shredder are shown in Figure 14–3. As the hammers reduce the size of the refuse components, they are themselves worn down. Typically, a set of hammers such as those shown can process 20,000 to 30,000 tons of refuse before having to be replaced.[1]

Shredder Performance and Design

The performance of a shredder is measured with respect to the degree of particle size reduction achieved and then compared with the cost in power and shredder wear.

The particle size distribution of a nonhomogeneous material such as municipal solid waste (MSW) is conveniently expressed by the Rosin–Rammler distribution function,[2] defined as

$$Y = 1 - \exp(-x/x_c)^a$$

where Y = the cumulative fraction of particles (by weight) less than size x
  n = a constant
  x_c = the *characteristic particle size*, defined as the size at which 63.2% (1 − 1/e = 0.632) of the particles (by weight) are smaller

This function plots as a straight line on log-log coordinates if $\ln[1/(1-y)]$ is plotted against x, as in Figure 14–4. Often, a shredder performance is specified on the basis of "90 percent of the particles (by weight) passing a given size." The conversion from characteristic size to 90 percent passing is possible by recognizing that

$$x_c = \frac{x}{[\ln(1/1-Y)]^{1/n}}$$

If Y = 90 percent, or 0.90, this expression reduces to

$$x_c = \frac{x_{90}}{(2.3)^{1/n}}$$

where x_{90} is the size at which 90 percent pass. Commonly n = 1.0, in which case $x_{90} = 2.3\ x_c$. Some typical values of x_c and n are shown in Table 14–1.

In addition to particle size reduction, shredder performance must be measured in terms of power usage, expressed as specific energy (kilowatt hours/ton of refuse processed). Figure 14–5 illustrates how solids feed rate, shredder speed, moisture content, and size of particles in the feed are interrelated.

A partially empirical expression often used to estimate the power requirements for shredders was developed by Bond.[3] The specific energy, W, required to reduce a unit weight of material 80 percent finer than some diameter L_F in

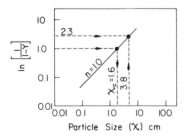

Figure 14–4. Rosin–Rammler particle size distribution equation. In this sample, n = 1.0 and x_c = 1.6.

Table 14–1.   Roisin–Rammler Exponents for Typical Shredder Installation*

Location	n (dimensionless)	x_c (cm)
Washington, DC	0.689	2.77
Wilmington, DE	0.629	4.16
Charleston, SC	0.823	4.03
St. Louis, MO	0.995	1.61
Pompano Beach, FL	0.587	0.67

* *Source*: Stratton, F.E., and H. Alter, "Application of Bond Theory to Solid Waste Shredding," *Journal of Environmental Engineering Division of the American Society of Civil Engineers* 104 (EE1)(1978).

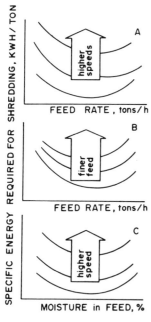

Figure 14–5.   (A) The effect of feed rate and rotor speed on specific energy used in shredding. (B) The effect of feed rate and feed particle size on specific energy. (C) The correlation between moisture content in the refuse and rotor speed.

micrometers, to a product 80 percent finer than some diameter L_p in micrometers, is expressed as

$$W = 10\, W_i\left(\frac{1}{\sqrt{L_p}} - \frac{1}{\sqrt{L_F}}\right)$$

where W_i is the *Bond work index*, a factor that is a function of the material processed for a given shredder and a function of shredder efficiency for a given material. W has the dimension of kilowatt hours/ton if L_p and L_F are in

Table 14–2. Typical Bond Work Indexes*

Shredder Location	Material	Bond Work Index (kWh/ton)
Washington, DC	MSW	463
Wilmington, DE	MSW	451
Charleston, SC	MSW	400
St. Louis, MO	MSW	434
Pompano Beach, FL	MSW	405
Washington, DC	Paper	194
Washington, DC	Steel Cans	262
Washington, DC	Alluminum cans	654
Washington, DC	Glass	8

* *Source*: Stratton, F.E., and H. Alter, "Application of Bond Theory to Solid Waste Shredding," *Journal of the Environmental Engineering Division of the American Society of Civil Engineers* 104 (EE1)(1978).

micrometers. The factor 10 also corrects for the dimensions. Typical values of the Bond work index for MSW and other specific materials are listed in Table 14–2.

It is convenient to combine the Rosin–Rammler particle size distribution equation with the Bond work index concept, by first noting that

$$(1 - Y) = \exp(-x/x_c)^n$$

where $(1 - Y)$ is the cumulative fraction *larger* than some size x. If $(1 - Y) = 0.2$, $x = L_p$ the screen size through which 80 percent of the product passes. Solving for L_p

$$0.2 = \exp\left(\frac{L_p}{x_c}\right)^a$$

$$L_p = x_c(1.61)^{1/n}$$

Substituting this back into the Bond work index equation,

$$W = \frac{10 \, W_i}{[x_c(1.61)^{1/n}]^{1/2}} - \frac{10 \, W_i}{[L_f]^{1/2}}$$

Example 14.1

Assuming $W_i = 400$ and x_c of the product is 1.62 cm, $n = 1$ and $L_F = 25$ cm (about 10 in., a realistic estimate of raw refuse), find the expected power requirements for a shredder processing 10 ton/hr.

$$W = \frac{10\,W_i}{[x_c(1.61)^{1/n}]^{1/2}} - \frac{10\,W_i}{[L_F]^{1/2}}$$

$$= \frac{10(400)}{[16,200(1.61)^{1/1}]^{1/2}} - \frac{10(400)}{[250,000]^{1/2}} = 16.7\ \text{kWh/ton}$$

(Note: x_c and L_F must be in micrometers, μm.)
The power requirement is $16.7\ \text{kWh/ton} \times 10\ \text{tons/hr} = 167\ \text{kW}$.

General Expressions for Materials Recovery

In the separation of any one material from a mixture, the separation is termed *binary*, in that only two outputs are required. When a device is to separate more than one material from a mixture, the process is termed *polynary* separation.

Figure 14–6 shows a binary separator receiving a mixed feed of x_0 and y_0. The objective is to separate out the x fraction. The first exit stream is to have the x component, but since the separation is not perfect, it also contains contamination in the amount of y_1. This stream is called the *product* or *extract*, while the second stream, containing mostly the y, but also some x, is known as the *reject*. The recovery of x can be expressed as

$$R_{(x_1)} = \left(\frac{x_1}{x_0}\right)100$$

where $R_{(x_1)}$ is the recovery of x in the first output stream, as percent.

This expression, however, is not an adequate descriptor of the performance of the binary separator. Imagine a situation in which the separator is turned *off*, i.e., all of the feed stream goes to the first output (extract). That is, $x_0 = x_1$, since there is no x_2. This would make $R_{(x_1)} = 100\%$. But obviously nothing has been done to the feed. Accordingly, a second requirement is the *purity* of the extract stream, defined as

$$P_{(x_1)} = \left(\frac{x_1}{x_1 + y_1}\right)100$$

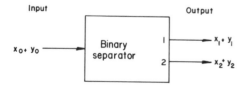

Figure 14–6. Definition sketch of a binary separator.

Similarly, purity is not an adequate descriptor of binary separator performance, since it might be possible to extract only a very small amount of x, in a pure state, but have the recovery $(R_{(x_1)})$ be very small, a clearly unacceptable condition from the process engineering standpoint. It is thus necessary to describe a materials separation device by *both* the recovery and purity.

This requirement, of course, makes it difficult to express the performance of a separation device as a single-valued parameter. This problem may be resolved by defining a term for binary separator *efficiency* as[4]

$$E_{(x,y)} = \left(\frac{x_1}{x_0}\right)\left(\frac{y_2}{y_0}\right)100$$

Example 14.2

A binary separator (magnet) is to separate a product (ferrous materials) from a feed stream (shredded refuse). The feed rate to the magnet is 1,000 kg/hr, and it contains 50 kg of ferrous (product x) and 950 kg of other material (reject y). Of the 40 kg in the product stream, 35 kg are ferrous. What is the percent recovery of ferrous, the purity of the product, and the overall efficiency?

$$R_{(x_1)} = \left(\frac{x_1}{x_0}\right)100 = \left(\frac{35}{50}\right)100 = 70\%$$

$$P_{(x_1)} = \left(\frac{x_1}{x_1 + y_1}\right)100 = \left(\frac{35}{40}\right)100 = 88\%$$

$$E_{(x,y)} = \left(\frac{x_1}{x_0}\right)\left(\frac{y_2}{y_0}\right)100 = \left(\frac{35}{50}\right)\left(\frac{[1,000 - 40] - [50 - 35]}{950}\right)100 = 70\%$$

Screens

Screens separate materials solely by size and do not identify the material by any other property. Consequently, screens are most often used in resource recovery as a classification step before a materials separation process. For example, it is possible technically (if not economically) to sort glass into clear and colored fractions by optical coding. This process, however, requires that the glass be of a given size, and screens may be used to produce such a feed to an optical sorter.

The most widely used screen in resource recovery is the *trommel*, pictured in Figure 14-7. The charge inside a trommel behaves in three distinctly different ways, depending on the speed of rotation. At slow speeds, the trommel material is *cascading*, i.e., not being lifted, but simply rolling back. At higher speed, *cataracting* occurs, in which the centrifugal force carries the material up to the side and then it falls back. At even higher speeds, *centrifuging* occurs, in which the material adheres to the inside of the trommel. Obviously, the efficiency of a

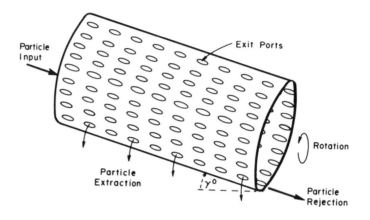

Figure 14–7. Trommel screen.

trommel is enhanced when the particles have the greatest opportunity to drop through the holes, and this occurs during the cataracting type of operation. Trommel speed is thus often designed as some fraction of the critical speed, defined as that rotational speed at which the materials will just begin centrifuging, and calculated as

$$n_c = \sqrt{\frac{g}{4\pi^2 r}}$$

where n_c = critical speed, rotations/sec
g = gravitational acceleration, cm/sec^2
r = radius of drum, cm

Example 14.3
    Find the critical speed for a 3-m-diameter trommel.

$$n_c = \left(\frac{g}{4\pi^2 r}\right)^{1/2}$$

$$= \left(\frac{980}{4(3.14)^2(150)}\right)^{1/2} = 0.407 \text{ rotation/sec}$$

Air Classifiers

Materials may be separated by their aerodynamic properties. In shredded MSW, most of the aerodynamically light materials are organic, and most of the heavy materials are inorganic; thus air classification can produce a refuse-derived fuel (RDF) superior to unclassified shredded refuse.

Most air classifiers are similar to the unit pictured in Figure 14–8. The fraction escaping with the air stream is the product (or the overflow), the fraction falling out the bottom is the reject (or underflow). The recovery of organics by air classification is adversely influenced by two factors:

1. Not all organics are aerodynamically light, and not all inorganics are aerodynamically heavy.
2. Perfect separation of heavy and light materials is difficult because of the stochastic nature of material movement in the classifier.

The first factor is illustrated in Figure 14–9, in which the terminal settling velocity (that air velocity at which the particle will just begin to rise with the air stream) is plotted against the fraction of particles of various materials. Regardless of the air velocity chosen, there can never be a complete separation of the organics (plastics and paper) from the inorganics (aluminum and steel).

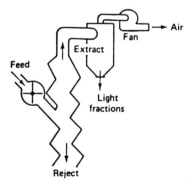

Figure 14–8.   Air classifier.

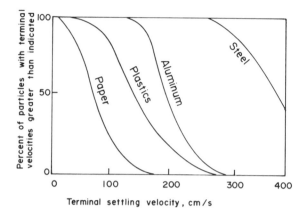

Figure 14–9. Terminal settling velocities of various components of MSW. Note that no single air speed will result in complete separation of organics (plastics and paper) from inorganics (steel and aluminum). Idealized curves.

The second factor is illustrated in Figure 14–10, which shows the efficiency (as defined above) versus the feed rate to the classifier. As solids loading increases, more particles that would have exited with the light fraction might get caught in the underflow stream, and vice versa.

Recovery of organics in the product stream may be plotted against the recovery of inorganics in the reject stream. The ideal performance curve may be obtained from terminal settling plots such as Figure 14–9 and by calculating $R_{(x_1)}$ and $R_{(y_2)}$ at different air velocities. The actual performance curve is obtained by measuring $R_{(x_1)}$ and $R_{(y_2)}$ for a continuous classifier at different air speeds. The closer the two curves are, the more effective is the air classifier. In addition, as noted in Figure 14–10, the performance deteriorates as the loading on the air classifier is increased. With high feed rates, the throat of the classifier becomes clogged and a clean separation is no longer possible.

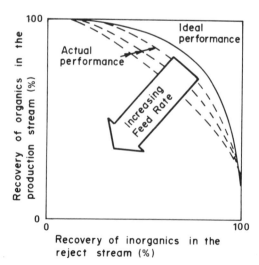

Figure 14–10. Actual and ideal performance of air classifiers. Idealized curves.

Example 14.4

If an air classifier operates with an air velocity of 200 cm/sec, and the feed contains equal amounts of paper, plastics, aluminum, and steel having terminal settling curves as shown in Figure 14–9, what would be the recovery of organics and the purity of the product?

At 200 cm/sec, the fractions of the components that report as the overflow (product) are (from Figure 14–9):

Paper	100%
Plastics	90%
Aluminum	50%
Steel	0%

The total organics in the product is

$$(100)(1/4) + (80)(1/4) = 45\% \text{ of feed}$$

The total inorganics in the product is

$$(50)(1/4) + (0)(1/4) = 12\% \text{ of the feed}$$

The recovery of organics is thus

$$R_{org} = \frac{45}{25 + 25} \times 100 = 90\%$$

The purity is

$$P_{product} = \frac{45}{45 + 12} \times 100 = 79\%$$

Magnets

Ferrous material is removed by using magnets, which continually extract the ferrous material and reject the remainder. Two types of magnets are shown in Figure 14–11. With the belt magnet, recovery of ferrous material is enhanced by placing the belt close to the refuse, but this also decreases the purity of the product. A major problem in using belt magnets is the depth of the refuse on the conveyor belt. The heavy ferrous particles tend to settle to the bottom of the refuse carried on a conveyor, and these are then the farthest away from the magnet.

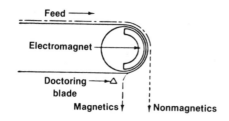

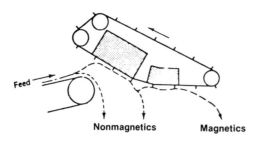

Figure 14–11. Two types of magnet used for resource recovery.

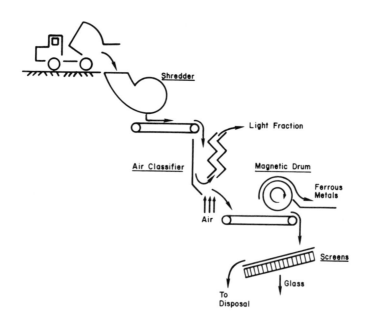

Figure 14–12. Schematic of typical materials separation facility for refuse processing.

Other Separation Equipment

Countless other unit operations for materials handling and storage have been tried. Jigs have been used for removing glass, froth rotation has been successfully employed for separating glass from ceramics, eddy current devices have recovered aluminum in commercial quantities, etc. As resource recovery operations evolve, more and better materials separation and handling equipment will be introduced. A schematic of a typical materials separation facility is presented in Figure 14–12.

Energy Recovery from the Organic Fraction of Municipal Solid Waste

The organic fraction of refuse is, as mentioned earlier, a useful secondary fuel. This shredded and classified product may be used in existing power plants either as a supplemental fuel with coal or fired as the sole fuel in separate boilers. A cross section of a typical heat recovery boiler is shown in the previous chapter (Figure 13–7). In smaller installations, modular heat recovery boilers, also described in Chapter 13, have found wide use.

In the combustion of an organic material, the presumed reaction is

$$(HC)_x + O_2 \rightarrow CO_2 + H_2O + \text{heat}$$

The process is much more complex, however, since not all the hydrocarbons are converted to carbon dioxide and water, and other components of the fuel such as nitrogen and sulfur also are oxidized. For nitrogen, the product of combustion is nitrogen oxides

$$N_2 + O_2 \rightarrow 2NO$$

$$2NO + O_2 \rightarrow 2NO_2$$

Nitrogen dioxide is an important component in the formation of photochemical smog.

Sulfur is also oxidized to sulfur oxides

$$S + O_2 \rightarrow SO_2$$

$$2SO_2 + O_2 \rightarrow 2SO_3$$

and SO_3 combines with water to form sulfuric acid,

$$SO_3 + H_2O \rightarrow H_2SO_4$$

which is, of course, highly corrosive and a hazard to health and vegetation. (See Chapter 18.)

The theoretical amount of oxygen required for combustion is known as *stoichiometric oxygen* (in terms of air it is *stoichiometric air*) and is calculated from the chemical reaction, as in the following example.

Example 14.5

If carbon is combusted, as

$$C + O_2 \rightarrow CO_2 + \text{heat}$$

how much air is required? One mole of oxygen is required for each mole of carbon used. The molecular weight of carbon is 14 and that of oxygen is $2 \times 16 = 32$. Hence, 1 g of C requires

$$\frac{32}{14} = 2.28 \text{ g of oxygen}$$

Since air is about 21 percent oxygen by volume and 23.15 percent by weight, the total amount of air required to combust 1 g of carbon is

$$\frac{2.28}{0.2315} = 9.87 \text{ g of air}$$

The yield of energy from combustion is measured as the calories of heat liberated per unit weight of material burned. This is known as the *heat of combustion*, or in engineering shorthand as the *heat value*, and is measured using *calorimeters*, in which a small sample of the fuel is placed in a stainless-steel bomb under a high pressure of pure oxygen and then fired. The heat is transferred to the water in which the bomb is placed, and the rise in temperature is measured. With knowledge of the mass of the water, the energy liberated during combustion may be calculated. In SI units the heat value is expressed as kilojoules/kilogram (kJ/kg); common engineering units express it as British thermal units/pound (Btu/lb). Table 14–3 lists some heats of combustion for common hydrocarbons and gives some typical values for refuse and RDF.

Table 14–3. Typical Values of Heats of Combustion

Fuel	Heat of Combustion (kJ/kg)	(Btu/lb)
Carbon (to CO_2)	32,800	14,100
Hydrogen	142,000	61,100
Sulfur (to SO_2)	9,300	3,980
Methane (CH_4)	55,500	23,875
Raw refuse	9,300	4,000
RDF (air classified)	18,600	8,000

In the design of any combustion operation, the system needs to be analyzed for thermal as well as materials balance. The latter requires simply that the mass of various inputs (air and fuel) must equal the mass of the outputs (stack emissions and bottom ashes). For a thermal balance, the heat input must equal the output plus losses. Typical thermal balance calculations are shown in Example 14.6.

Example 14.6

Assume a processed refuse containing 20 percent moisture and 60 percent organics. It is to be fed to a boiler at a rate of 1,000 kg/hr. From a calorimetric analysis of the refuse, it was determined that a dry sample has a heat value of 19,000 kJ/kg.

The heat from the combustion of the RDF is

$$H_{comb} = 19,000 \text{ kJ/kg} \times 1,000 \text{ kg/hr}$$
$$= 19 \times 10^6 \text{ kJ/hr}$$

The RDF organic fraction contains hydrogen, which becomes water. The heat of combustion value above includes the latent heat of vaporization (the hydrogen in the sample all condenses to water, thus yielding heat). Since this does not occur in full-scale combustion (i.e., the water exists as steam), the latent heat of vaporization must be considered a heat loss. Assuming the organics had 50 percent hydrogen, the latent heat loss resulting from vaporization is

$$H_{vap} = 1,000 \text{ kg/hr} \times 0.6 \text{ (organics)} \times 0.5 \text{ (hydrogen)}$$
$$\times 2,420 \text{ kJ/kg (latent heat of vaporization for water)}$$
$$= 0.726 \times 10^6 \text{ kJ/hr}$$

The RDF also contains moisture that is vaporized.

$$H_{mois} = 1,000 \text{ kg/hr} \times 0.2 \text{ (moisture)} \times 2,420 \text{ kJ/kg} = 0.484 \times 10^6 \text{ kJ/hr}$$

There is heat loss secondary to radiation, usually assumed as 5 percent of the heat input, or

$$H_{rad} = 19 \times 10^6 \text{ kJ/hr} \times 0.05 = 0.95 \times 10^6 \text{ kJ/hr}$$

Not all of the organics will combust. Assume that the ashes contain 10 percent of the organics, so that the heat loss is

$$H_{noncom} = (10/60) \times 19 \times 10^6 \text{ kJ/hr} = 3.17 \times 10^6 \text{ kJ/hr}$$

The stack gases also contain heat, and since the amount is unknown, it is usually calculated as the difference, as

heat input = heat output

$$H_{comb} = (H_{vap} + H_{mois} + H_{rad} + H_{noncom}) + H_{stack}$$

$$19 \times 10^6 = (0.726 + 0.484 + 0.95 + 3.17) \times 10^6 + H_{stack}$$

$$H_{stack} = 13.67 \text{ kJ/hr}$$

So the heat lost in the stack gases is 13.67 kJ/hr. This heat may be recovered by running cold water into the boiler (the water wall tubes) and producing steam. If 2,000 kg/hr of steam at a temperature of 300°C and a pressure of 4×10^3 kPa is required and the temperature of the boiler water is 80°C, calculate the heat loss in the stack gases. Heat in the boiler water is

$$H_{wat} = 2,000 \text{ kg/hr (water)} \times (80 + 273)°K \times 0.00418$$

$$\text{kJ/kg} \cdot °K = 2,951 \text{ kJ/hr}$$

(Note: the specific heat of water is 0.00418 kJ/kg·°K.)
The heat in the stream, at 300°C and 4×10^3 kPa (see steam tables in any thermodynamics textbook) is 2,975 kJ/kg

$$H_{steam} = 2,000 \text{ kg/hr} \times 2,975 \text{ kJ/kg} = 5.95 \times 10^6 \text{ kJ/hr}$$

The heat balance then yields

$$\text{In} = \text{Out}$$

$$\left.\begin{array}{l}\text{Heat in combustion} \\ \text{Heat in boiler water}\end{array}\right\} = \left\{\begin{array}{l}\text{Latent heat from vaporization} \\ \text{Heat loss from radiation} \\ \text{Heat in ashes} \\ \text{Heat in steam} \\ \text{Heat in stack gases}\end{array}\right.$$

By subtraction, the heat in the stack gases is

$$H_{comb} + H_{wat} = (H_{vap} + H_{mois} + H_{rad} + H_{noncom} + H_{steam}) + H_{stack}$$

$$(19 + 0.0002) \times 10^6 = (0.726 + 0.484 + 0.95 + 3.17 + 5.95) \times 10^6 + \begin{array}{l}\text{heat in} \\ \text{stack} \\ \text{gases}\end{array}$$

$$\text{Heat in stack gases} = 7.72 \times 10^6 \text{ kJ/hr}$$

Combustion of the organic fraction of refuse is not the only means of extracting useful energy from it. This extraction may also be by chemical or biochemical means.

Using a process known as *acid hydrolysis,* the cellulose fraction of RDF may be treated so as to produce methane gas. Other chemical processes are presently being developed for producing alcohol from RDF. Alcohol may be added to gasoline and thus serve as substitute liquid fuel. A mixture of 80 percent gasoline and 20 percent alcohol will not require readjustments in an internal combustion engine.

The two types of biochemical process used for extracting useful products from refuse are anaerobic and aerobic decomposition. In the anaerobic system, refuse is mixed with sewage sludge and the mixture is digested. Operational problems have made this process impractical on a large scale, although single household units that combine human excreta with refuse have been used.

Aerobic decomposition of refuse is better known as *composting* and results in the production of a useful soil conditioner that has moderate fertilizer value. The process is exothermic, and at a household level has been used as a means of producing hot water for heating homes. On a community scale, composting may be a mechanized operation, using an aerobic digester (Figures 14–13 and 14–14), or it may be a low-technology operation, using long rows of shredded refuse known as *windrows*.

Windrows are normally 3 m (10 ft) wide at the base and 1.5 m (4 to 6 ft) high. Under these conditions, sufficient moisture and oxygen are available to support aerobic life. The piles must be turned periodically to allow sufficient oxygen to penetrate to all parts of the pile. A modification of this system is to blow air into the piles. This is commonly known as *static pile composting* (Figure 14–15).

Temperatures within a windrow approach 140°F, entirely due to biological activity. The pH will approach neutrality after an initial drop. With most

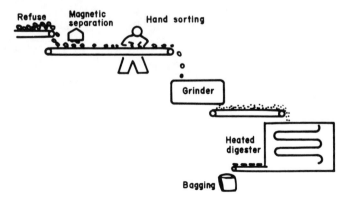

Figure 14–13. Schematic of a mechanical composting operation.

Figure 14–14. A community composting operation in Auckland, New Zealand, showing the rotary digester and windrows.

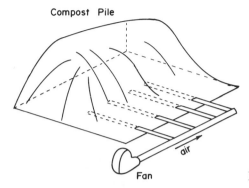

Figure 14–15.   Static pile composting.

wastes, additional nutrients are not needed. The composting of bark and other materials, however, is successful only with the addition of nitrogen and phosphorous nutrients.

Moisture must usually be controlled. Excessive moisture makes if difficult to maintain aerobic conditions, whereas a dearth inhibits biological life. A 40 to 60 percent moisture content is considered desirable.

There has been some controversy over the use of inoculants, freeze-dried cultures, used to speed up the process. Once the composting pile is established, which requires about 2 weeks, the inoculants have not proven to be of any significant value. Most municipal refuse contains all the organisms required for successful composting, and "mystery cultures" are thus not needed.

The endpoint of a composting operation may be measured by noting a drop in temperature. The compost should have an earthy smell, similar to peat moss, and should have a dark brown color.

Although compost is an excellent soil conditioner, it is not widely used by U.S. farmers. Inorganic fertilizers are cheap and easy to apply, and most farms are located where soil conditions are good. The plentiful food supply in most developed countries does not dictate the use of marginal lands where compost would be of real value.

CONCLUSION

It is obvious that we must attack the solid waste problem from both ends—from the source as well as from the disposal methods. Solid wastes, often called the "third pollution," are only now being considered a problem equal in magnitude to that of air and water pollution.

It has been suggested that one solution to the solid waste problem is the development of truly biodegradable forms of materials such as plastics and glass. A scientist in England has developed a plastic that will disintegrate when exposed to ultraviolet light. Direct sunlight will thus cause disintegration, whereas light filtered through a window pane will not. The materials thus broken down will degrade biologically.

We are still many years away from the development and use of fully recyclable or biodegradable materials. The only truly disposable package available today is the ice cream cone.

PROBLEMS

14.1   Estimate the critical speed for a trommel screen, 3 m in diameter, processing refuse so as to run 25 percent full.

14.2   What air speed is required in an air classifier to suspend a spherical piece of glass 1 mm in diameter? (See p. 149, assume $\rho_s = 2.65$ g/cm^3, $C_D = 2.5$, $\mu = 2 \times 10^{-4}$ poise, and $\rho = 0.0012$ g/cm^3.)

14.3   Refer to the curves shown in Figure 14–16. By replotting the curves, estimate Rosin–Rammler constants n and x_c. Compare these with the data in Table 14–1.

14.4   If the Bond work index is 400 kWh/ton, what is the power requirement to process 100 ton/hr of the refuse described in Figure 14–16?

14.5   An air classifier performance is shown below

	Organics (kg/hr)	Inorganics (kg/hr)
Feed	80	20
Product	60	10
Reject	20	10

Calculate the recovery, purity, and efficiency.

14.6   Estimate the heating value of MSW based on the ultimate analysis shown in Table 12–1.

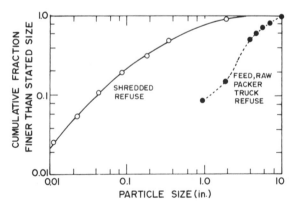

Figure 14–16. Feed and product curves in shredding, for Problem 14.3.

14.7 A power plant burns 100 ton/hr of coal. How much air is needed if 50 percent excess air is used?

14.8 Using the data in Example 14.6, what percentage of the heating value can be saved by reducing the moisture content of refuse to zero?

14.9 In Example 14.6, what fraction of the coal is "wasted" owing to the loss of heat in the stack gases (if steam is produced)?

14.10 What is the "wasted" coal in Problem 14.9 heat equivalent in barrels of oil? (The heating value of a barrel of oil is 6×10^6 kJ.)

LIST OF SYMBOLS

E = efficiency of materials separation
L_F = feed particle size, 80 percent finer than, μm
L_p = product size, 80 percent finer than, μm
MSW = municipal solid waste
n_c = critical speed of a trommel, rotations/sec
P_x = purity of a product x, percent
r = radius, cm
RDF = refuse-derived fuel
R_x = recovery of a product x, percent
W = specific energy, kWh/ton
W_i = Bond work index, kWh/ton
x = particle size, m
x_c = characteristic particle size, m
x_0, y_0 = mass per time of feed to a materials separation device
x_1, y_1 = mass per time of components x and y existing from a materials separation device through exit stream 1
x_2, y_2 = mass per time of components x and y exiting from a materials separation device through exit stream 2
Y = cumulative fraction of particles (by weight) less than some specified size

REFERENCES

1. Vesilind, P.A., A.E. Rimer, and W.A. Worrell. "Performance Testing of a Vertical Hammermill Shredder," *Proceedings of the National Conference on Solid Waste Processing* (New York: American Society of Mechanical Engineers, 1980).
2. Rosin, P., and E. Rammler. "Laws Covering the Fineness of Powdered Coal," *Journal of the Institute for Fuel* 7:29(1933).
3. Bond, F.C., "The Third Theory of Comminution," *Transactions of the American Institute of Mining Engineers* 193:484(1952).
4. Worrell, W.A., and P.A. Vesilind. "Testing and Evaluation of Air Classifier Performance," *Resource Recovery and Conservation* 4:247 (1979).

Chapter 15
Hazardous Waste

For centuries, chemical wastes have been the by-products of developing societies. Here a disposal site, there a disposal site—all with little or no attention to potential impacts on groundwater quality, runoff to streams and lakes, and skin contact as children played hide-and-seek in a forest of abandoned 55-gallon drums. Engineering decisions here historically were made by default; little or no handling/processing/disposal planning at the corporate or plant level necessitated quick and dirty decision by mid- and entry-level engineers at the end of production processes. These production engineers solved disposal problems by simply piling or dumping these waste products "out back."

Attitudes began to change in the 1960s and 1970s. As other chapters of this text indicate, air, water, and land were no longer viewed as commodities to be polluted with the problems of cleanup freely passed to neighboring towns or future generations. Individuals responded with court actions against pollution, and governments responded with revised local zoning ordinances, updated public health laws, and new major Federal Clean Air and Clean Water Acts. In 1976, the Federal Resource Conservation and Recovery Act (RCRA) was enacted to give the U.S. Environmental Protection Agency specific authority to regulate the generation, transport, and disposal of hazardous waste. In the 1980s we find that engineering knowledge and expertise has not kept pace with this awakening to the necessity to manage hazardous wastes adequately. This chapter discusses the state of knowledge in the field of hazardous waste engineering, tracing the quantities of wastes generated in the nation from handling and processing options through transportation controls to resource recovery and ultimate disposal alternatives.

MAGNITUDE OF THE PROBLEM

Over the years, the term "hazardous" has evolved in a confusing setting as different groups advocate many criteria for classifying a waste as "hazardous."

Within the federal government, different agencies used such descriptions as toxic, explosive and radioactive to label a waste as hazardous. Different states had other classification systems, as did the National Academy of Sciences and the National Cancer Institute. These past systems are displayed in Table 15–1. Selected classification criteria are described in more detail in Table 15–2.

The federal government attempted to impose a nationwide classification system under the implementation of RCRA, in which a hazardous waste is defined by the degree of ignitability, corrosivity, reactivity, or toxicity. This definition includes acids, toxic chemicals, explosives, and other harmful or potentially harmful waste. In this chapter, this will be the applicable definition of hazardous waste. Radioactive wastes are excluded. Such wastes obviously are

Table 15–1. Historical Definitions of Hazardous Waste

System	Toxicological	Flammability	Explosives	Corrosive	Reactivity	Oxidizing material	Radioactive	Irritant	Strong sanitizer	Bioconcentration	Carcinogenic, mutagenic, teratogenic	Sufficient quantity
Title 15, U.S. Code, Sec. 1261	×	×		×		×	×	×				×
CPSC-Title 16, CFR, Part 1500	×	×		×		×	×	×				×
Food, Drug, and Cosmetic Act	×								×		×	×
DOT-Title 49, CFR, Parts 100–199	×	×	×	×		×	×	×				×
Pesticides-Title 40, CFR, Part 162	×	×							×	×		
Ocean Dumping-Title 40, CFR, Part 227	×								×	×	×	
NIOSH-Toxic Substances List	×										×	
Drinking Water Standards	×								×	×		
FWPCA Sec. 304 (a)(1)	×								×	×		
Sec. 307 (a)	×								×	×		×
Sec. 311 (b)(2)(A)	×											×
Clean Air Act-Sec. 112	×							×			×	
California State List	×	×	×	×				×	×			
National Academy of Sciences	×	×			×							
TRW Systems Group	×	×	×	×			×			×		×
Batelle Memorial Institute	×	×	×		×	×	×	×		×	×	
Booz-Allen Applied Research, Inc.	×	×	×		×							×
Dept. of the Army	×											
Dept. of the Navy	×	×	×	×	×	×	×					
National Cancer Institute	×									×	×	×

Table 15–2.   Classification Criteria for Hazardous Waste

Criteria	Description
1. Bioconcentration (bioaccumulation)	The process by which living organisms concentrate an element or compound in levels in excess of the surrounding environment.
2. LD_{50} (lethal dose 50)	A calculated dose of a chemical substance that is expected to kill 50 percent of a population exposed through a route other than respiration (mg/kg of body weight).
3. LDC_{50} (Lethal dose concentration 50)	A calculated concentration of a chemical substance that, when following the respiratory route, will kill 50 percent of a population during a 4-hour exposure period (ambient concentration in mg/L).
4. Phytotoxicity	The ability of a chemical substance to cause poisonous reactions in plants.

hazardous, but because their generation, handling, processing, and disposal differ so drastically from those of nonnuclear hazards, the radioactive waste problem is addressed separately in Chapter 16.

Given this somewhat limited definition, more than 60 million metric tons, by wet weight, of hazardous waste are generated annually throughout the United States. More than 60 percent is generated by the chemical and allied products industry, and the machinery, primary metals, paper, and glass products industries each generate between 3 and 10 percent of the nation's total. Approximately 60 percent of the hazardous waste is liquid or sludge. Major generating states, including New Jersey, Illinois, Ohio, California, Pennsylvania, Texas, New York, Michigan, Tennessee, and Indiana, contribute more than 80 percent of the nation's total production of hazardous waste, and most waste is disposed of on the generator's property. More than 80 percent of all disposal is in inadequately designed and operated pits, ponds, landfills, and incinerators.

A hasty reading of these hazardous waste facts points to several interesting, through shocking, conclusions. First, most hazardous waste is generated and inadequately disposed of in the eastern portion of the country. In this region, the climate is wet with patterns of rainfall that permit infiltration or runoff to occur. Infiltration permits the transport of hazardous waste into groundwater supplies, and surface runoff leads to the contamination of streams and lakes. Second, most hazardous waste is generated and disposed of in areas where people rely on aquifers for drinking water. Major aquifers and well withdrawals underlie areas where the wastes are generated. Thus, the hazardous waste problem is compounded by two considerations: the wastes are generated and disposed of in areas where it rains and in areas where people rely on aquifers for supplies of drinking water.

WASTE PROCESSING AND HANDLING

Waste processing and handling are key concerns as a hazardous waste begins its journey from the generator site to a secure long-term storage facility. Ideally, the waste can be stabilized or detoxified or somehow rendered harmless in a treatment process similar to those outlined briefly below.

Chemical Stabilization/Fixation. In these processes, chemicals are mixed with waste sludges, the mixture is pumped onto land, and solidification occurs in several days or weeks. The result is a chemical nest that entraps the waste, and pollutants such as heavy metals may be chemically bound in insoluble complexes. Proponents of these processes have argued for building roadways, dams, and bridges with a selected cement as the fixing agent. The environmental adequacy of the processes has not been documented, however, as long-term leaching and defixation potentials are not well understood.

Volume Reduction. Volume reduction is usually achieved in an incineration process. This process takes advantage of the large organic fraction of waste being generated by many industries, but may lead to secondary problems for hazardous waste engineers: air emissions in the stack of the incinerator and ash production in the base of the incinerator. Both by-products of incineration must be addressed in terms of legal, cost, and ethical constraints. Because incineration is often considered a very good method for the ultimate disposal of hazardous waste, we discuss it in some detail later in this chapter.

Waste Segregation. Before shipment to a processing or long-term storage facility, wastes are segregated by type and chemical characteristics. Similar wastes are grouped in a 55-gallon drum or group of drums, segregating liquids such as acids from solids such as contaminated laboratory clothing and animal carcasses. Waste segregation is generally practiced to prevent undesirable reactions at disposal sites and may lead to economies of scale in the design of detoxification or resource recovery facilities.

Detoxification. Numerous thermal, chemical, and biological processes are available to detoxify chemical wastes. Options include:

- neutralization
- ion exchange
- incineration
- pyrolysis
- aerated lagoons
- waste stabilization ponds

These technologies are extremely waste specific; ion exchange obviously doesn't work for every chemical, and some forms of heat treatment may be prohibitively expensive for sludges that have a high water content. It's time to call in the chemical engineers whenever detoxification technologies are being considered.

Degradation. Methods exist that chemically degrade some hazardous wastes and render them safer, if not completely safe. Chemical degradation

processes, which are very waste specific, include hydrolysis, to destroy or-
ganophosphorus and carbonate pesticides, and chemical dechlorination, to
destroy some polychlorinated pesticides. Biological degradation generally involves incorporating the waste into the soil. Landfarming, as it has been termed,
relies on healthy soil microorganisms to metabolize the waste components. Such
landfarming sites must be strictly controlled for possible water and air pollution
that results from overactive or underactive organism populations. For the most
part, degradation of hazardous waste is in the research and development stages.

Encapsulation. A wide range of material is available to encapsulate hazardous waste. Ranging from the basic 55-gallon steel drums utilized throughout the
nation, options include:

- concrete
- asphalt
- plastics

Several layers of different materials are often recommended, such as a steel
drum coated with an inch or more of polyurethane foam to prevent rust.

TRANSPORTATION OF HAZARDOUS WASTES

Hazardous wastes are transported across the nation on trucks, rail flatcars, and
barges. Because many hazardous wastes are often generated in relatively small
quantities, truck transportation, and often small-truck transportation, is a
highly visible and constant threat to public safety and the environment. There
are four basic elements in the control strategy for the movement of hazardous
waste from a generator.

1. *Haulers.* Major concerns over hazardous waste haulers include
operator training, insurance coverage, and special registration of transport
vehicles. Handling precautions include gloves, face masks, and coveralls for
workers, as well as registration of handling equipment to control future use of
the equipment and avoid situations in which hazardous waste trucks today are
used to carry produce to market tomorrow. Schedules for relicensing haulers
and checking equipment are part of an overall program for ensuring proper
transport of hazardous wastes.

2. *Hazardous Waste Manifest.* The concept of a cradle-to-grave tracking
system has long been considered key to proper management of hazardous waste.
This "bill of lading" or "trip ticket" ideally accompanies each barrel of waste
and describes the content of each barrel to its recipient. Copies of the manifest
are submitted to generators and state officials so all parties know that each
waste has reached its desired destination in a timely manner. This system serves
four major purposes: (1) it provides the government with a means of tracking
waste within a given state and determining quantities, types, and locations

where the waste originates and is ultimately disposed; (2) it certifies that wastes being hauled are accurately described to the manager of the processing/ disposing facility; (3) it provides information for recommended emergency response if a copy of the manifest is not returned to the generator; and (4) it provides a data base for future planning within a state. Figure 15–1 illustrates one possible routing of copies of a selected manifest. In this example, the original manifest and five copies are passed from the state regulatory agency to the generator of the waste. Copies accompany each barrel of waste that leaves the generating site and are signed and mailed to the respective locations to indicate the transfer of the waste from one location to another.

3.   *Labeling and Placarding.* Before a waste is transported from a generating site, each container is labeled and the transportation vehicle is placarded. Announcements that are appropriate include warnings for explosives, flammable liquids, corrosive material, strong oxidizers, compressed gases, and poisonous or toxic substances. Multiple labeling is desirable if, for example, a waste is both explosive and flammable. These labels and placards warn the general public of possible dangers and assist emergency response teams as they react in the event of a spill or accident along a transportation route.

4.   *Accident and Incident Reporting.* Accidents involving hazardous wastes must be reported immediately to state regulatory agencies and local health officials. Accident reports that are submitted immediately and indicate the amount of materials released, the hazards of these materials, and the nature of the failure that caused the accident may be instrumental in containing a waste and cleaning the site. For example, if liquid waste can be contained, groundwater and surface water pollution may be avoided.

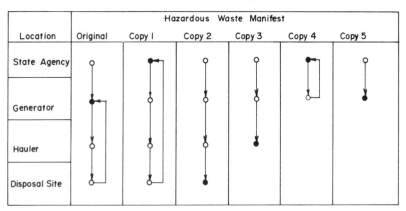

Figure 15–1.   Possible routing of copies of a hazardous waste manifest.

RESOURCE RECOVERY ALTERNATIVES

Resource recovery alternatives are based on the premise that one man's waste is another man's prize. What may be a worthless drum of electroplating sludge to the plating engineer may be a silver mine to an engineer skilled in metals recovery. In hazardous waste management, two types of system exist for transferring this waste to a location where it is viewed as a resource: *hazardous waste materials transfers* and *hazardous waste information clearinghouses*. In practice, one organization may display characteristics of both of these pure systems.

The rationale behind both transfer mechanisms is illustrated in Figure 15–2. An industrial process typically has three outputs: (1) a principal product, which is sold to a consumer; (2) a useful by-product available for sale to another industry; and (3) waste, historically destined for ultimate disposal. Waste transfers and clearinghouses act to minimize this flow of waste to a landfill or to ocean burial by directing it to a heretofore unidentified industry or firm that perceives the waste as a resource. As the regulatory and economic climate of the nation evolves, these perceptions may continue to change and more and more waste may be economically recycled.

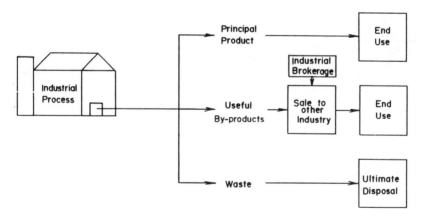

Figure 15–2. Rationale for hazardous waste clearinghouses and exchanges.

Information Clearinghouses

The pure clearinghouse has limited functions. Basically, these institutions offer a central point for collecting and displaying information about industrial wastes. Their goal is to introduce interested potential trading partners to each other through the use of anonymous advertisements and contacts. Clearinghouses generally do not seek customers, negotiate transfers, set prices, process materials, or provide legal advice to interested parties. One major function of a clearinghouse is to keep all data and transactions confidential so trade secrets are not compromised.

Clearinghouses are also generally subsidized by sponsors, either trade or governmental. Small clerical staffs are organized in a single office or offices spread throughout a region. Little capital is required to get these operations off the ground, and annual operation expenses may range from $40,000 to $150,000.

The value of clearinghouse operations should not be overemphasized. Often they are only able to operate in the short term; they evolve from an organization with many listings and active trading to rapid drop-offs in activity as plant managers make their contacts with waste suppliers and short-circuit the system by eliminating the clearinghouse.

Materials Exchanges

In comparison with the clearinghouse concept, a pure materials exchange has many complex functions. A transfer agent within the exchange typically identifies generators of waste and potential users of the waste. The exchange will buy or accept waste, analyze its chemical and physical properties, identify buyers, reprocess the waste as needed, and sell it at a profit.

The success of an exchange depends on several factors. Initially, a highly competent technical staff is required to analyze waste flows and design and prescribe methods for processing the waste into a saleable resource. Capitalization may easily exceed $500,000 in laboratory equipment and supplies, and a yearly cash flow, including debt service, may exceed $200,000 for even small operations. The ability to diversify is critical to the success of an exchange. Its management must be able to identify local suppliers and buyers of their products. Additionally, an exchange may even enter the disposal business and incinerate or landfill waste.

Clearinghouses and exchanges have been attempted with some success in the United States. However, a longer track record exists in Europe. Belgium, Switzerland, West Germany, most of the Scandinavian countries, and the United Kingdom all have experienced some success with exchanges. The general characteristics of European waste exchanges include:

- operation by the national industrial associations
- services offered without charge
- waste availability made known through published advertisements
- advertisements discussing chemical and physical properties, as well as quantities, of waste
- advertisements coded to keep the identity of offerer confidential

Much can be learned in the United States from these experiences.

Five wastes are generally recognized as having transfer value: (1) wastes having high concentration of metals, (2) solvents, (3) concentrated acids, (4) oils, and (5) combustibles for fuel. That is not to say these wastes are the only transferable items. Transformed from waste to resource in one European exchange were:

- foundry slag, 50 to 60 percent metallic Al, 400 ton/yr
- methanol, 90 percent with trace mineral acids, 150 m³/yr
- cherries, deep frozen, 4 tons

One man's waste may truly be another man's valued resource.

HAZARDOUS WASTE MANAGEMENT FACILITIES

Siting Considerations

A wide range of factors must be considered in siting hazardous waste management facilities. Some of these are determined by law: for example, RCRA prohibits landfilling of flammable or pyrophoric liquids. Socioeconomic factors are often the key to siting. Joseph Koppel[1] has coined the acronym LULU (for locally undesirable land use) for a facility that no one wants nearby but that is going to be put somewhere. Certainly hazardous waste facilities are LULUs.

In selecting a site, all of the relevant "-ologies" must be considered: hydrology, climatology, geology, and ecology, as well as current land use, environmental health, and transportation. Risk analysis is usually required also. (See Chapter 2.)

Hydrology. Hazardous waste landfills should be located well above historically high groundwater tables. Care should be taken to ensure that a location has no surface or subsurface connection, such as a crack in confining strata, between the site and a water course. Hydrologic considerations limit direct discharge of wastes into groundwater or surface water supplies.

Climatology. Hazardous waste management facilities should be located outside the paths of recurring severe storms. Hurricanes and tornadoes disrupt the integrity of landfills and incinerators and cause immediate catastrophic effects on the surrounding environment and public health in the region of the facility. In addition, areas of high air pollution potential should be avoided in site selection processes. These areas include valleys where winds or inversions act to hold pollutants close to the surface of the earth, as well as areas on the windward side of mountain ranges, i.e., areas similar to the Los Angeles area where long-term inversions are prevalent.

Geology. A disposal or processing facility should be located only on stable geologic formations. Impervious rock, which is not littered with cracks and fissures, is an ideal final liner for hazardous waste landfills.

Ecology. The ecological balance must be considered as hazardous waste management facilities are located in a region. Ideal sites in this respect include areas of low fauna and flora density, and efforts should be made to avoid wilderness areas, wildlife refuges, and migration routes. Areas with unique plants and animals, especially endangered species, should also be avoided.

Alternative Land Use. Areas with low ultimate land use should receive prime consideration as facilities are sited in a region. Areas with high recreational use potential should be avoided because of the increased possibility of direct human contact with the wastes.

Environmental Health. Landfills and processing facilities should be located away from private wells, away from municipal water supplies, and away from high population densities. Flood plains should be avoided, at least up to the 100-year storm level.

Transportation. Transportation routes to facilities are a major consideration in siting hazardous waste management facilities. Such facilities should be accessible by all-weather highways to avoid spills and accidents during periods of rain and snowfall. Ideally, the closer a facility is to the generators of the waste, the less likely are spills and accidents as the wastes move along the countryside.

Socioeconomic Factors. Factors that could make or break an effort to site a hazardous waste management facility fall under this major heading. Such factors, which range from citizen acceptance to long-term care and monitoring of the facility, are:

1. Citizen acceptance and public education programs: Will local townspeople permit it?
2. Land use changes and industrial development trends: Does the region wish to experience the industrial growth that is induced by such facilities?
3. User fee structures and recovery of project costs: Who will pay for the facility; can user changes be used to induce industry to reuse, reduce, or recover the resources in the waste?
4. Long-term care and monitoring: How will postclosure maintenance be guaranteed and who will pay?

All are critical concerns in a hazardous waste management scheme.

The term "mixed waste" refers to mixtures of hazardous and radioactive wastes (organic solvents used in liquid scintillation counting are an excellent example). Siting a mixed waste facility is virtually impossible at present, because in many cases the laws governing handling of chemically hazardous waste conflict with those governing handling of radioactive waste.

Incinerators

Incineration is a controlled process that uses combustion to convert a waste to a less bulky, less toxic, or less noxious material. The principal products of incineration from a volume standpoint are carbon dioxide, water, and ash, but the products of primary concern because of their environmental effects are compounds containing sulfur, nitrogen, and halogens. When the gaseous combustion products from an incineration process contain undesirable compounds, a secondary treatment such as afterburning, scrubbing, or filtration is required to lower concentrations to acceptable levels before atmospheric release. The solid ash products from the incineration process are a major concern and must reach adequate ultimate disposal.

Figure 15–3 shows the schematic of a generalized incineration system that includes the components of a waste incineration system. The actual system, which may contain one or more of the components, is usually dependent on the individual waste's application requirement. The advantages of incineration as a means of disposal for hazardous waste are:

1. Burning wastes and fuels in a controlled manner has been carried on for many years and the basic process technology is available and reasonably well developed. This is not the case for some of the more exotic chemical degradation processes.
2. Incineration is broadly applicable to most organic wastes and can be scaled to handle large volumes of liquid waste.
3. Large expensive land areas are not required.

The disadvantages of incineration include:

1. The equipment tends to be more costly to operate than many other alternatives.

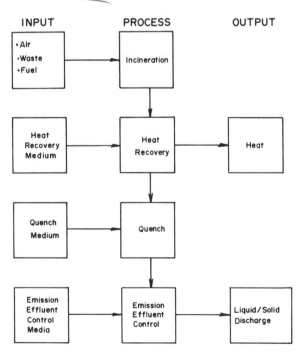

Figure 15–3.   Waste incineration system.

2. It is not always a means of ultimate disposal in that normally an ash remains that may or may not be toxic but that in any case must be disposed of properly.
3. Unless controlled by applications of air pollution control technology, the gaseous and particulate products of combustion may be hazardous to health or damaging to property.

The decision to incinerate a specific waste will therefore depend, first, on the environmental adequacy of incineration as compared with other alternatives, and second, on the relative costs of incineration and the environmentally sound alternatives.

The variables that have the greatest effect on the completion of the oxidation of wastes are waste combustibility, dwell time in the combustor, flame temperature, and the turbulence present in the reaction zone of the incinerator. The combustibility is a measure of the ease with which a material may be oxidized in a combustion environment. Materials with a low flammability limit, low flash point, and low ignition and autoignition temperatures may be combusted in a less severe oxidation environment, i.e., at a lower temperature and with less excess oxygen.

Of the three "T's" of good combustion—time, temperature, and turbulence—only the *temperature* may be readily controlled after the incinerator unit is constructed. This may be done by varying the air-to-fuel ratio. If solid carbonaceous waste is to be burned without smoke, a minimum temperature of 760°C (1,400°F) must be maintained in the combustion chamber. Upper temperature limits in the incinerator are dictated by the refractory materials available. Above 1,300°C (2,400°F) special refractories are needed.

The degree of *turbulence* of the air for oxidation with the waste fuel will affect the incinerator performance significantly. In general, both mechanical and aerodynamic means are utilized to achieve mixing of the air and fuel. The completeness of combustion and the time required for complete combustion are significantly affected by the amount and the effectiveness of the turbulence.

The third major requirement for good combustion is *time*. Sufficient time must be provided to the combustion process to allow slow-burning particles or droplets to burn completely before they are chilled by contact with cold surfaces or the atmosphere. The amount of time required depends on the temperature, fuel size, and degree of turbulence achieved.

The type and form of waste will dictate the type of combustion unit required. A number of control methods have been successfully developed for applications in which the pollutants are in the form of fumes or gas. If the waste gas contains organic materials that are combustible, then incineration should be considered as a final method of disposal. When the amount of combustible material in the mixture is below the lower flammable limit, it may be necessary to add small quantities of natural gas or other auxiliary fuel to sustain combustion in the burner. Thus economic considerations are critical in the selection of incinerator systems because of the high costs of these additional fuels.

Boilers for some high-temperature industrial processes may serve as incinerators for toxic or hazardous carbonaceous waste. Cement kilns, which must operate at temperatures in excess of 1,400°C (2,500°F) to produce cement clinker, can use organic solvents as fuel, and this provides an acceptable method of waste solvent and waste oil disposal.

Incineration is also a possibility for the destruction of liquid wastes. Liquid wastes may be classified into two types from a combustion standpoint: combustible liquids and partially combustible liquids. Combustible liquids include all materials having sufficient calorific value to support combustion in a conventional combustor or burner. Noncombustible liquids cannot be treated or disposed of by incineration and include materials that would not support combustion without the addition of auxiliary fuel and would have a high percentage of noncombustible constituents such as water.

When starting with a waste in liquid form, it is necessary to supply sufficient heat for vaporization in addition to raising it to its ignition temperature. For a waste to be considered combustible, several rules of thumb should be used. The waste should be pumpable at ambient temperature or capable of being pumped after heating to some reasonable temperature level. Since liquids vaporize and react more rapidly when finely divided in the form of a spray, atomizing nozzles are usually used to inject waste liquids into incineration equipment whenever the viscosity of the waste permits atomization. If the waste cannot be pumped or atomized, it cannot be burned as a liquid but must be handled as a sludge or solid.

To support combustion in air without the assistance of an auxiliary fuel, the waste must generally have a calorific value of 18,500 to 23,000 kJ/kg (8,000–10,000 Btu/lb) or higher. Liquid waste having a heating value below 18,500 kJ/kg (8,000 Btu/lb) is considered a partially combustible material and requires special treatment.

Several basic considerations are important in the design of an incinerator for a partially combustible waste. First, the waste material must be atomized as finely as possible to present the greatest surface area for mixing with combustion air. Second, adequate combustion air to supply all the oxygen required for oxidation or incineration of the organics present should be provided in accordance with carefully calculated requirements. Third, the heat from the auxiliary fuel must be sufficient to raise the temperature of the waste and the combustion air to a point above the ignition temperature of the organic material in the waste.

Incineration of wastes that are not pure liquids but that might be considered sludges or slurries is also an important waste disposal problem. Incinerator types applicable for this kind of waste would be fluidized bed incinerators, rotary kiln incinerators, and multiple hearth incinerators.

Incineration is not a total disposal method for many solids and sludges because most of these materials contain noncombustibles and have residual ash.

Complications develop with the wide variety of materials that must be burned. Controlling the proper amount of air to give combustion of both solids and sludges is difficult, and with most currently available incinerator designs this is impossible.

The types of incinerator that are applicable to solid wastes are open pit incinerators and closed incinerators such as rotary kilns and multiple hearth incinerators. Generally, the incinerator design does not have to be limited to a single combustible or partially combustible waste. Often it is both economical and feasible to utilize a combustible waste, either liquid or gas, as the heat source for the incineration of a partially combustible waste that may be either liquid or gas.

Experience indicates that wastes that contain only carbon, hydrogen, and oxygen and that may be handled in power generation systems may be destroyed in a way that reclaims some of their energy content. These types of waste may also be judiciously blended with wastes having low energy content, such as the highly chlorinated organics, to minimize the use of purchased fossil fuel. On the other hand, rising energy costs will not be a significant deterrent to the use of thermal destruction methods when they are clearly indicated to be the most desirable method on an environmental basis.

Air emissions from hazardous waste incinerators include the common air pollutants, discussed in Chapter 18. In addition, inadequate incineration may result in emission of some of the hazardous materials that the incineration was intended to destroy. Incomplete combustion, particularly at relatively low temperatures, may also result in production of a class of compounds known collectively as *dioxins*, including both polychlorinated dibenzodioxins (PCDD) and polychlorinated dibenzofurans (PCDF). The compound in this class that has been identified as a carcinogen and teratogen is 2,3,7,8-tetrachlorodibenzo-*p*-dioxin (2,3,7,8-TCDD), shown in Figure 15–4. There is growing public concern about TCDD emissions from incinerators.

TCDD was first recognized as an oxidation product of trichlorophenol herbicides (2,4-D and 2,4,5-T, one of the ingredients of Agent Orange).[2] In 1977, it was one of the PCDDs found present in municipal incinerator fly ash and air emissions, and it has subsequently been found to be a constituent of gaseous emissions from virtually all combustion processes, including trash fires and barbecues. TCDD is degraded by sunlight in the presence of water.

The acute toxicity of TCDD in animals is extremely high (LD_{50} in hamsters of 3.0 mg/kg); carcinogenesis and genetic effects (teratogenesis) have also been observed in chronic exposure to high doses in experimental animals. In humans,

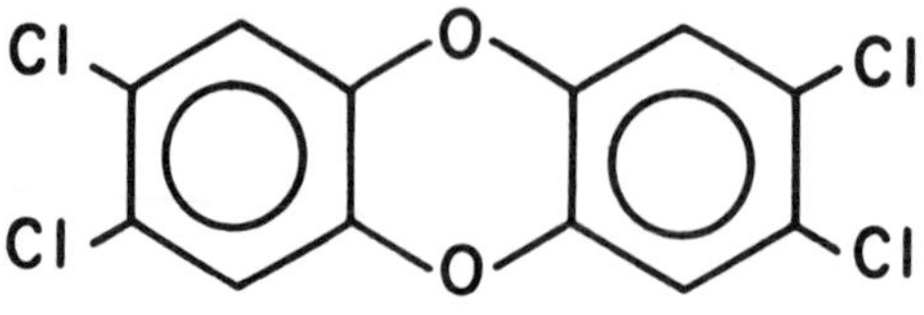

Figure 15–4. 2,3,7,8-Dichlorodibenzo-*p*-dioxin.

the evidence for these adverse effects is mixed. Although acute effects such as skin rashes and digestive difficulties have been observed on high accidental exposure, these are transitory. Public concern has focused on chronic effects, but existing evidence for either carcinogenesis or birth defects in humans from chronic TCDD exposure is inconsistent. Regulations governing incineration are designed to limit TCDD emission to below measureable quantities; these limits may usually be achieved by the proper combination of temperature and residence time.

Engineers should understand, however, that public concern about TCDD (and dioxins in general) has occasionally reached hysterical proportions and is a major factor in opposition to incinerator siting.

Landfills

Landfills must be adequately designed and operated if public health and the environment are to be protected. The general components that go into the design of these facilities, as well as the correct procedure to follow during the operation and postclosure phase of the facility's life, are discussed below.

Design

Three levels of safeguard must be incorporated into the design of a hazardous landfill. These levels are displayed in Figure 15–5. The primary system is an impermeable liner, either clay or synthetic material, coupled with a leachate collection and treatment system. Infiltration may be minimized with a cap of impervious material overlaying the landfill and sloped to permit adequate runoff and to discourage pooling of the water. The objectives are to prevent rainwater and snow melt from entering the soil and percolating to the waste

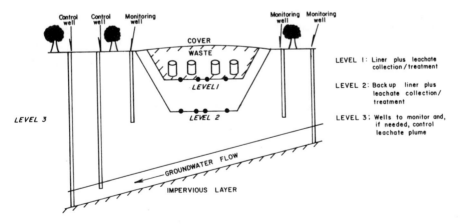

Figure 15–5. Three levels of safeguard in hazardous waste landfills.

containers and, in case water does enter the disposal cells, to collect and treat it as quickly as possible. Side slopes of the landfill should be a maximum of 3 : 1 to reduce stress on the liner material. Research and testing of the range of synthetic liners must be viewed with respect to a liner's strength, compatibility with wastes, costs, and life expectancy. Rubber, asphalt, concrete, and a variety of plastics are available, and such combinations as polyvinyl chloride overlaying clay may prove useful on a site-specific basis.

A leachate collection system must be designed by contours to promote movement of the waste to pumps for extraction to the surface and subsequent treatment. Plastic pipes, or sand and gravel, similar to systems in municipal landfills and used on golf courses around the country, are adequate to channel the leachate to a pumping station below the landfill. One or more pumps direct the collected leachate to the surface, where a wide range of waste-specific treatment technologies are available, including:

- sorbent material: carbon and fly ash arranged in a column through which the leachate is passed.
- packaged physical-chemical units, including chemical addition and flash mixing, controlled flocculation, sedimentation, pressure filtration, pH adjustment, and reverse osmosis.

The effectiveness of each method is highly waste specific, and tests must be conducted on a site-by-site basis before a reliable leachate treatment system can be designed. All methods produce waste sludges which must reach ultimate disposal.

A secondary safeguard system consists of another barrier contoured to provide a backup leachate collection system. In the event of failure of the primary system, the secondary collection system conveys the leachate to a pumping station, which in turn relays the wastewater to the surface for treatment.

A final safeguard system is also advisable. This system consists of a series of discharge wells up-gradient and down-gradient to monitor groundwater quality in the area and to control leachate plumes if the primary and secondary systems fail. Up-gradient wells act to define the background levels of selected chemicals in the groundwater and to serve as a basis for comparing the concentrations of these chemicals in the discharge from the down-gradient wells. This system thus provides an alarm mechanism if the primary and secondary systems fail.

If methane generation is possible in a hazardous waste landfill, a gas collection system must be designed into the landfill. Sufficient vent points must be allowed so that the methane generated may be burned off continuously.

Operation

As waste containers are brought to a landfill site for burial, specific precautions should be taken to ensure the protection of public health, worker safety, and the environment. Wastes should be segregated by physical and chemical characteristics and buried in the same cells of the landfill. Three-dimensional mapping of the site is useful for future mining of these cells for resource recovery purposes. Observation wells with continuous monitoring should be maintained, and regular core soil samples should be taken around the perimeter of the site to verify the integrity of the liner materials.

Site Closure

Once a site is closed and does not accept more waste, the operation and maintenance of the site must continue. The impervious cap on top of the landfill must be inspected and maintained to minimize infiltration. Surface water runoff must be managed, collected, and possibly treated. Continuous monitoring of surface water, groundwater, soil, and air quality is necessary, as ballooning and rupture of the cover material may occur if gases produced or released from the waste rise to the surface. Waste inventories and burial maps must be maintained for future land use and waste reclamation. A major component of postclosure management is maintaining limited access to the area.

CONCLUSION

Hazardous waste is a relatively new concern of environmental engineers. For years, the necessary by-products of an industrialized society were piled "out back" on land that had little value. As time passed and the rains came and went, the migration of harmful chemicals moved hazardous waste to the front page of the newspaper and into the classroom. Engineers employed in all public and private sectors must now face head-on the processing, transport, and disposal of these wastes.

Hazardous waste is appropriately addressed at the "front end" of the generation process: either in maximizing resource recovery or in detoxification at the site of generation. Storage—landfilling in particular—is at best a last resort measure for hazardous waste handling.

PROBLEMS

15.1 Assume you are an engineer working for a hazardous waste processing firm. Your vice president thinks it would be profitable to locate a new regional facility near the state capitol. Given what you know about that region, rank the factors that distinguish a good site from a bad site. Discuss the reasons for this ranking; i.e., why, for example, are hydrologic considerations more critical in that region than, for example, the geology.

15.2 If you were a town engineer just informed of a chemical spill on Main Street, sequence your responses. List and describe the actions your town should take for the next 48 hours if the spill is relatively small (100 to 500 gallons) and is confined to a small plot of land.

15.3 The manifest system, through which hazardous waste must be tracked from generator to disposal site, is expensive for industry. Make the following assumptions about a simple electroplating operation: 50 barrels of a waste per day, 1 "trip ticket" per barrel, $25/hr labor charge. Assume the generator's laboratory technician can identify the contents of each barrel at zero additional time for no additional cost to the company because he routinely has done that for years anyway. What is the cost in man-hours and in dollars for this generator to comply with the manifest system illustrated in Figure 15–1? Document your necessary assumptions about the time required for the generator to complete each step of each trip ticket.

15.4 Compare and contrast the design considerations of the hazardous waste landfill with the design considerations of a conventional municipal refuse landfill.

15.5 As town engineer, design a system to detect and stop the movement of hazardous wastes into your municipal refuse landfill.

LIST OF SYMBOLS

LD_{50} = Lethal dose, at which 50 percent of the subjects are killed
LDC_{50} = Lethal dose concentration at which 50 percent of the subjects are killed
RCRA = Resource Conservation and Recovery Act

REFERENCES

1. Koppel, J. *Environment* (March 1985).
2. Tschirley, F.H. *Scientific American* 254:29(1986).

Chapter 16
Radioactive Waste

In the late 1800s, French scientists determined that uranium minerals routinely emitted invisible radiation capable of passing through apparently solid objects. Building on the research, Pierre and Marie Curie were able to isolate two new chemical elements from the uranium minerals: polonium and radium, and these newly discovered elements were observed to produce radiation that was even more intense than uranium emissions. Subsequently, it was determined that the radiation was a result of an automatic disintegration, as atoms of the minerals somehow exploded spontaneously through time.

This chapter presents a general background discussion of the interaction of ionizing radiation with matter and of the environmental effects of radionuclides. The chapter highlights radioactive waste as a pollutant, discusses the impacts of ionizing radiation on the natural environment and on public health, and summarizes options available to environmental engineers for the disposal of radioactive waste.

RADIATION

The Curies and their contemporaries classified the radiation from uranium minerals (pitchblende) into three types, according to the direction of deflection in a magnetic field. These three types of radiation were called alpha (α), beta (β), and gamma (γ) radiation. Becquerel recognized that gamma radiation was the equivalent of the "x-rays" discovered by Roentgen. In 1932, Chadwick identified the neutron as the highly penetrating radiation which results when beryllium is bombarded with alpha particles. Modern physics has subsequently identified emission of positrons, muons, and pions, but not all are of equal concern to the environmental engineer. The significant problems associated with the management of radioactive wastes require a basic understanding of alpha, beta, and gamma emissions and emitters and an understanding of the effect of neutrons.

Radioactive Decay

An atom that is radioactive has an unstable nucleus. The nucleus moves to a more stable condition by emitting an alpha or beta particle; this emission is frequently accompanied by emission of additional energy in the form of gamma radiation. This emission is called radioactive decay. Radioactive decay follows the laws of chance, and the rate of decay, or rate of decrease of the number of radioactive nuclei, can be expressed as an ordinary first-order reaction:

$$dN = -K_b N dt$$

where N = the number of radioactive nuclei
  K_b = a proportionality factor called the disintegration constant; K_b has the units $1/time$

Integrating this equation, letting $N = N_0$ at time $= 0$, we get the classical exponential radioactive decay equation:

$$\int_{N_0}^{N} \frac{dN}{N} = -\int_{0}^{t} K_b dt$$

$$\ln \frac{N}{N_0} = -K_b t$$

$$\frac{N}{N_0} = \exp(-K_b t)$$

The data points in Figure 16–1 correspond to this equation.

After a specific time period, $t = {}_{1/2}L$, the value equals one-half of the previous N. That is, at the end of some time period defined as ${}_{1/2}L$, half of the radioactive atoms have disintegrated. This half-life is determined from the above equation:

$$_{1/2}L = \frac{\ln 2}{K_b} = \frac{0.693}{K_b}$$

The half-lives of selected radioactive products are presented in Table 16–1.

Looking at Figure 16–1, we see that N does not become zero at $t = 2 \cdot {}_{1/2}L$. In fact, N becomes $\frac{1}{4}N_0$. We therefore see that the equation

$$\frac{N}{N_0} = \exp(-K_b t)$$

is so constructed that N never becomes zero in any finite time period; for every half-life that passes, the number of atoms is halved.

Table 16–1.   Some Important Radionuclides

Product	Type of Radiation	Half-life
Krypton-85	Beta and gamma	10 years
Strontium-90	Beta	29 years
Iodine-131	Beta and gamma	8 days
Cesium-137	Beta and gamma	30 years
Tritium	Beta	12 years
Cobalt-60	Beta and gamma	5 years
Carbon-14	Beta	5,770 years

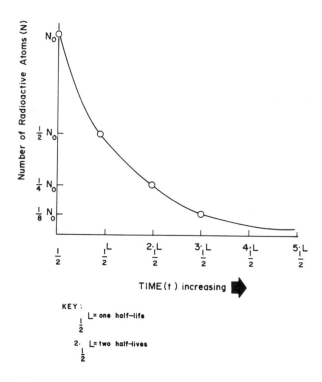

Figure 16–1.   General description of radioactive decay.

Example 16.1

An engineer, with the assistance of a chemist, prepares 10.0 g of pure $_6C^{11}$. This notation represents carbon with 6 protons and a total mass of 11 atomic mass units (amu, the number of protons plus the number of neutrons). (Conversion: 1 amu = 1.66×10^{-24} g.) The radioactive decay of this atom is

$$_6C^{11} \rightarrow {}_1e^0 + {}_5B^{11}$$

If the half-life is 21 minutes, how many grams of carbon are left 24 hours after the preparation?

The equations above describing the half-life refer to number of atoms, so we must initially calculate the number of atoms in 1 g of $_6C^{11}$

$$\frac{1 \text{ atom of } _6C^{11}}{11.0 \text{ amu}} \times \frac{1 \text{ amu}}{1.66 \times 10^{-24} \text{ g of } _6C^{11}} = \frac{1 \text{ atom of } _6C^{11}}{18.3 \times 10^{-24} \text{ g of } _6C^{11}}$$

Now, compute the number of atoms of $_6C^{11}$ in 10 g:

$$\frac{1 \text{ atom of } _6C^{11}}{18.3 \times 10^{-24} \text{ g of } _6C^{11}} \times 10 \text{ g of } _6C^{11} = 55 \times 10^{22} \text{ atoms of } _6C^{11}$$

Now, apply the equation for $_{1/2}L$:

$$K_b = \frac{0.693}{_{1/2}L} = \frac{0.693}{21 \text{ min}} = 33 \times 10^{-3} \text{ min}^{-1}$$

Following convention to avoid dealing with negative logarithms, we can rearrange our half-life equation as:

$$-K_b t = \ln \frac{N}{N_0} = -\ln \frac{N_0}{N}$$

or

$$K_b t = \ln \frac{N_0}{N}$$

Now, realizing that we must express t and K_b so that $K_b t$ is unitless, we must include the number of minutes in a day in our calculations:

$$K_b t = (33 \times 10^{-3} \text{ min}^{-1})(1{,}440 \text{ min}) = 47.5$$

and

$$\ln \frac{N_0}{N} = 2.303 \log \frac{N_0}{N}$$

So we have

$$2.303 \log \frac{N_0}{N} = 47.5$$

or

$$\log \frac{N_0}{N} = 20.6$$

Applying the properties of logarithms:

$$\frac{N_0}{N} = (\text{antilog } 0.6) \times (\text{antilog } 20)$$

we get:

$$\frac{N_0}{N} = 4.0 \times 10^{20} \quad \text{or} \quad N = \frac{N_0}{4.0 \times 10^{20}}$$

This gives us

$$N = \frac{55 \times 10^{22}}{4.0 \times 10^{20}} = 14 \times 10^2 \text{ atoms of } {}_6C^{11}$$

After 1 day there are approximately 1,400 atoms of ${}_6C^{11}$ remaining. To convert this to grams:

$$(14 \times 10^2 \text{ atoms of } {}_6C^{11}) \times \left(\frac{18.3 \times 10^{-24} \text{ g}}{\text{atoms of } {}_6C^{11}} \right) = 250 \times 10^{-22} \text{ g of } {}_6C^{11}$$

A useful figure of merit for radioactive decay is the following:

- After 10 half-lives, 10^{-3}, or 0.1 percent, of the original quantity of radioactive material is left.
- After 20 half-lives, 10^{-6}, or 0.0001 percent, of the original quantity of radioactive material is left.

Alpha, Beta, and Gamma Radiation

Emissions from radioactive nuclei are characterized, in general, as ionizing radiation, because collision between these emissions and an atom or molecule ionizes that atom or molecule. Alpha, beta, and gamma rays or particles may be characterized further by their movement in an electric or magnetic field. Apparatus for such a characterization is shown in Figure 16–2. A beam of

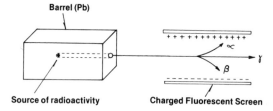

Figure 16–2. Controlled measurement of alpha (α), beta (β), and gamma (γ) radiation.

disintegrating radioactive atoms is aimed with a lead barrel at a fluorescent screen, which is designed to glow when hit by the radiation. Two alternately charged probes direct the positively charged α radiation and negatively charged β radiation accordingly. The γ radiation is seen to be "invisible light," a stream of neutral particles that passes through the electromagnetic force field. Alpha and beta emissions are typically classified as particles, whereas gamma emissions consist of electromagnetic radiation (waves).

Alpha radiations, identified as being physically identical to the nuclei of helium atoms, stripped of their planetary electrons with only two protons and two neutrons remaining, are exploded from the nucleus of selected radioactive atoms with a kinetic energy of somewhere between 4 and 10 MeV.* The particles have a mass of about 4 amu (6.642×10^{-4} g) and a positive charge of 2.** As these charged particles travel along at approximately 10,000 miles/sec, they interact with other atoms. Each interaction transfers a portion of the energy to the electrons of these other atoms and results in the production of an ion pair; i.e., a negative electron with an associated positive ion. The huge number of interactions produces between 30,000 and 100,000 of these ion pairs per centimeter of air traveled, resulting in rapid depletion of the alpha energy. Subsequently, the range of the alpha particle is only 1 to 8 cm in air. If such a particle runs into a solid object, notably human skin cells, the energy is rapidly dissipated. Thus, alpha particles present no direct problem of external radiation damage to humans, but could cause health problems when also emitted inside the body where protective layers are not present to diffuse the energy. Even the strongest alpha particles are stopped by the epidermal layer of the skin and rarely reach the sensitive layers. People are typically contaminated with material that emits alpha particles only if the material is inhaled, ingested, or absorbed in a skin wound.

Beta radiation is electrons that are also emitted from the nucleus of a radioactive atom at a velocity approaching the speed of light, with kinetic energy between 0.2 and 3.2 MeV. Given their lower mass of approximately 5.5×10^{-4} amu (9.130×10^{-28} g), the interactions between beta particles and the atoms of pass-through materials are much less frequent than alpha particle interactions. Fewer than 200 ion pairs are formed in each centimeter of flight through air, and the resulting slower rate of energy loss enables beta particles to travel several meters in air and several centimeters through human tissue. Thus, although beta radiation can damage tissue under the human skin, internal organs are generally protected. However, exposed organs such as eyes are sensitive to beta damage.

Gamma radiation is identified as invisible, electromagnetic rays emitted from the nucleus of radioactive atoms. These rays are like medical x-rays, in

---

*Mass is designated as energy in the equation $E = mc^2$. One MeV $= 10^6$ electron volts (eV). One electron volt $= 1.603 \times 10^{-12}$ erg.

**This charge, in atomic mass units (amu), is expressed in units relative to an electron with a negative charge of 1.

that they are composed of photons. Because of their neutral charge, gamma photons collide randomly with the atoms of the material as they pass through. Thus, the *relaxation length*, the distance to decrease the intensity of the ray by a factor of 1/e, is much greater than for α and β particles. A typical gamma ray, with an energy of about 0.7 MeV, has a unique relaxation length for different pass-through materials: for example, lead, water, and air have relaxation lengths of 5, 50, and 10,000 cm, respectively. The dose of gamma radiation received by unprotected human tissue may be significant because the dose is not greatly impacted by air molecules.

The properties of the more common radioactive emissions are summarized in Table 16–2.

When ionizing radiation is emitted from a nucleus, the nature of that nucleus changes: another element is formed, and there is a change in mass as well. This process may be written as a nuclear reaction. In such a reaction, both mass and charge must balance for reactants and products. For example, the beta decay of C-14 may be written:

$$_6C^{14} = {}_{-1}e^0 + {}_7N^{14}$$

The mass balance for this nuclear equation is

$$14 = 0 + 14$$

The charge balance is

$$6 = -1 + 7$$

A typical reaction for alpha decay is the first step in the uranium-238 decay chain:

$$_{92}U^{238} = {}_2He^4 + {}_{90}Th^{234}$$

Thus, when a nuclide emits a beta particle, the mass number remains unchanged

Table 16–2. Properties of Ionizing Radiation

Particle or Wave	Mass (amu)	Charge
Alpha ($_2He^4$)	4	+2
Beta (electron)	5.5×10^{-4}	−1
Gamma (X-ray)	approx. 0	0
Neutron	1	0
Positron	5.5×10^{-4}	+1

and the atomic number increases by one. Beta decay is thought to be the decay of a neutron in the nucleus to a proton and an electron (beta particle), with emission of the beta particle from the decaying nucleus. When a nuclide emits an alpha particle, the atomic mass decreases by 4 and the atomic number decreases by 2. Gamma emission does not yield a change in atomic mass or atomic number.

Nuclear reactions may also be written for bombardment of nuclei with subatomic particles. For example, tritium (H-3) is produced by bombarding a lithium target with neutrons:

$$_0n^1 + {_3}Li^6 = {_1}H^3 + {_2}He^4$$

The reactions written above tell us nothing about the energy with which ionizing radiation is emitted. However, the emitted particle always has a certain amount of kinetic energy. This energy is lost in collisions with target atoms as the alpha, beta, neutron, and gamma radiations pass through different materials. If the material is human tissue, the energy gained (by the tissue) may cause a disordering in the chemical or biological structure of the tissue and subsequent dysfunction of the cells. These effects are discussed later in this chapter.

Units for Measuring Radiation

Standard units have evolved that are used to measure radiation and its impacts on material. As we see above, the damage to human tissue is directly related to the amounts of energy deposited in the tissue by the alpha, beta, and gamma particles. This energy, in the form of ionization and excitation of molecules, results in heat damage to the tissue or even radiation burn. For this reason, many of the units are related to energy deposits.

A *curie* is a measure of total radioactivity or source strength, and is equal to 3.7×10^{10} disintegrations/sec, the radioactivity of 1 g of the element radium. The decay rate

$$dN/dt$$

is measured in curies. The source strength in curies is not sufficient for a complete characterization of a source; the nature of the element (e.g., Pu-139, U-138, Sr-90) and the type of emission (e.g., alpha) are also necessary.

The mass of material that disintegrates 3.7×10^{10} times per second varies widely from material to material. Note the data in Table 16–8. The relationship between the curies and the mass of a given radionuclide is given by:

$$Q = \frac{K_T M N^0}{W(3.7 \times 10^{10})}$$

where   Q = number of curies
        K_T = disintegration constant = $0.693/_{1/2}L$; i.e., the fraction of atoms
             that decay per second
        $_{1/2}L$ = half-life of the radionuclide, in seconds
        M = mass of the radionuclide, in grams
        N^0 = Avogadro's number, 6.03×10^{23} g/g of atomic weight
        W = atomic weight of the radionuclide

A complex relationship exists between the quantity of radiation, the curie, and
the radiation dose rate, rads per second. This relationship depends on the energy
of the radiation, the type of radiation, the path of the radiation, and the amount
of absorbing material between the emitter and the receptor.

A *rad* (for "radiation absorbed dose") is that quantity of ionizing radiation
which leads to absorption of 100 ergs/g of absorbing material.

The *roentgen* is a standard unit of exposure corresponding to a quantity of
gamma rays that deposit 87.7 erg/g of air at standard temperature and pressure.
This unit is equivalent to the ejection of 1 electrostatic unit (esu) of charge in
1 cm^3 of air and describes the electromagnetic field associated with radioactive
decay. However, the unit does not indicate where the gamma rays deposit their
energy. For the environmental engineer interested in health effects of radiation,
the location of the energy deposit is also of concern.

The *dose equivalent,* measured in *rem* (for "roentgen equivalent man"),
addresses the following issue: all types of ionizing radiation do not produce
identical biological effects for a given amount of energy delivered to human
tissue. Radiation yielding higher specific ionization along its track will generally
produce a greater effect, but the quantitative difference will depend on the tissue
or organ and biological change selected for study. The dose equivalent is the
product of the dose in rads and a *quality factor* (sometimes called *relative
biological effectiveness,* or RBE) that describes the biological effect being
considered.

Different types of ionizing radiation, even for a given amount of energy,
produce different effects on living tissue. The quality factor takes into account
the differing biological effects of alpha, beta, and gamma radiation. The
standard for comparison is gamma radiation having a linear energy transfer
(LET) in water of 3.5 keV/μm and a rate of 10 rads/min. Together with this
standard, the quality factor can be defined as:

$$QF = rem/rad$$

Table 16–3 gives sample quality factors for internal dose: the dose from
radionuclides incorporated in human tissue.

Quality factors are determined for chronic, low-level doses of radiation,
using effects that occur in an individual during a lifetime of exposure. Acute,
high-level doses produce different effects and would have different quality
factors. The quality factor also takes into account the pathway of the radionuclide into the human body: ingestion, inhalation, or immersion.

Table 16–3. Sample Quality Factors for Internal Radiation

Internal Radiation	Quality Factor
α	10.0
β and γ	
$E_{max} > 0.03$ Mev*	1.0
$E_{max} < 0.03$ Mev*	1.7

*E_{max} refers to the maximum level of emissions by the β and γ source.

Doses are also often expressed in terms of *population dose*, which is measured in *man-rem*. The population dose is the product of the number of people affected and the average dose in rems. That is, if a population of 100,000 individuals receives an average whole-body dose of 0.5 rem, the population dose is 0.5×10^5 man-rems, or 50,000 man-rems. The utility of the population dose concept will become evident in the section dealing with health effects.

Table 16–4 gives some average radiation doses in the United States. It should be remembered that these are rather imprecise estimates.

Table 16–4. Estimates of Annual Whole-Body Dose Rates in the United States

Source	Average Whole-Body Dose (mrems/yr)	Annual Population Dose (10^6 man-rems)
Natural sources		
Cosmic radiation	44	
Internal (K-40)	18	
External alpha (minerals)	30	
External beta	10	
Subtotal	102	21
Anthropogenic sources		
Diagnostic x-ray	72	15
Radiopharmaceuticals	1	0.2
Global fallout	9	2
Occupational exposure	1	0.2
Nuclear power	0.03	0.01
Miscellaneous	2	0.5
Subtotal	85	18
Total	187	39

Measuring Radiation

Devices have been developed to measure the radiation dose, dose rate, or the quantity of active material that is present. The particle counter, the ionization chamber, photographic film, and the thermoluminescent detector are four methods widely used in the field.

Counters are designed to note the movement of single particles through a defined volume. Gas-filled counters collect the ionization produced by the radiation as it passes through the gas and amplify it to produce an audible pulse. Counters are typically used to determine the radioactivity present by measuring the number of particles of protons that are emitted.

Ionization chambers basicaly consist of a pair of charged electrodes that collect ions formed within their respective electrical fields. Chambers are generally designed to determine dose or dose-rate measurements because they provide an indirect representation of the energy deposited in the chamber.

Photographic film darkens if exposed to radiation and is a useful indicator of the presence of radioactivity. Such film is often used for determining personnel exposure and making other dose measurements for which a long record of dose is necessary or a permanent record of dose is required.

Thermoluminescent detectors (*TLD*) are crystals, such as NaI, that can be excited to high electronic energy levels by ionizing radiation. The energy is then released as a short burst or flash of light, which can be detected by a photocell or photomultiplier. TLD systems are replacing photographic film in personnel dosimeters, because they are more sensitive and consistent. Liquid scintillators (organic phosphors) are used in biomedical applications.

HEALTH EFFECTS

When alpha and beta particles and gamma radiation penetrate living cells, or any other matter, they transfer their energy through a series of collisions with the atoms or nuclei of the receiving material. Many molecules are damaged in the process, as chemical bonds are broken and electrons are lost (ionization). In fact, energy is lost all along the path of the radiation, and a measure of this rate of *linear energy transfer* (LET) is the density of ionization activity along this path. The more ionization that is observed, the higher the intensity of biological damage to the cells. Table 16–5 gives some typical LET values. LET is a measure of the energy lost by the ionizing radiation with each successive collision. Thus, charged particles may have a higher LET than uncharged particles of the same size, and may penetrate further. Values given for the dose equivalent vary widely with different measures of biological damage.

Biological effects from these penetrations may be grouped as *somatic* and *genetic*. Somatic effects are the impacts on individuals directly exposed to the radiation and include damage to the circulatory system, carcinogenesis, and decrease in organ function because of cell killing. Genetic effects occur because

Table 16–5.   Representative LET Values

Radiation	Kinetic Energy (MeV)	Average LET (kEV/μm)	Dose Equivalent
X-rays	0.01–0.2	3.0	1.00
Gamma rays	1.25	0.3	0.7
Electrons (beta)	0.1	0.42	1.0
	1.0	0.25	1.4
Alpha particles	0.1	260	
	5.0	95	10
Neutrons	Thermal		4–5
	1.0	20	2–10
Protons	2.0	16	2
	5.0	8	2

ionizing radiation damages the genetic material of the cell and can cause chromosome breakage. Genetic effects are not evident in the individual receiving the radiation, but are transferred to that individual's offspring and descendants.

Radiation sickness (circulatory system breakdown, nausea, hair loss) and resulting death are acute somatic effects occurring after very high exposure, as from a nuclear bomb or intense radiation therapy. To date, there has been no radiation-related death from a nuclear electric generating plant accident, although such an accident is possible. For humans, it is estimated that a gamma radiation dose of 400 to 500 rads delivered to a major portion of the body in a short time will result in death within a few months.[1] Most mammals appear to have the same degree of sensitivity to the LD_{50} (lethal dose resulting in a 50 percent kill within 30 days). Lower doses, but above 100 rads, will lead to vomiting, diarrhea, and nausea in humans; hair loss is generally observed within 2 weeks after an individual has been exposed to 300 rads or more.[2,3]

Ionizing radiation presents a good example of the different effects of acute high doses and chronic low doses. Construction of dose-response curves depends on retrospective studies of atom bomb survivors, industrial and occupational exposures, and a few studies of very high medical exposures. Since low doses produce effects (cancer, genetic effects) that have very long latency periods, good data do not exist for low doses, and most dose-response curves have been extrapolated from high-dose data.

There is no threshold for cell damage from ionizing radiation. Thus, for a long time, low-dose responses were linear extrapolations of high-dose data. Recent thorough re-examination of all data on radiation effects indicates that the dose-response curve is probably linear at low doses and quadratic at high doses. This linear-quadratic dose-response relationship is shown in Figure 16–3.

Unless there is a nuclear war, only low-LET radiation need be considered as an environmental pollutant. Since such radiation is a carcinogen, there is

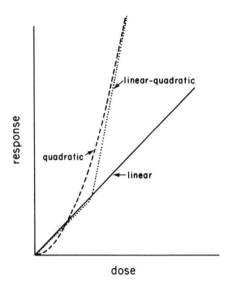

response

dose

quadratic

linear-quadratic

linear

Figure 16–3. Linear-quadratic dose-response relationship.

assumed to be no threshold for damage, and risk analysis is necessary. As effects of exposure have been expressed over the years since the first atomic bomb, risk estimates have become more refined. Tables 16–6 and 16–7 give a variety of risk estimates for effects from exposure to low-LET radiation, including the estimates presently used by EPA.

The uncertainty in these numbers is estimated to be ±250 percent. As we have seen in Chapter 2, the number of cancer fatalities in the United States per year per 10^6 persons is 1,930. In other words, exposure to an additional rad per year would increase the number of fatal cancers per million population by about 15 percent.

The uncertainty in these numbers is estimated to be ±300 percent. By comparison, the current incidence of genetic defects in the United States per 10^6 liveborn children is 107,100.

There is a documented risk from ionizing radiation, but it is apparently small and clearly uncertain. The health effects of a given dose of radiation

Table 16–6. Fatal Somatic Risk from Lifetime Exposure to 1 rad/yr Low-LET Radiation

Year	Dose Model	Effects per 10^6 Person-Rad
1972		667
1980	Linear	403
1980	Quadratic	169
1977	Quadratic	75–175
1985	Linear	280

Table 6-7. Genetic Risk per 10^6 Liveborn from Exposure to 1 Rem/Generation Low-LET Radiation

Year	First Generation	All Generations
1956		500
1972	12–200	60–1,500
1977	63	300
1980	89	320
1982	22	149
1985	20–90	260–1,110

depends on a large number of factors, including

- magnitude of the absorbed dose
- type of radiation
- penetrating power of the radiation
- sensitivity of the receiving cells and organs
- rate at which the dose is delivered
- proportion of the cell, organ, or human body exposed

To protect public health, the environmental engineer must focus on the factors that are most in his control: the magnitude of potential doses, the type of radiation available to impact public health in an area, and the rate at which doses are delivered. The environmental engineer must, in short, act to minimize unnecessary radioactive exposure.

SOURCES OF RADIOACTIVE WASTE

There are a number of sources of radioactive waste: the nuclear fuel cycle, radiopharmaceutical manufacture and use, biomedical research and applications, and a number of industrial uses. The behavior of radionuclides is determined by their physical and chemical properties; radionuclides may exist as gases, liquids, or solids and may be soluble or insoluble in water or other solvents.

Until 1980, there was no classification of radioactive wastes. In that year, the U.S. Nuclear Regulatory Commission (NRC) classified radioactive wastes into the following categories:

- *High-Level Waste (HLW)*. HLW includes two categories: spent nuclear fuel from nuclear reactors and the solid and liquid waste from reprocessing of spent fuel. In the United States, only military irradiated fuel is reprocessed, to recover plutonium and fissile uranium. The NRC reserves the right to classify additional materials as HLW as necessary.

- *Uranium Mining and Mill Tailings.* The pulverized rock and leachate from uranium mining and milling operations.
- *Transuranic (TRU) Waste.* Radioactive waste that is not HLW but contains more than 100 nanocuries per gram of elements heavier than uranium (the elements with atomic number higher than 92). Most TRU waste in the United States is the product of defense reprocessing.
- *By-product Material.* Any radioactive material, except fissile nuclides, that is produced as waste during plutonium production or fabrication.
- *Low-Level Waste (LLW).* LLW includes everything that is not included in one of the other four categories. To assure appropriate disposal, the NRC has designated several classes of LLW:

    Class A contains only short-lived radionuclides or extremely low concentrations of longer-lived radionuclides and must be chemically stable. Class A waste may be disposed of in landfills without particular stabilization, as long as it is not mixed with other hazardous or flammable waste. Most trash is Class A waste.

    Class B contains higher levels of radioactivity and must be physically stabilized before transportation or disposal. It must not contain free liquid.

    Class C is waste that will not decay to acceptable levels in 100 years and must be isolated from the environment for 300 years or more. Power plant LLW is in this category.

    Formerly Used Sites Remedial Action Program (FUSRAP) waste is contaminated soil from radium and uranium refining and atomic bomb research and development that took place during World War II. Little was known about the long-term effects of ionizing radiation at the time, and there was widespread contamination of the structures in which this work was done and of the land surrounding those structures. FUSRAP waste contains exceedingly low concentrations of radionuclides, but there is a large amount of this waste.

LLW is not necessarily less radioactive than HLW and may even have a higher specific activity (curies per gram). The distinguishing feature of LLW is that it contains virtually no alpha emitters.

The Nuclear Fuel Cycle

The nuclear fuel cycle (shown in Figure 16–4) generates radioactive waste at every stage. Mining and milling generate the same sort of waste that any mining and milling operations generate, except that this waste is radioactive. Mining and milling dust must be stabilized to keep it from dispersing, and leachate must be prevented from contaminating waterways and groundwater.

Partially refined uranium ore ("yellowcake" because of its bright yellow color) is then enriched in the fissile isotope U-235, and nuclear fuel is fabricated. Mined uranium is more than 99 percent U-238, which is not fissile, and

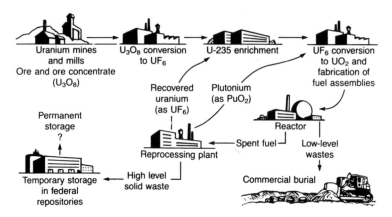

Figure 16–4.   The nuclear fuel cycle.

approximately 0.7 percent U-235. The concentration of U-235 is increased to about 3 percent by converting to UF_6; the lighter isotope is then separated by gaseous diffusion. UF_6 enriched in U-235 is then converted to UO_3 and fabricated into fuel. Enrichment and fabrication produce TRU waste.

The nuclear fuel is then inserted into a nuclear reactor core, where a controlled fission reaction produces heat, which in turn produces pressurized steam for electric power generation. The steam system, turbines, and generators in a nuclear power plant are essentially the same as those in any thermal electric power plant, such as a coal plant. The difference between nuclear and fossil fuel electric generation is in the evolution of the heat that drives the plant.

Figure 16–5 is a diagram of a typical pressurized water nuclear reactor. Commercial reactors in the United States are either pressurized water reactors, in which the water that removes heat from the nuclear reactor core (the "primary coolant") is under pressure and does not boil, or boiling water reactors, in which the primary coolant is permitted to boil.* In any case, the primary coolant transfers heat to the steam system (the "secondary coolant") by a heat exchanger that ensures complete physical isolation of the primary coolant from the secondary coolant. A third cooling system provides water from external sources to condense the spent steam in the steam system.

All thermal electric power generation produces large quantities of waste heat. Fossil fuel electric-generating plants are, at best, about 42 percent thermally efficient: that is, 42 percent of the heat liberated by combustion in the boiler is converted to electricity, and 58 percent is simply dissipated in the environment. By comparison, nuclear plants are 32 percent efficient at best.

---

*There is one exception: the Fort St. Vrain reactor in Colorado, in which the primary coolant is helium gas.

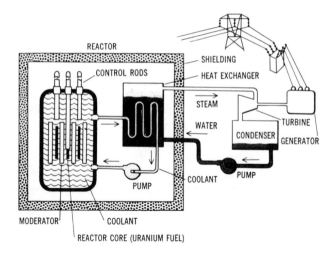

REACTOR

CONTROL RODS

SHIELDING

HEAT EXCHANGER

STEAM

WATER

CONDENSER

TURBINE

GENERATOR

PUMP

COOLANT

PUMP

MODERATOR

COOLANT

REACTOR CORE (URANIUM FUEL)

Figure 16–5. Typical pressurized water nuclear reactor.

The nuclear reactor is perhaps the key radioactive waste producer in the nuclear fuel cycle. The production of high-level radioactive waste is a direct consequence of the fission reaction. One expression of this reaction is

$$_{92}U^{235} + {}_0n^1 = {}_{42}Mo^{95} + {}_{57}La^{139} + 2{}_0n^1 + 204 \text{ MeV}$$

Several features of this reaction merit discussion.

- A large amount of energy is released in this reaction: 204 MeV per uranium atom, or 80 million BTU/g of uranium. Uranium fission releases about 100,000 times as much heat per gram of fuel as natural gas combustion; the development of nuclear power is based on this phenomenon.
- One neutron is required for the fission reaction, but the reaction itself produces two neutrons. Each of these neutrons can then initiate another fission reaction, which will produce another two neutrons, and so on, resulting in a *fission chain reaction*. However, there must also be a high enough concentration of U-235 so that the neutrons produced are likely to collide with U-235 nuclei. The mass of U-235 (and, in general, the mass of fissile material) needed to sustain a fission chain reaction is called the *critical mass*.
- Neutron flow in a reactor may be interrupted and the fission reaction stopped by inserting control rods that absorb neutrons (see Figure 16–5). Although inserting control rods stops the fission reaction, it does not stop heat generation within the core, since the many radioactive fission products continue to emit energy. Therefore, continued coolant flow is critical.

● In this particular reaction, Mo-95 and La-139 are the products of the fission. However, it is a characteristic of the fission reaction that the fissile nucleus does not split in the same way each time. In fact, in a nuclear reactor, there are about 80 different fission products. Although many of these have very short half-lives and decay very quickly, a number of fission products have long half-lives. It is these fission products that make up most of the high-level radioactive waste. Table 16–8 lists the relatively long-lived fission products from a typical reactor.

As is evident, this mixture of fission products has a very high specific radioactivity. Fission products are beta or gamma emitters, because they are simply too small to emit alpha particles. The longest-lived fission product has a 30-year half-life and thus remains significantly radioactive for 600 years.

Table 16–8. Long-Lived Reactor Fission Products

Isotope	Type of Emission	Half-life	Curies Produced/MW_t
Kr-85	Beta/gamma	10.6 yrs	500
Sr-89	Beta	53 days	23,000
Sr-90	Beta	29 yrs	2,600
Y-91	Beta/gamma	58 days	33,500
Zr-95	Beta	65 days	50,000
Nb-95	Beta/gamma	35 days	52,000
Ru-103	Beta	40 days	40,000
Ru-106	Beta	1 yr	19,000
Ag-111	Beta	7.6 days	1,300
Sn-125	Beta	9.4 days	420
Sb-125	Beta	2 yrs	200
Te-127	Gamma	105 days	580
Te-129	Beta	33.5 days	3,000
I-129	Beta	1.6×10^7 yr	< 200
I-131	Beta	8.3 days	28,000
Xe-131	Beta	12 days	270
Xe-133	Beta/gamma	5.3 days	56,500
Cs-134	Beta	2.1 yrs	2,000
Cs-137	Beta	30 yrs	3,600
Ba-140	Beta	13 days	45,000
Ce-141	Beta	33 days	48,000
Ce-144	Beta	286 days	38,000
Pr-143	Beta	13.7 days	46,000
Nd-147	Beta	11 days	19,000
Pm-147	Beta	2.6 yrs	7,600
Pm-148	Gamma	42 days	620
Eu-156	Beta	14 days	4,800

Moreover, spent fuel (HLW) contains other radionuclides that compound its polluting potential.

About one fission reaction in 10,000 yields three fission fragments instead of two, and the third fission fragment is tritium (H-3). Tritium has a 12.3-year half-life and, since it is chemically a form of hydrogen, exchanges freely with nonradioactive hydrogen gas in a reactor and with the H^+ ions in the reactor cooling water. Containment of tritium is thus difficult.

In a fission reaction, not all of the neutrons will collide with a fissile nucleus to initiate fission. Some neutrons will collide, and react, with the fuel container and the reactor vessel itself. There are several long-lived products of such neutron activation, notably Co-60 (half-life, 5.2 years) and Fe-59 (half-life, 45 days). Other neutrons will react with nonfissile isotopes of uranium, mostly U-238, as, for example,

$$_{92}U^{238} + _0n^1 = _{92}U^{239} = _{93}Np^{239} + _{-1}e^0$$

$$_{93}Np^{239} = _{94}Pu^{239} + _{-1}e^0$$

Pu-239, "weapons-grade" plutonium, has a half-life of 24,600 years.

Other isotopes of plutonium are formed also. One of these isotopes decays by beta emission to form another long-lived radionuclide, Am-241 (half life, 433 years), by the reaction

Table 16–9.   Uranium Decay Series

Isotope	Emission	Half-Life
U-238	Alpha	4.5×10^9 yrs
Th-234	Beta	24.1 days
Pa-234	Beta	1.18 min
U-234	Alpha	248,000 yrs
Th-230	Alpha	80,000 yrs
Ra-226	Alpha	1,620 yrs
Rn-222	Alpha	3.28 days
Po-218	Alpha/beta	3.05 min
Pb-214	Beta	26.8 min
At-218	Alpha	2 sec
Bi-214	Alpha/beta	19.7 min
Po-214	Alpha	0.00016 sec
Tl-210	Beta	1.32 min
Pb-210	Beta	19.4 yrs
Bi-210	Alpha/beta	5.0 days
Po-210	Alpha	138.4 days
Tl-206	Beta	4.2 min
Pb-206		Stable

Note: The chain branches where two emissions are listed; e.g., Pb-214 and At-218 are both daughters of Po-218.

$$_{94}Pu^{241} = {}_{95}Am^{241} + {}_{-1}e^0$$

Am-241 is commonly used as an ion producer in smoke detectors.

As has been mentioned, each uranium isotope is part of a decay series: when U-238 decays, the daughter element is also radioactive and decays, and so on until a stable element is reached. The U-238 decay series is shown in Table 16–9. The plutonium isotopes, which are formed by neutron reactions with uranium, also decay in a series of radionuclides. Spent nuclear fuel thus contains fission products, tritium, neutron activation products, plutonium, and plutonium and uranium daughter elements in a very radioactive, very long lived mixture that is chemically difficult to segregate.

Reprocessing Waste

In the United States, plutonium for weapons is produced by irradiating U-238 with neutrons (in the reactions given above) in military breeder reactors. The plutonium, along with uranium and neptunium, is then extracted by dissolving the entire irradiated fuel element in nitric acid, and then extracted with tributyl phosphate. Further partition and selective precipitation result in recovery of plutonium, uranium, neptunium, strontium, and cesium. The fissile isotopes of plutonium and uranium are categorized as *special nuclear material*; the remainder of these are considered by-product material.

Both the acid solvent (which is ultimately neutralized) and the organic extraction solvents contain high concentrations of radioisotopes and are classified as HLW. This process also yields TRU waste and LLW.

The only use for plutonium in the United States is in weapons manufacture; commercial spent fuel is not reprocessed. In France, however, plutonium is produced (also by neutron reaction with U-238) in the Superphenix breeder reactor, for use in nuclear electric power generation.

Other Reactor Waste

The primary and secondary coolants in a nuclear generating plant pick up considerable radioactive contamination through controlled leaks. Contaminants are removed from the cooling water by ion-exchange columns. The loaded columns are class C low-level radioactive waste. Class A and B wastes are also produced in routine cleanup activities in nuclear reactors.

Eventually, the reactor core and the structures immediately surrounding it become very radioactive, primarily by neutron activation, and the reactor must be shut down and decommissioned. Although the two small reactors now being decommissioned are being treated as LLW, the ultimate implications of decommissioning are not yet clear.

Other Sources of Radioactive Waste

The increasingly widespread use of radioisotopes in research, medicine and industry has created a lengthy list of potential sources of other radioactive waste. Sources range from a large number of laboratories using small quantities of a few isotopes to large medical and research laboratories where many different isotopes are produced, used, and wasted in large volumes.

Liquid scintillation counting has become an important biomedical tool and produces large volumes of waste organic solvents, such as toluene, which have a low specific activity. The long-lived contaminants of liquid scintillators are tritium and C-14, for the most part. These wastes are characteristic of "mixed wastes": mixtures of hazardous and radioactive waste. Their chemical nature as well as their radioactivity poses disposal problems.

Naturally occurring radionuclides, and those inadvertently released in times past, may also pose threats to public health (see Table 16–10). An outstanding example is the buildup of Rn-222 in homes and commercial buildings with restricted air circulation. Rn-222 is a member of the uranium decay series and is thus found ubiquitously in rock. Chemically, Rn-222 is an inert gas, like helium. When uranium-bearing minerals are crushed or machined, Rn-222 is released; there is even a steady release of Rn-222 from rock outcrops. Buildings that are

Table 16–10. Some Radionuclides of Concern

Isotope	Emission	Half-life
Cosmic Ray Production		
H-3	Beta	12 yrs
Be-10	Beta	2.5×10^6 yrs
C-14	Beta	5730 yrs
Si-32	Beta	650 yrs
Cl-36	Beta	3.1×10^5 yrs
Rock and Building Materials		
Rn-222	Alpha	3.8 days
Uranium and daughters	Alpha/beta/gamma	
Thorium and daughters	Alpha/beta/gamma	
K-40	Beta	1.3×10^9 yrs
Atmospheric Fallout		
C-14	Beta	5,730 yrs
Cs-137	Beta	30 yrs
Sr-90	Beta	29 yrs
I-131	Beta	8 days
H-3	Beta	12 yrs

insulated to prevent convective heat loss often have too little air circulation to keep the interior purged of Rn-222. Although Rn-222 has a short half-life, it decays to much longer-lived metallic radionuclides.

Coal combustion, copper mining, and phosphate mining release isotopes of uranium and thorium into the environment. K-40, C-14, and H-3 are found in many foodstuffs. Although atmospheric nuclear testing was discounted many years ago, fallout from past tests continues to enter the terrestrial environment.

MOVEMENT THROUGH THE ENVIRONMENT

A radionuclide is an environmental pollutant because of a combination of the following properties:

- Half-life.
- Chemical nature and properties: a radioactive isotope will behave chemically and biochemically exactly like stable (nonradioactive) isotopes of the same element.
- Abundance.

Radioactive wastes may thus contaminate air, water, and land. This contamination may have a significant impact on public health. Figure 16–6 pictures the pathways of contamination from a hypothetical source of radioactive waste.

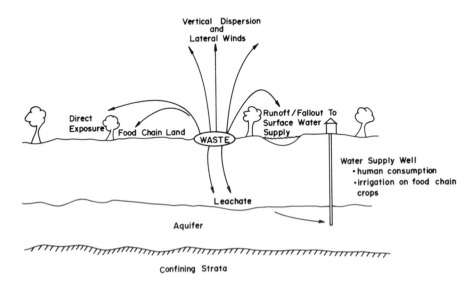

Figure 16–6. Potential movement of radioactivity from waste storage and "disposal" areas.

From an *air transport* standpoint, all nuclear waste handling and disposal facilities fall into two general categories: (1) those that have a planned and predictable discharge to the atmosphere and (2) those for which any discharge would be purely accidental. Boiling-water reactors, fuel-processing plants, and a large number of atomic experimental systems—particularly military—fit into the first category and, on occasion, also into the second category. Good examples of the second category are well-contained pressurized water reactors and long-term storage facilities.

The atmospheric pathways from both types of discharge have received a great deal of attention because of the analogies between radioactive emission and sulfur dioxides and inert gases generated by automobiles and smoke stacks across the nation. Airborne radioactivity is available for inhalation or disposition on water supplies or food chain land and thus impacts public health. Chapters 18 and 19 discuss air pollution and meteorology, and much of that discussion is applicable to the study of radioactive particles.

Water transport occurs whenever radionuclides in surface or subsurface soil erode or leach into a watercourse or whenever fallout occurs from the atmosphere. The radioactive waste is transferred from its solid or liquid state into groundwater or surface water supplies where the radionuclides may enter the human food chain or drinking water. However, little is known about the leachability of radionuclides, and only gross estimates can be made about surface water transport and disposition. As with any such radioactive material transport equations, the half-life of the isotopes must be considered. Chapters 5 and 10 address water transport considerations that are somewhat applicable to the problems associated with radioactive waste.

Radioactive isotopes also move *from the land* into the human food chain. Plant uptake mechanisms have virtually no way of screening an element that is radioactive; thus, plants generally collect a deposit of radionuclides in their structure if the radionuclides are present with the water, phosphorus, nitrogen, and trace elements necessary for plant growth. Rain and wind also act to deposit the hazardous material on the roots and leaves of the plants, all to be passed along to whatever or whomever eats the plants. Cows eating contaminated corn, for example, may pass radioactivity along to milk drinkers.

Transuranic wastes last for long periods of time within the environment, but for the most part are strongly held by soil particles. They are not easily translocated through most food chains, although some concentration does pass through certain aquatic food systems. Additionally, they pose only a slight biological hazard to humans because adsorption in the gastrointestinal track is limited. The greater potential hazard from TRU wastes is from inhalation of dust particles containing the materials since a large fraction is generally retained in the human lung.

In summary, very little is known or understood about the transport of radioactive materials through the environment, making the management of such wastes especially troublesome.

RADIOACTIVE WASTE MANAGEMENT

The objective of the environmental engineer is to prevent the introduction of radioactive materials into the biosphere during the effective lifetime (about 20 half-lives) of these materials. Control of the potential direct impact on the human environment is necessary but not sufficient, because radionuclides may be transmitted through water, air, and land pathways for many years and, in some cases, for many generations. Some radionuclides that are currently treated as waste may be retrieved and recycled by future reprocessing, but such recycling creates its own waste stream. Most radioactive waste is not amenable to recycling. It may be treated only by isolating it from pathways to the human food chain and environment until its radioactivity no longer poses a threat. Isolation requirements differ for the different classes of radioactive waste.

Engineers charged with radioactive waste control must, because of current technical, economic, and political factors outside their control, focus on long-term storage technologies, i.e., *disposal*. With the given half-lives of many radioisotopes, it is difficult to imagine any technology that truly offers ultimate disposal for these wastes; thus, we will think in terms of long-term storage: 10, 100, or 10,000 years or longer. Many of the issues discussed in Chapter 15 are applicable to the radioactive waste problem as well.

High-Level Radioactive Waste

The following options have been considered for long-term disposal of HLW:

1. land disposal
    *   burial in very deep holes
    *   burial on an inaccessible island
    *   deep (mined) geologic disposal
    *   liquid injection into geologic formations
    *   rock melting
2. sub-seabed disposal
3. disposal in polar ice sheets
4. disposal in space
5. transmutation into shorter-lived or stable nuclides

Only two of these options appeared to warrant further investigation: mined geologic disposal and sub-seabed disposal. After preliminary investigation, the sub-seabed option has been discontinued in favor of mined geologic disposal.

To store radioactive waste with a reasonable degree of assurance that it will not be dispersed into the environment, a two-stage barrier has been proposed. The first barrier would be provided by the waste form itself: the optimum waste form would be radioactive material dispersed in a glass matrix, or vitrified. The second barrier would be the geologic (rock) matrix.

This scheme is planned for defense HLW. However, since commercial spent fuel is not reprocessed, it will be stored in the geologic repository in the form in which it leaves the reactor: as spent fuel rods, which are combined in heavy stainless steel casks.

When the fissile uranium in a fuel rod is about 75 percent used up, the rods are ejected into a very large pool of water, where they remain until the short-lived nuclides have decayed away and until the rods are thermally cooler. Initially, this phase was intended to be about 6 months long, but the spent fuel has remained in storage pools as long as 10 years in some cases because there is no repository for it. After sufficient decay and cooling, the spent fuel will be loaded into casks and placed in a repository.

Investigations for a geologic repository began in 1972. Salt is the most thoroughly investigated, for the following reasons:

- Salt-mining technology is well developed, and storage sites could be constructed.
- Salt deposits tend to have a high plasticity and thus have a tendency to seal themselves if fractures are created by major movements in the earth's crust.
- Salt deposits have low permeability and are essentially sealed from groundwater and surface water supplies.
- Salt has a high thermal conductivity, which helps dissipate heat that builds in waste containers.
- Salt formations have a high structural strength, with the ability to withstand effects of heat and radiation.

Salt has some disadvantages, however. No salt deposit is free of brine inclusions, and these tend to migrate toward heat sources, which the emplaced radioactive waste would provide. Salt may also contain some fossil water.

The Nuclear Waste Policy Act of 1982 mandates two geologic repositories and further mandates geological diversity, since the behavior of any single medium over such long time periods may only be guessed at. Accordingly, three rock types—salt, volcanic tuff, and basalt—are presently being investigated for suitability for the first respository. Granite will be added to this list when investigations on a second repository begin. The first repository is scheduled to begin accepting waste in 1998, and to close, sealing the waste permanently and irretrievably, in 2098.

Except for salt, the media being investigated are all hard rock, which cracks and fissures under thermal and chemical stress. Radioactive emissions may even assist in the cracking. Repository integrity thus depends on keeping radionuclides out of groundwater aquifers, if any are present; this poses potential problems for the hard rock repositories. Figure 16–7 shows the geologic formations under investigation.

SCALE

500 miles

300 km

KEY

⬭ Area over rocksalt deposits (beds)

⬭ Area over salt domes

⊗ Area over granite

⬤ Volcanic tuff

⬭ Area over basalt

Figure 16–7. Salt storage potential in the United States.

TRU Waste

Since 1979, a salt formation near Carlsbad, New Mexico, has been under investigation as a radioactive waste repository. The Waste Isolation Pilot Project was found to be unable to withstand the thermal stress from spent fuel and will serve, instead, as the TRU waste repository for the United States.

Low-Level Radioactive Waste

Unlike HLW, commercial LLW has been the responsibility of the private sector since 1960. Low-level waste has been disposed of, without any treatment at all, in shallow burial pits: trenches that are about 60 feet deep and with an area of about 25,000 ft². Three of the six existing commercial sites were shut down between 1975 and 1978 because of leaks of radionuclides into surface water and drinking water. Three sites remain open today, at Hanford, Washington, Beatty, Nevada, and Barnwell, South Carolina, but the last two of these are scheduled to be closed by 1992.

Newer techniques being applied to LLW include packaging, compaction, and incineration. Compaction can achieve volume reduction of 8:1, and incineration can reduce waste volumes by factors of 30 or more. For wastes such as research animal carcasses, liquid scintillators, and other mixed radioactive and organic solvent waste, incineration appears to provide an excellent treatment option. Low-level waste incinerator emissions must be very carefully controlled, however: decontamination factors of 10,000 are usually required. Air emission controls are discussed further in Chapter 20.

CONCLUSIONS

Radiation is energy traveling in the form of particles or waves. Examples include microwave ovens, X-rays in medicine, and sunlight. This energy flows from a natural and spontaneous process in which unstable atoms of an element emit excess energy from their nuclei; it also results from human-induced chain reactions in nuclear power facilities.

This energy flux may have adverse impacts on biological matter. Radiation sickness, cancer, shortened life, or immediate death may result from varying exposures. The most useful units for measuring radiation doses to humans are the *rem* and *millirem* (1/1,000 of a rem, abbreviated *mrem*). These units take into account the effects on living tissue of the three types of radiation: alpha, beta, and gamma radiations.

The general concern of the environmental engineer is radiation from man-made sources, particularly the wastes from nuclear power plants and laboratory research studies. These wastes must be handled with care exceeding the care given the hazardous wastes discussed in Chapter 15, but in much the same manner.

There are now about 100 operating nuclear electric generating plants in the United States, and no more are being ordered or designed. Plutonium production reactors are nearing the end of their useful lives, and no more are planned at present. However, even if the entire nuclear industry were to be shut down, the waste would still be with us. We have just embarked on thorough investigations of disposal options for radioactive waste, and we may anticipate steps toward resolution in the next decade.

Should the nuclear industry shut down? Should the currently operating plants finish their useful lives of 30 to 40 years and not be replaced? In this context, we must realize that 15 percent of the electricity in the United States today is nuclear generated. In some states, the share is as high as 30 percent (Massachusetts) and, in one instance, better than 60 percent (Illinois). It is unlikely that we will be able to replace this generating capacity by solar-, wind-, and geothermally generated electricity in the next 30 years, if indeed we can do it at all. Energy conservation may increase supplies by 5 percent.

The other future option for electric generation is coal combustion. We must make a thorough and objective comparison of the environmental and public health costs and benefits of these options before making any decisions that apply unilaterally to nuclear power. It is wise to remember that generating electricity, no matter how it is done, causes permanent, irrevocable environmental damage, on a scale directly proportional to the power produced.

Military plutonium production may be curtailed by the mid-1990s. Open debate is needed on the future of the nuclear arsenal.

In the past 40 years, the pendulum of public opinion has swung from the "pro-nuclear" extreme to the "anti-nuclear" extreme. Politics follows public opinion, but political decisions about nuclear power and nuclear weapons are highly emotion-laden and not always wise ones. The role of engineers, now more than ever, is to inform and influence political decisions.

PROBLEMS

16.1   Show that after 10 half-lives, about 0.1 percent of the initial radioactive material is left and that after 20 half-lives, 10^{-4} percent is left.

16.2   Fe-55 has a half-life of 2.4 years. Calculate the disintegration constant. If there are 16,000 curies of Fe-55 in a reactor core vessel, how much will be left after 100 years?

16.3   I-131 is used in diagnosis and treatment of thyroid diseases. How should a hospital dispose of waste I-131? Should this waste be disposed of in a shallow land burial site? Are other disposal methods available and adequate?

16.4   Discuss the *environmental* impacts of a nuclear power plant, including water, air, and land quality considerations. Include discussions of thermal pollution and impacts on the aquatic environment.

16.5   Write nuclear reactions for the following, identifying the product elements and particles:
   a. Fusion of two deuterium atoms to form He-3.
   b. Beta decay of Sr-90.
   c. Beta decay of Kr-85.
   d. Neutron emission from Kr-87.
   e. Decay of Th-230 to Ra-226.

16.6   How many grams of Cs-137 are produced in a 750 MW_t power plant? How many grams would be left if this amount were allowed to decay for 100 years?

16.7   From Table 16-4, calculate the average energy absorbed per year by an individual in the United States. Assume that all beta and gamma radiation is low energy (≈ 0.03 MeV).

16.8   The mixture of uranium and its daughter elements that is found in old uranium deposits is called *secular equilibrium*: the equilibrium amounts of U-238 and each daughter element. If you begin with pure U-238, how long does it take until secular equilibrium is reached?

16.9   The environmental impact statement for radioactive waste transportation limits the dose that a truck driver may receive to 4.3 mrem. For how many hours would a truck driver receive this dose until his or her total dose would equal the annual dose from natural radiation?

16.10 The Environmental Protection Agency limits the allowed dose from airborne radioactivity to 25 mrem/yr. Estimate the number of somatic effects that might occur in a population of 50,000 who receive this dose over a 70-year period, in addition to the background dose. Make your assumptions explicit.

16.11 What should be the upper limit for annual occupational radiation dose? Why?

LIST OF SYMBOLS

α = alpha radiation
β = beta radiation
γ = gamma radiation
amu = atomic mass unit
esu = electrostatic unit
K_b = fraction of atoms that disintegrate per unit time
K_T = fraction of atoms that disintegrate per second = $0.693/_{1/2}L$
$_{1/2}L$ = half-life of a radionuclide, in seconds
LD_{50} = lethal dose 50
LDC_{50} = lethal dose concentration 50
LET = linear energy transfer
M = mass of a radionuclide, grams
MeV = 10^6 electron volts
N_0 = number of radioactive atoms within a material
N^0 = Avogadro's number, 6.02×10^{23} grams/gram of atom weight
Q = number of curies
rem = roentgen equivalent man
t = time
TRU = transuranic (long-lived isotopes)
W = atomic weight of a radionuclide

REFERENCES

1. Braestrup, C.G., and H.O. Wyckoff. *Radiation Protection* (Springfield, IL: Charles C. Thomas, 1958).
2. Glasstone, S. *Effects of Nuclear Weapons* (U.S. Government Printing Office, 1962).
3. Nickson, J.J., and H.N. Baine. "Physiological Effects of Radiation," in *Radiation Hygiene Handbook* (New York: McGraw-Hill, 1959).

Chapter 17

Solid and Hazardous Waste Law

Laws controlling environmental pollution are discussed in this text in terms of their evolution from the courtroom through Congressional committees to administrative agencies. The gaps in common law are filled by statutory laws adopted by Congress and state legislatures and implemented by administrative agencies such as the U.S. Environmental Protection Agency (EPA) and state departments of natural resources. For several reasons, this evolutionary process was particularly rapid in the area of solid waste law.

For decades, and in fact centuries, solid waste was disposed of on land no one really cared about. Municipal refuse was historically trucked to a landfill in the middle of a woodland, and industrial waste—often hazardous—was generally "piled out back" on land owned by the industry itself. In both cases, environmental protection and public health were not perceived as issues. Solid wastes were definitely out of sight and neatly out of mind.

Only relatively recently has public interest in solid waste disposal sites reached the level of concern that equals that for water and air pollution. Some local courtrooms and zoning commissions have dealt with siting selected disposal facilities over the years, but most decisions simply resulted in the city or industry hauling the waste a little farther away from the complaining public. At these remote locations, solid waste was not visible like the smokestacks that emitted pollutants into the atmosphere and the pipes discharging wastewater into the rivers. Pollution of the land from solid waste disposal facilities was and is a much more subtle type of pollution.

The passage of federal and state environmental statutes reflects this naiveté. Initially, the highly visible problem of air pollution was addressed in a series of clean air acts. Then, the next most visible pollution was confronted in the water pollution control laws. Finally, as researchers dug deeper into environmental and public health concerns, it was realized that even obscure landfills and holding lagoons had the potential to pollute the land significantly and impact public health. Subsurface and surface water supplies, and even local air quality, are threatened by solid waste.

This chapter addresses solid waste law in two major sections: nonhazardous waste and hazardous waste. The division reflects the regulatory philosophy for dealing with these two very distinct problems. Realizing the general lack of common law in this area, we move directly into statutory controls of solid waste.

NONHAZARDOUS SOLID WASTE

The most significant solid waste disposal regulations were developed under the Resource Conservation and Recovery Act (RCRA) of 1976. This federal statute, which amended the elementary Solid Waste Disposal Act of 1965, reflected the concerns of the public in general and Congress in particular about: (1) protecting public health and the environment from solid waste disposal, (2) filling the loopholes in existing surface water and air quality laws, (3) ensuring adequate land disposal of residues from air pollution technologies and sludge from wastewater treatment processes, and (4) promoting resource conservation and recovery.

The EPA implemented RCRA in a manner that reflected these concerns. Disposal sites were cataloged as either landfills, lagoons, or land spreading operations, and the adverse effects of improper disposal were grouped into eight categories of impacts:

- *Floodplains* were historically prime locations for industrial disposal facilities because many firms elected to locate along rivers for power generation and/or transportation of process inputs or production outputs. When the rivers flood, the disposed wastes are washed downstream, impacting water quality.
- *Endangered species* may be threatened directly by habitat destruction or kills as the disposal site is being developed and operated.
- *Surface water quality* may also be impacted by certain disposal practices. Without proper controls of runoff and leachate, rainwater has the potential to transport pollutants from the disposal site to nearby lakes and streams.
- *Groundwater* impacts are of great concern because about half of the nation's population depends on groundwater for water supply.
- *Food chain crops* may be adversely affected by landspreading or solid waste that may impact on public health and agricultural productivity. Food chain crops often accumulate trace chemicals and increase the dangers to public health.
- *Air quality* may be degraded by pollutants emitted from waste decomposition, such as methane, and may cause serious pollution problems downwind from a disposal site.
- *Health* and *safety* of onsite workers and the nearby public may also be directly impacted by explosions, fire, and bird hazards to aircraft.

The EPA guidelines spell out the operational and performance requirements to eliminate these eight types of impact from solid waste disposal.

Under operational standards, technologies, designs, or operating methods are specified to a degree that theoretically ensures the protection of public health and the environment. Any or all of a long list of operational considerations could go into a plan to build and operate a disposal facility: type of waste to be handled, facility location, facility design, operating parameters, and monitoring and testing procedures. The advantage of operational standards is that the best practical technologies can be utilized for solid waste disposal, and the state agency mandated with environmental protection can easily determine compliance with a specified operating standard. The major drawback is that compliance is generally not measured by monitoring actual effects on the environment, but rather as an either/or situation in which the facility either does or does not meet the required operational requirement.

On the other hand, performance standards are developed to provide given levels of protection to land/air/water quality around the disposal site. Determining compliance with performance standards is not easy because the actual monitoring and testing of groundwater, surface water, and land and air quality are costly, complex undertakings.

Because solid wastes are heterogeneous and because site-specific considerations are all-important in their adequate disposal, the federal regulatory effort realized the necessity of allowing state and local discretion in protecting the environmental quality and public health. Therefore, the EPA developed both operational and performance standards to minimize the effect of each of the eight potential impacts of solid waste disposals.

Floodplains. The protection for floodplains focuses on limiting disposal facilities in such areas unless the local area has been protected against being washed out by the base flood. A base flood is defined as a flood that has a 1 percent or greater chance of occurring in any year, or a flood equaled or exceeded only once in every 100 years on the average.

Endangered and Threatened Species. Solid waste disposal facilities must be constructed and operated so they will not contribute to the "taking" of endangered and threatened species. "Taking" means the harming, pursuing, hunting, wounding, killing, trapping, or capturing of species so listed by the U.S. Department of the Interior. Operators of a solid waste disposal facility may be required to modify its operation to comply with these rules if nearby areas are designated as critical habitats.

Surface Water. Solid waste disposal leads to the pollution of surface waters whenever rainwater percolates through the refuse and runoff occurs, or whenever spills take place during solid waste shipments. Point source discharges from disposal sites are regulated by the NPDES permit system discussed in detail in Chapter 11. Nonpoint source, or diffused water, movement is regulated by the requirements implementing an areawide or statewide water quality management plan approved by the EPA. Generally, these plans regulate the degradation of surface water quality by facility design, operation, and maintenance. Artificial and natural runoff barriers such as liners, levees, and dikes are often required by states. If runoff waters are collected, the site becomes by definition a point source of pollution, and the facility must have a NPDES permit if this new point source discharges into surface watercourses.

Groundwater. To prevent contamination of groundwater supplies, the owners and operators of a solid waste disposal facility must comply with one or more of five design and operational requirements: (1) utilize natural hydrogeologic conditions like underlying confining strata to block flow to aquifers, (2) collect and properly dispose of leachate by installing natural or synthetic liners, (3) reduce the infiltration of rainwater into the solid waste with correct cover material, (4) divert contaminated groundwater or leachate from groundwater supplies, and (5) conduct groundwater monitoring and testing procedures.

Land Used for Food Chain Crops. The federal guidelines attempt to regulate the movement of heavy metals and synthetic organics, principally polychlorinated biphenyls (PCB), from solid waste into soils used to produce human food chain crops. Typically, the solid waste of concern here is sludge from municipal wastewater treatment facilities. To control heavy metals, the regulations give the operator of such disposal facilities two options to follow: the controlled application approach and the dedicated facility approach. If the operator elects the controlled application approach, he must conform with an annual application rate of cadmium from solid waste. Cadmium is controlled basically as a function of time and the type of crop to be grown. A cadmium addition of 2.0 kg/ha (1.8 lb/acre) is allowed for all food chain crops other than tobacco, leafy vegetables, and root crops grown for human consumption. An application rate of 0.5 kg/ha (0.45 lb/acre) is in force for these accumulator crops. Under this option, the operator must conform to a cumulative limit on the amount of cadmium that may be disposed of on food chain land. This cumulative limit is a function of the cation exchange capacity (CEC) of the soil and the background soil pH. At low pH and low CEC, a maximum cumulative application rate of 5.0 kg/ha (4.50 lb/acre) is allowed; whereas at high CEC and soils with high or near neutral pH levels, a 20 kg/ha (18 lb/acre) amount is allowed. The theory is that in soils of high CEC and high pH, the cadmium remains bound in the soil and is not taken up by the plants. All operators electing this option must ensure that the pH of the solid waste and soil mixture is 6.5 or greater at the time of application.

The second option open to food chain landspreaders is the dedicated facility approach. This option is distinguished from the first option by its reliance on output control, or crop management as opposed to input control, or limiting the amounts of cadmium that may be applied to the soil. This second cadmium control option is designed specifically for facilities with the resources and the capabilities to closely manage and monitor the performance of their respective operations. The requirements under this option include: (1) only growing animal feed crops, (2) the pH of the solid waste and soil mixture is 6.5 or greater at the time of application, (3) there is a facility operating plan that demonstrates how the animal feed will be distributed to preclude direct ingestion by humans, and (4) there are deed notifications for future property owners that the property has received solid wastes at high cadmium application rates and that food chain crops should not be grown owing to a possible health hazard.

Solid waste with PCB concentrations greater than or equal to 10 mg/kg of dry weight must be incorporated into the soil when applied to land used for producing animal feed, unless the PCB concentration is found to be less than 1.5 mg/kg fat basis in the produced milk. By requiring soil incorporation, the solid waste regulations attempt to provide that the Food and Drug Administration levels for PCB in milk and animal feed will not be exceeded.

Health. The federal rules require that the operation of a solid waste disposal facility protect the public health from disease vectors. Disease vectors are any routes by which disease is transmitted to humans (birds, rodents, flies, and mosquitoes are good examples). This protection must be achieved by minimizing the availability of food for the vectors. At landfills, an effective means to control vectors, especially rodents, is the application of cover material at the end of each day of operation. Other techniques include poisons, repellents, and natural controls such as the supply predators. Treating sewage sludge with pathogen-reduction processes serves to control the spread of disease from landscaping activities.

Air Quality. A solid waste disposal facility must comply with the Clean Air Act and the relevant State Implementation Plans developed by state and local air emission control boards. Open burning of residential, commercial, institutional, and industrial solid waste is generally prohibited. However, several special wastes are excluded from this prohibition:. diseased trees, land-clearing debris, debris from emergency cleanup operations, and wastes from silviculture and agriculture operations. Open burning is defined as the combustion of solid waste without: (1) control of combustion in the air to maintain adequate temperatures for efficient combustion, (2) containment of the combustion reaction in an enclosed device to provide adequate residence time, or (3) control of the emissions from the combustion process.

Safety. Fires at solid waste disposal facilities are a constant threat to public safety. Death, injury, and vast property damage have resulted in the past from fires breaking out in open dumps. The ban on open burning also reduces the chance of accidental fires. In terms of aircraft safety, facilities should not be located between airports and bird feeding, roosting, or watering sites. This hazard is important when birds fly across an airport or landing/takeoff pattern.

HAZARDOUS WASTE

Common law is particularly not well developed in the hazardous waste area. The issues discussed in Chapter 15 are newly recognized by society, and little case law has had a chance to develop. Since the public health impacts of solid waste disposal in general and hazardous waste disposal in particular were so poorly understood, few plaintiffs ever bothered to take a defendant into the common law courtroom to seek payment for damages or injunctions against such activities. Hazardous waste law is therefore discussed here as statutory law, specifically in terms of the compensation for victims of improper hazardous waste disposal and efforts to regulate the generation, transport, and disposal of hazardous waste.

Historically, federal statutory law was generally lacking in describing how victims should be compensated for improper hazardous waste disposal. A complex, repetitious, confusing list of federal statutes was the only recourse for victims.

The federal Clean Water Act only covers wastes that are discharged to navigable waterways. Only surface water and ocean waters within 200 miles from the coast are considered, and a $5 million revolving fund is set up by the act and administered by the Coast Guard. Fines and charges are deposited into the fund to compensate victims of discharges, but the funds are only available if the discharge of the waste is clearly identifiable. The fund is used most often for compensating states for cleanup of spills from large tanker ships.

The Outer Continental Shelf (OCS) Lands Act sets up two funds to help pay for hazardous waste cleanup and to compensate victims. Under the act, OCS leaseholders are required to report spills and leaks from petroleum-producing sites, and the Offshore Oil Spill Pollution Fund exists to finance cleanup costs and compensate injured parties for loss of the use of their property, loss of the use of natural resources, loss of profits, and a state or local government's loss of tax revenue. The U.S. Department of Transportation (DOT) is responsible for the administration of this fund, with a balance maintained between $100 and $200 million provided by a 3 percent tax on oil produced at the OSC sites. If the operator of the site cooperates with the DOT once a spill occurs, his liability is limited to a maximum of $35 million for a facility spill and $250,000 for a ship spill at the site.

The Fisherman's Contingency Fund also exists under the OCS Lands Act to repay fishermen for loss of profit and equipment owing to oil and gas exploration, development, and production. The Secretary of the U.S. Department of Commerce is responsible for the administration of this fund. The balance of under $1 million is funded by assessments collected by the U.S. Department of the Interior from holders of permits and pipeline easements. If a fisherman cannot pinpoint the site responsible for the discharge of hazardous waste, his claim against the fund may still be acceptable if his boat was operated within the vicinity of OCS activity and if he filed his claim within 5 days of his injury. The fact that two agencies are involved in the administration makes this fund especially confusing to the victims of hazardous waste incidents.

The Price-Anderson Act was passed to provide compensation to victims of discharge from nuclear facilities. Nuclear incidents including radioactive material/toxic spills and emissions are covered, as are explosions. The licensees of a nuclear facility are required by the Nuclear Regulatory Commission (NRC) to obtain insurance protection. If damages from an accident are greater than this insurance coverage, the federal government indemnifies the licensee up to $500 million. In no case does financial liability exceed $560 million. If a major meltdown had occurred at the Three Mile Island in Pennsylvania, the inadequacy of this fund would have been painfully obvious.

Under the Deepwater Ports Act, the Coast Guard acts to remove oil from deepwater ports. The DOT administers a liability fund that helps pay for the cleanup and that compensates injured parties. Financed by a 2¢ per barrel tax on oil loaded or unloaded at such a facility, the fund takes effect once the required insurance coverage is exceeded. The deepwater port itself must hold insurance for $50 million for claims against its waste discharges, and vessels using the port must hold insurance for $20 million for claims against their waste spills. Once these limits are exceeded, the fund established by the Deepwater Ports Act takes over. Again, the usefulness of the fund is dependent on how well two agencies work together and how well insurance claims are administered. This situation is again confusing at best to injured parties.

Statutes at the state level have generally paralleled these federal efforts. New Jersey has a Spill Compensation and Control Act for hazardous wastes listed by the EPA. The fund covers cleanup costs, losses of income, losses of tax revenues, and the restoration of damaged property and natural resources. A tax of 1¢ per barrel of hazardous substance finances a fund that pays injured parties if they file a claim within 6 years of the hazardous waste discharge and within 1 year of the discovery of the damage.

The New York Environmental Protection and Spill Compensation Fund is similar to this New Jersey fund, but with two major differences. In New York, only petroleum discharges are covered, and the generator may not blame a third party for a spill or discharge incident. Thus, if a trucker or handler accidentally spills hazardous waste, the generator of the waste may still be held liable for damages.

Other states have limited efforts in the compensation of victims of hazardous waste incidents. Florida has a Coastal Protection Trust Fund of $35 million that compensates victims for spills, leaks, and dumping of waste. However, the coverage is limited to injury resulting from oil, pesticides, ammonia, and chlorine, which provides a loophole for many types of hazardous waste discussed in Chapter 15. Other states that do have compensation funds generally spell out limited compensatable injuries and provide limited funds, often less than $100,000.

In the past, the value of these federal and state statutes has been questioned. Even when taken as a group, they do not provide for complete compensation strategy for dealing with hazardous waste. Few funds address personal injuries, and abandoned disposal sites are not considered. Huge administrative problems also exist as several agencies attempt to respond to a hazardous waste spill. Questions of which fund applies, which agency has jurisdiction, and what injuries may receive compensation linger.

To clean up their acts, the federal government passed a law labeled the Superfund Act. This act evolved out of several years of battles between congressional committees, the EPA staff, and special interest groups. Originally, the Superfund Act was to serve two purposes: (1) provide money for the cleanup of abandoned hazardous waste disposal sites, and (2) establish liability so discharges could be made to pay for injuries and damages. Three considerations went into the development of this act:

- *What type of incidents should be covered?* The focus here was on decisions to include spills and abandoned sites under the Superfund Act, as well as onsite toxic pollutants in harbors and rivers across the country. Fires and explosions caused by hazardous materials were also a focus of this consideration as the Superfund concept evolved through the year.
- *What type of damages should be compensated?* As alluded to above, three types of damage could be covered by a superfund: environmental cleanup costs; economic losses associated with property use, income and tax revenues; and personal injury in the form of medical costs for acute injuries, chronic illness, death, and general pain and suffering.
- *Who pays for the compensation?* Choices here include federal appropriations into the fund, industry contributions from sales tax, or income tax surcharges. Federal and state cost sharing is also possible, as are fees for disposal of hazardous waste at permitted disposal facilities.

The version adopted by Congress addressed each consideration and reflected the mood of compromise that surrounded the Superfund's evolution. A fund of between $1 billion and $4 billion was established that is financed 87 percent by a tax on the chemical industry and 13 percent by general revenues of the federal government. Small payments are permitted out of the fund, but only for out-of-pocket medical costs and partial payment for diagnostic services. The active Superfund Act does not set strict liability for spills and abandoned hazardous wastes.

Compensation for damages is one concern; regulations that control generators, transporters, and disposers of hazardous waste are the other side of the hazardous waste law coin. The federal RCRA, which is discussed at length in this chapter as it relates to solid waste, also is the principal statute that deals with hazardous waste. The RCRA requires the EPA to establish a comprehensive regulatory program to control hazardous waste. This solid and hazardous waste act offers a good example of the types of reporting and recordkeeping requirement that are mandated in federal environmental laws. The Clean Water Act and Clean Air Act place requirements on water and air polluters similar to the type of requirement discussed below for generators of hazardous waste.

A generator of hazardous waste must meet these EPA requirements:

1. determine if the waste is hazardous, as defined by an EPA listing of hazardous waste, by EPA testing procedures, or as indicated by the materials and processes used in production

2. obtain an EPA general identification number
3. obtain a facility permit if hazardous waste is stored at the generating site for 90 days or longer
4. use appropriate containers and labels before shipment offsite
5. prepare a manifest for tracking the waste shipment offsite, as described in Chapter 15 (see Figure 15-3)
6. assure arrival of the waste at the disposal site
7. submit annual summaries of activities to federal or state regulatory agencies or both

The key to the regulatory system is this manifest system of "cradle to grave" tracking (see Figure 15-3).

A transporter of hazardous waste must meet several requirements under RCRA:

1. obtain an EPA Transporter Identification Number
2. comply with the provisions of the manifest system
3. deliver the entire quantity of waste to the disposal/processing site
4. keep a copy of the manifest for 3 years
5. follow DOT rules for responding to spills of hazardous waste

Again, the key element of transportation controls is the implementation and administration of the manifest system.

Facilities which treat, store, or dispose of hazardous waste also have requirements placed on them by RCRA:

1. The owner of the facility must apply for an operator's permit and supply data on the proposed site and waste to be handled. This permit will spell out the terms of compliance: construction and operating schedules, as well as monitoring and recordkeeping procedures.
2. As permits are granted by the federal or state agency, minimum operating standards will be placed on the facility. Design and engineering standards for containing, neutralizing, and destroying the wastes are addressed in these permits, as are safety and emergency measures in the event of accident. Personnel training is included in each permit's requirements.
3. The financial responsibility of the facility is defined, and a trust fund generally must be established at site closure. This fund enables the monitoring of groundwater and surface discharges to continue after the site is closed and enables the executor of the trust to maintain the site for years to come.

The key to the successful regulation of hazardous waste generation, transportation, and disposal is the manifest system and the manner in which the disposal processing facilities are regulated. The responsibilities for administering RCRA are passed from the EPA to state agencies as these agencies show they have the authority and expertise to regulate hazardous waste effectively.

In addition to the control of hazardous waste, Congress has passed the Toxic Substances Control Act (TOSCA) in an attempt to control hazardous

Table 17–1. Checklist to Comply with TOSCA's Premanufacture Notification (PMN) System

Step	Requirement
1	Inspect EPA inventory of toxic substances to see whether new product is listed.
2	Assess EPA rules for testing procedures to determine whether substances not listed by the EPA are, in fact, "toxic."
3	Prepare and submit the PMN form at least 90 days before manufacture.
4	Obtain an EPA ruling: • *No ruling*: begin manufacturing at end of 90-day period • *EPA order*: respond to EPA order for more data • *EPA Ruling*: EPA may prohibit or limit the proposed manufacturing, processing, use, or disposal of the substance.

substances before they become hazardous waste. One facet of this regulatory program is known as the Premanufacturing Notification (PMN) System. Under the system, all manufacturers must notify the EPA before marketing any substance not included in the EPA's 1979 inventory of toxic substances. In this notification, the manufacturer must analyze the predicted effect of the substance of workers, on the environment, and on consumers. This analysis must be based on test data and all relevant literature. A compliance checklist is presented in Table 17–1 in which specific steps are spelled out for manufacturers who wish to produce a new product.

CONCLUSION

Solid waste statutory law developed rather rapidly once scientists and engineers realized the real and potential impacts of improper disposal. Air, land, and water quality from such disposal became a key concern to local health officials and federal and state regulators. Hazardous and nonhazardous solid wastes are now regulated under a complex system of federal and state statutes that place operating requirements on facility operators.

The newness of these statutes leaves their effectiveness in doubt. How will the EPA implement RCRA? If states are given the responsibility to control hazardous and nonhazardous waste within their boundaries, how will they respond? In the end, what will be the public health, environmental quality, and economic impacts of RCRA and the Superfund?

PROBLEMS

17.1   Local land use ordinances often play key roles in limiting the number of sites available for a solid or hazardous waste disposal facility. Discuss the types of these zoning restrictions that apply in your home town.

17.2   Assume you work for a firm that contracts to clean the inside and outside of factories and office buildings in the state capitol. Your boss thinks he could make more money by expanding his business to include the handling and transportation of hazardous wastes generated by his current clients. Outline the types of data you must collect to advise him to expand/not expand.

17.3   Laws that govern refuse collection often begin with controls on the generator; i.e., rules that each household must follow if city trucks are to pick up their solid waste. If you were asked by a town council to develop a set of such "household rules," what controls would you include? Emphasize public health concerns, minimizing the cost of collection, and even resource recovery considerations.

17.4   The oceans have long been viewed as a bottomless pit into which the solid wastes of the world may be dumped. Engineers, scientists, and politicians are divided on the issue. Should ocean disposal be banned? Develop arguments pro and con.

LIST OF SYMBOLS

DOT = U.S. Department of Transportation
EPA = U.S. Environmental Protection Agency
NRC = Nuclear Regulatory Commission
OCS = outer continental shelf
PMN = Premanufacture Notification
RCRA = Resource Conservation and Recovery Act
TOSCA = Toxic Substances Control Act

Chapter 18
Air Pollution

Be it known to all within the sound
of my voice, whoever shall be
found guilty of burning coal shall
suffer the loss of his head.
King Edward II, ca. 1300

Obviously, air pollution is not a new problem. King Edward II (1307–1327) tried to solve the problem of what Eleanor of Aquitaine called "the unendurable smoke" by prohibiting the burning of coal while Parliament was in session. His successors, Richard II (1377–1399) and Henry V (1413–1422) both took action against smoke, the former by taxation and the latter by forming a commission to study the problem. This was the first of a plethora of commissions, none of which helped reduce the level of air pollution in England. Under Charles II (1630–1685) a pamphlet was authored in 1661 by John Evelyn, entitled "Fumifugium or the Inconvenience of Aer and Smoak of London Dissipated, together with some Remedies Humbly Proposed." His suggestions included moving industry to the outskirts of town and establishing green belts around the city. None of his proposals were implemented. A subsequent commission in 1845 suggested, among other solutions to the smoke problems, that locomotives "consume their own smoke." In 1847 this requirement was extended to chimneys, and in 1853 it was decreed that offending chimneys be torn down.

In fact, for all the rhetoric and commissions, there was little action in England, or anywhere else in the industrially developing world, until after World War II. Action was finally prompted in part by two major air pollution episodes, in which human deaths were directly attributed to high levels of pollutants.

The first of these occurred in Donora, a small steel town (pop. 14,000) in western Pennsylvania. Donora is located in a bend of the Monongahela River, and in 1948 had three main industrial plants—a steel mill, a wire mill, and a zinc plating plant.

363

During the last week of October 1948, a heavy smog settled in the area, and a weather inversion prevented the movement of pollutants out of the valley. (See Chapter 19 for a definition of inversion.) On Wednesday, the smog became especially intense. It was reported that streamers of carbon appeared to hang motionless in the air and visibility was so poor that even natives of the area became lost.[1] By Friday, the doctors' offices and hospitals were flooded with calls for medical help. Yet no alarm had been sounded. The Friday Halloween parade was well attended, and a large crowd watched the Saturday afternoon football game between Donora and Monongahela High Schools.*

The first death had, however, occurred at 2 A.M. More followed in quick succession, and by midnight 17 persons were dead. Four more died before the effects of the smog abated. By this time the emergency had been recognized and special medical help was rushed in.

Although the Donora episode helped focus attention on air pollution problems in the United States, it took four more years before England suffered a similar disaster and action was finally taken. The "Killer Smog" of 1952 occurred in London, with meteorological conditions similar to those during the Donora episode. A dense fog at ground level coupled with bitter cold and the smoke from coal burners caused the formation of another infamous "pea souper." This one was more severe than the usual smog, however, and lasted for over a week. The smog was so heavy that visibility during daylight hours was cut to only a few meters. Bus conductors had to walk in front of their vehicles to guide them through the streets (Figure 18–1). Two days after the fog set in, the death rate in London began to soar (Figure 18–2). Sulfur dioxide concentrations increased to nearly seven times their normal levels, and carbon monoxide was twice the normal. It is important, however, not to conclude from the apparent relationship in Figure 18–2 that sulfur dioxides *caused* the deaths. Many other factors may have been responsible, and the various air pollutants probably acted synergistically to affect the death rate.

Primarily owing to these acute episodes in London and Donora, public opinion and concern forced the initial attempts to clean up urban air.

Interestingly, we have difficulty in defining what *is* clean air. From the scientific standpoint, clean air is composed of the constituents listed in Table 18–1. If we accept this as a definition of clean air, however, we are in trouble, since any naturally occurring suspended material may be called a pollutant, and one never finds such "clean air" in nature. It may thus be more appropriate to define air pollutants as those substances that exist in such concentrations as to cause an unwanted effect. These pollutants may be natural (the blue haze over the Blue Ridge Mountains) or anthropogenic ("human-made"), such as auto-

---

* It has been reported that the Monongahela coach protested the game and asked that it not be recorded in the books, because he alleged that the Donora coach contrived to have a pall of smog hanging over the field so that on kicking and passing plays, the ball would disappear and his players did not know where it would reappear.[2]

Figure 18–1. London smog during the 1952 episode. [Courtesy of BBC Hulton Picture Library.]

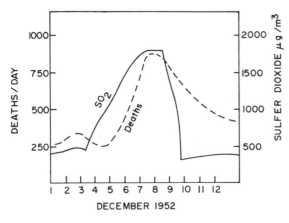

Figure 18–2. Deaths and pollutant concentration during the London 1952 air pollution episode. [Source: Wilkins, E.T., *Journal of the Royal San. Inst.* 74(1)(1954).]

Table 18–1.   Components of Normal Dry Air

	Concentration (ppm)
Nitrogen	780,900
Oxygen	209,400
Argon	9,300
Carbon Dioxide	315
Neon	18
Helium	5.2
Methane	1.0–1.2
Krypton	1.0
Nitrous Oxide	0.5
Hydrogen	0.5
Xenon	0.008
Nitrogen Dioxide	0.02
Ozone	0.01–0.04

mobile exhaust. Air pollutants may be in the form of gases or suspended particulate matter (liquid or solid particles greater than 1 μm in diameter).

TYPES OF AIR POLLUTANTS

Gaseous Pollutants

In the United States, five gaseous pollutants are recognized as being air pollutants of national concern. These are:

> sulfur dioxide
> oxides of nitrogen (NO and NO_2)
> carbon monoxide
> hydrocarbons
> ozone and other oxidants

Some gaseous pollutants of particular concern near major industrial sources are hydrogen sulfide, ammonia, and hydrogen fluoride. These substances exist in the ambient air as gases; under commonly found temperature and atmospheric pressure conditions, they usually behave like ideal gases. Several of these pollutants, notably H_2S, NH_4OH, HF, CO, and NO undergo relatively rapid chemical reaction in the atmosphere. Table 18–2 summarizes the properties of some gaseous air pollutants.

Table 18–2. Gaseous Air Pollutants

Name	Formula	Properties of Importance	Significance as Air Pollutant
Sulfur dioxide	SO_2	Colorless gas, intense choking odor, highly soluble in water to form sulfurous acid, H_2SO_3	Damage to vegetation, property, and health
Sulfur trioxide	SO_3	Soluble in water to form sulfuric acid, H_2SO_4	Highly corrosive
Hydrogen sulfide	H_2S	Rotten egg odor at low concentrations, odorless at high concentrations	Highly poisonous
Nitrous oxide	N_2O	Colorless gas, used as carrier gas in aerosol bottles	Relatively inert. Not produced in combustion
Nitric oxide	NO	Colorless gas	Produced during high-temperature, high-pressure combustion. Oxides to NO_2
Nitrogen dioxide	NO_2	Brown to orange gas	Major component in the formation of photochemical smog
Carbon monoxide	CO	Colorless and odorless	Product of incomplete combustion. Poisonous
Carbon dioxide	CO_2	Colorless and odorless	Formed during complete combustion. Possible effects in producing changes in global climate
Ozone	O_3	Highly reactive	Damage to vegetation and property. Produced mainly during the formation of photochemical smog
Hydrocarbons	C_xH_y	Many	Some hydrocarbons are emitted from automobiles and industries; others are formed in the atmosphere
Hydrogen fluoride	HF	Colorless, acrid gas	Product of aluminum refining; causes fluorosis in cattle

Pollutant concentrations are commonly expressed as micrograms per cubic meter ($\mu g/m^3$). An older and still used method is to express gaseous pollutant concentrations as parts per million, where

$$1 \text{ ppm} = \frac{1 \text{ volume of gaseous pollutant}}{10^6 \text{ volumes, pollutant} + \text{air}}$$

and 1 ppm = 0.0001 percent by volume

Conversion between $\mu g/m^3$ and ppm is done by using the ideal gas law

$$PV = nRT$$

where   P = pressure of gas
V = volume of gas
n = number of moles
R = gas constant
T = temperature of gas in °K

Example 18.1

The exhaust of an automobile is found to contain 2 percent CO at a temperature of 80°C. Express the concentration of CO in the exhaust in $\mu g/m^3$.

$$2\% = 20{,}000 \text{ ppm} = 20{,}000 \text{ L of CO}/10^6 \text{ L of exhaust}$$

T = 273° + 80° = 353°K
P = 1 atmosphere
R = 0.082 L-atm/mol-°K
molecular weight of CO = 12 + 16 = 28 g/mole

From the ideal gas law

$$PV = nRT = (\text{weight of CO/mol. wt.})RT$$

Or, solving for the weight of CO:

$$\text{weight of CO} = PV(\text{mol. wt})/RT$$

$$= \frac{(1 \text{ atm})(20{,}000 \text{ L})(28 \text{ g/mol})}{(0.082 \text{ L-atm/mol-°K})(353°\text{K})}$$

$$= 1.93 \times 10^4 \text{ g}$$

Thus

$$20{,}000 \text{ L of } CO/10^6 \text{ L of exhaust} = 1.93 \times 10^4 \text{ g}/10^6 \text{ L}$$

$$= 1.93 \times 10^4 \text{ g}/10^3 \text{ m}^3$$

$$= 19.3 \ \mu\text{g/m}^3$$

We can see from the ideal gas law that, at constant temperature, the pressure-volume product is a constant. This can be expressed as Boyle's Law

$$\text{At constant temperature,} \quad P_1V_1 = P_2V_2$$

At constant pressure, the relationship between volume and temperature can be expressed as Charles' Law

$$\text{At constant pressure,} \quad \frac{V_1}{T_1} = \frac{V_2}{T_2}$$

If we apply Charles' Law to Example 18–1, we have

$$\frac{V_1}{T_1} = \frac{V_2}{T_2}$$

$$\frac{24.5}{273° + 25°} = \frac{V_2}{273° + 80°}$$

$$V_2 = 29.0 \text{ liters}$$

This 1 g-mole of a gas at 80°C occupies 29.0 liters. The molecular weight of CO is $12 + 16 = 28$, and

$$\mu\text{g/m}^3 = \frac{20{,}000 \text{ ppm} \times 28 \times 10^3}{29.0}$$

$$= 19.3 \times 10^6 \ \mu\text{g/m}^3$$

Particulate Pollutants

Particulate pollutants are classified as follows:

1. *Dust*—solid particles that are: (a) entrained by process gases directly from the material being handled or processed, e.g., coal, ash, and cement; (b) direct offspring of a parent material undergoing a mechanical operation, e.g., sawdust from woodworking; (c) entrained materials used in a mechanical operation, e.g., sand from sandblasting. Dusts from grain elevators and coal-cleaning plants typify this class of particulate. Dust consists of relatively large particles. Cement dust, for example, is about 100 μm in diameter.

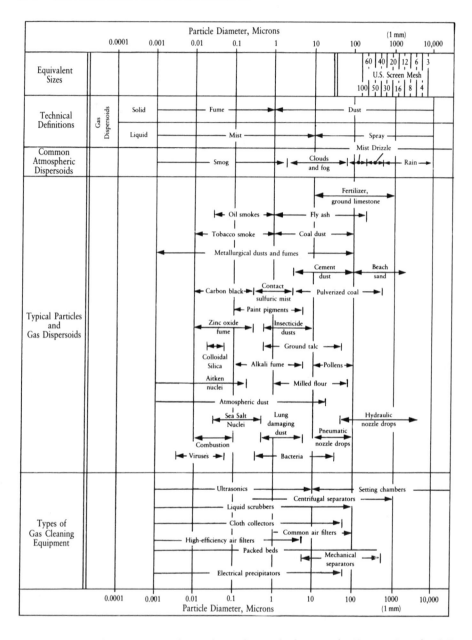

Figure 18–3. Characteristics of particles and particle dispersoids. [Source: Lapple, C.E., Stanford Research Institute Journal 5, 1961.]

2. *Fume*—a solid particle, frequently a metallic oxide, formed by the condensation of vapors by sublimation, distillation, calcination, or chemical reaction processes. Examples of fumes are zinc and lead oxide resulting from the condensation and oxidation of metal volatilized in a high-temperature process. The particles in fumes are quite small, with diameters from 0.03 to 0.3 μm.

3. *Mist*—a liquid particle formed by the condensation of a vapor and perhaps by chemical reaction. An illustration of this process is the formation of sulfuric acid mist:

$$SO_3 \text{ (gas) } 22°C \rightarrow SO_3 \text{ (liquid)}$$

$$SO_3 \text{ (liquid)} + H_2O \rightarrow H_2SO_4$$

Sulfur trioxide gas becomes a liquid since its dew point is 22°C, and SO_3 particles are hydroscopic. Mists typically range from 0.5 to 3.0 μm in diameter.

4. *Smoke*—solid particles formed as a result of incomplete combustion of carbonaceous materials. Although hydrocarbons, organic acids, sulfur oxides, and nitrogen oxides are also produced in combustion processes, only the solid particles resulting from the incomplete combustion of carbonaceous materials are smoke. Smoke particles have diameters from 0.05 to approximately 1 μm.

5. *Spray*—a liquid particle formed by the atomization of a parent liquid.

Approximate size ranges of the various types of air pollutants are shown in Figure 18–3. The concentration of particulates is always expressed as $\mu g/m^3$.

There are several particulate air pollutants that are of particular concern. These are particles smaller than 10 μm in diameter ("respirable particulate matter"), lead, asbestos, heavy metal compounds, and radioactive substances.

The Greenhouse Effect

Carbon dioxide and water vapor are also constituents of the atmosphere but are not presently considered pollutants. There is some question that the atmosphere is accumulating an excess of CO_2 sufficient to bring about climate changes. Internal vibration and rotation of molecular CO_2 cause it to absorb infrared radiation. Carbon dioxide in the air will thus absorb some of the heat that the earth normally radiates into space, and reradiate that heat back to the earth's surface. A canopy of CO_2 has an effect similar to glass in a greenhouse: it traps heat that would otherwise be lost by radiation. It is possible that an accumulation of atmospheric CO_2 could trap enough heat to warm the earth considerably and eventually cause some melting of the polar ice caps. There is not, at present, enough evidence to state conclusively whether this is occurring or not.

SOURCES OF AIR POLLUTION

Many of the pollutants of concern are formed and emitted through natural processes. For example, naturally occurring particulates include pollen grains, fungus spores, salt spray, smoke particles from forest fires, and dust from volcanic eruptions. Gaseous pollutants from natural sources include carbon monoxide as a product of animal and plant respiration, hydrocarbons (terpenes) from conifers, sulfur dioxide from geysers, hydrogen sulfide from the breakdown of cysteine and other sulfur-containing amino acids by bacterial action, nitrogen oxides, and methane.

Near cities and in populated areas, more than 90% of the volume of air pollutants is the result of human activity. Anthropogenic sources of air pollution may be broadly classified as stationary or mobile sources. Stationary sources include combustion processes, solid waste disposal, industrial processes, and construction and demolition. The principal pollutants from stationary combustion processes are particulate matter, in the form of fly ash and smoke, and the gaseous pollutants sulfur dioxide, nitric oxide, and nitrogen dioxide.

Sulfur Dioxide

For uncontrolled sources, sulfur dioxide emissions are a function of the amount of sulfur present in the fuel. Thus, combustion of fossil fuels that contain sulfur may yield significant quantities of SO_2.

The amount of sulfur oxide produced can be calculated if the sulfur content of the fuel is known, as the example below illustrates.

Example 18.2

During the 1952 London episode, it is estimated that 25,000 metric tons of coal was burned that had an average sulfur content of about 4 percent. The mixing depth (height of the inversion layer or cap over the city that prevented the pollutants from escaping) was about 150 m, over an area of 1,200 km^2. What is the expected SO_2 concentration at the end of this time?

Amount of sulfur emitted = 25,000 metric tons × 0.04 = 1,000 metric tons. Sulfur has an atomic weight of 32. Sulfur dioxide has a molecular weight of 64. The amount of sulfur dioxide emitted was thus = (64/32) × 1,000 metric tons = 2,000 metric tons.

The volume into which this is mixed is 150 m × 1,200 km^2 × 10^6 m^2/km^2 = 180,000 × 10^6 m^3.

The concentration would thus be =

$$\frac{2,000 \text{ metric tons} \times 10^6 \text{ g/metric ton} \times 10^6 \text{ } \mu g/g}{180,000 \times 10^6 m^3} = 11,000 \text{ } \mu g/m^3$$

The actual peak concentration of SO_2 during the London episode was less than 2,000 $\mu g/m^3$. Where did all of the SO_2 go? This question is addressed in Chapter 19.

In the United States, the largest single source of atmospheric sulfur dioxide is fossil fuel combustion, both for electric power generation and for process heat. In any given community, however, an industrial plant may be the principal source, since many industrial operations emit significant quantities of SO_2 as well. Some significant industrial emitters are:

- *Nonferrous smelters.* With the exception of iron and aluminum, metal ores are sulfur compounds. When the ore is reduced to the pure metal, the sulfur in the ore is ultimately oxidized to SO_2. Copper ore, for example, is usually a form of CuS. Since the atomic weight of copper is 64 g/mol, the same as the molecular weight of SO_2, every pound of copper that is refined yields a pound of sulfur dioxide, which goes into the air unless it is trapped.
- *Oil refining.* Sulfur and hydrogen sulfide are constituents of crude oil, and H_2S is released as a gas during catalytic cracking. Since hydrogen sulfide is considerably more toxic than sulfur dioxide, it is flared to SO_2 before release to the ambient air.
- *Pulp and paper manufacture.* The sulfite process for wood pulping uses hot H_2SO_3 and thus emits SO_2 into the air. The kraft pulping process produces H_2S, which is then flared to SO_2.

Oxides of Nitrogen

Nitric oxide is formed by the thermal fixation of atmospheric nitrogen. The reaction is strongly temperature dependent: the equilibrium constant for the reaction

$$N_2 + O_2 = 2NO$$

is approximately proportional to the fourth power of the absolute temperature at which the reaction takes place. Thus, all high-temperature processes produce NO, which is then oxidized further to NO_2 in the ambient air.

In the United States, about half of the atmospheric nitrogen oxides are produced by stationary sources. Fossil fuels contain nitrogen compounds, which are also oxidized to NO on combustion.

Carbon Monoxide

CO is a product of incomplete combustion of carbon-containing compounds. Stationary combustion sources produce CO, which is usually transformed to CO_2 rapidly enough so that there is little significant dispersion of CO from the stationary source in the ambient air.

Most of the CO in the ambient air comes from vehicle exhaust. Internal combustion engines do not burn fuel completely to CO_2 and water; some unburned fuel will always be exhausted, with CO as a component. Various afterburner devices may be used to reduce CO. CO is a localized pollutant and tends to build up in areas of concentrated vehicle traffic, in parking garages, and under building overhangs.

Hydrocarbons

Vehicles are also a major source of atmospheric hydrocarbons. Figure 18–4 shows some of the points of hydrocarbon emission in an automobile engine. Stationary sources of hydrocarbons include petrochemical manufacture, oil refining, incineration, paint manufacture and use, and dry cleaning.

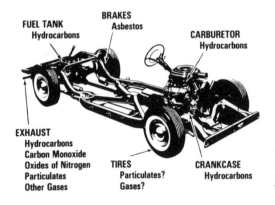

FUEL TANK
Hydrocarbons

BRAKES
Asbestos

CARBURETOR
Hydrocarbons

EXHAUST
Hydrocarbons
Carbon Monoxide
Oxides of Nitrogen
Particulates
Other Gases

TIRES
Particulates?
Gases?

CRANKCASE
Hydrocarbons

Figure 18–4. Sources of automotive air pollution. [Courtesy of General Motors.]

PRIMARY AND SECONDARY SOURCES

Ozone (Photochemical Oxidant)

Photochemically formed organic oxidants, which are classified as ozone, are a secondary air pollutant. That is, ozone is not emitted directly into the air, but is the result of chemical reactions in the ambient air. The components of automobile exhaust are particularly important in the formation of atmospheric ozone and are the primary contributors to Los Angeles smog, which is mostly ozone. Table 18–3 lists in simplified form some of the key reactions in the formation of photochemical smog.

Table 18–3. Simplified Reaction Scheme for Photochemical Smog

NO_2 Nitrogen dioxide	+Light	→NO Nitric oxide	+O Atomic oxygen
O	+O_2 Molecular oxygen	→O_3 Ozone	
O_3	+NO	→NO_2	+O_2
O	+HC Hydrocarbon	→HCO˙ Radical	
HCO˙	+O_2	→$HCO_3˙$ Radical	
$HCO_3˙$	+HC	→Aldehydes, ketones, etc.	
$HCO_3˙$	+NO	→$HCO_2˙$ Radical	+NO_2
$HCO_3˙$	+O_2	→O_3	+$HCO_2˙$
$HCO_x˙$ Radical	+NO_2	→Peroxyacetyl nitrates	

The reaction sequence illustrates how nitrogen oxides formed in the combustion of gasoline and other fuels and emitted to the atmosphere are acted upon by sunlight to yield ozone (O_3), a compound not emitted as such from a source and hence considered a secondary pollutant. Ozone in turn reacts with hydrocarbons to form a series of compounds that includes aldehydes, organic acids, and epoxy compounds. Thus, the atmosphere may be viewed as a huge reaction vessel in which new compounds are being formed while others are being destroyed.

The formation of photochemical smog is a dynamic process. Figure 18–5 is an illustration of how the concentrations of some of the components vary during the day. Note that as the morning rush hour begins the NO levels increase, followed quickly by NO_2. As the latter reacts with sunlight, O_3 and other oxidants are produced. The hydrocarbon level similarly increases at the beginning of the day and then drops off in the evening.

The reactions involved in photochemical smog remained a mystery for many years. Particularly baffling was the formation of high ozone levels. As seen from the first three reactions in Table 18–3, for every mole of NO_2 reacting to make atomic oxygen and hence ozone, one mole of NO_2 was created from reaction with the ozone. All of these reactions are fast. How, then, could the ozone concentrations build to such high levels?

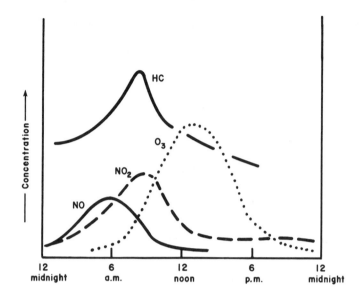

Figure 18–5. Formation of photochemical smog.

One answer is that NO enters into other reactions, especially with various hydrocarbon radicals, and thus allows excess ozone to accumulate in the atmosphere (the seventh reaction in Table 18–3). In addition, some hydrocarbon radicals react with molecular oxygen and also produce ozone.

The chemistry of photochemical smog is still not clearly understood. This was witnessed by the recent attempt to reduce hydrocarbon emissions to control ozone levels. The thinking was that if HC is not available, the O_3 will be used to oxidize NO to NO_2, thus using the available ozone. Unfortunately, this control strategy was a failure, and the answer seems to be that all primary pollutants involved in photochemical smog formation must be controlled.

Acid Rain

Acidic precipitation is another pollutant formed by reactions in the air. Sulfur dioxide and nitrogen dioxide react with water and atmospheric oxygen to produce sulfuric and nitric acid. Normal, uncontaminated rain has a pH of about 5.6, but acid rain can be as low as pH 2 or even below. Hundreds of lakes in North America and Scandinavia have become so acidic that they no longer can support fish life. In a recent study of Norwegian lakes, more than 70 percent of the lakes having a pH of less than 4.5 contained no fish, and nearly all lakes with a pH of 5.5 and above contained fish. The low pH not only affects fish directly, but contributes to the release of potentially toxic metals such as aluminum, thus magnifying the problem. In Norway, storms that travel over the

	pH of rain water at sampling site
"Normal"	5.8
Storm of 25 Aug.	4.2
Storm of 26 Aug.	4.0
Storm of 27 Aug.	4.1

Figure 18–6. Paths of three storms sampled in Kristiansand, Norway.

industrial areas of Great Britain and continental Europe can be tracked and have been found to dump especially destructive precipitation. Figure 18–6 shows three such storms and the corresponding pH values of the rainwater. The recognition of this problem makes the use of the "tall stack" method of air pollution control highly questionable (see Chapter 21).

In North America, acid rain has already wiped out all fish and many plants in 50 percent of the high mountain lakes in the Adirondacks. The pH in many of these lakes has reached such levels of acidity as to replace the trout and native plants with acid-tolerant mats of algae.[3]

Suspended Particulate Matter

Virtually every industrial process is a potential source of dust, smoke, or aerosol emissions. Waste incineration, coal combustion, and the combustion of heavy (bunker-grade) oil are sources. Agricultural operations are a major source of dust, especially in dry land farming. Traffic on roads, even on completely paved roads, is also a major source of dust. In 1985, 40 percent of the total suspended particulate matter in the air of Seattle's downtown industrial center was identified as "crustal dust": dust raised by traffic on paved streets.[4] Construction and demolition may be particularly troublesome sources of dust in urban areas.

Fires are a major source of airborne particulate matter as well as of hydrocarbon emissions and carbon monoxide. Forest fires are usually considered a natural (nonanthropogenic) source, but fires for land clearing, slash burns, agricultural burns, and trash fires contribute considerably. Since 1970, most communities in the United States have prohibited open burning of trash and dead leaves.

Wood-burning stoves and fireplaces also produce smoke that contains partly burned hydrocarbons, aromatic compounds, tars, and aldehydes as well as smoke and ash. Several cities (the largest is Portland, Oregon) now prohibit the use of wood stoves and fireplaces during unfavorable weather conditions.

Lead

Airborne lead is particulate matter and is chemically generally lead oxide or lead chloride. Vehicle exhaust is virtually the only source of airborne lead except in the vicinity of nonferrous smelters. The Bunker Hill lead smelter at Kellogg, Idaho, the largest in the United States, was shut down in 1980, eliminating a significant source of lead pollution in Kellogg.

Lead from vehicle exhaust is deposited for several hundred yards downwind of highways. In urban areas where there is a concentration of freeway intersections, this deposited lead has been identified as a source of lead intoxication.

HAZARDOUS SUBSTANCES

There are a number of hazardous substances emitted into the air that pose particular control problems. Some of these, together with several sources which have been identified, are:

- Asbestos: construction, demolition, and remodeling of existing structures, replacement of pipes, furnaces, etc., asbestos refining and fabrication, dust from soil erosion
- Mercury: chlor-alkali manufacture, battery manufacture
- Vinyl chloride monomer: manufacture and fabrication of polyvinyl chloride products
- Fluorides: primary aluminum smelting, phosphate fertilizer manufacture
- Hydrogen sulfide: crude oil transportation, oil desulfurizing and refining, kraft paper manufacture
- Benzene: petrochemical manufacture, industrial solvent use, pharmaceutical manufacture

HEALTH EFFECTS

Much of our knowledge of the effects of air pollution on people comes from the study of acute air pollution episodes. As previously noted, the two most famous episodes occurred in Donora, Pennsylvania, and in London, England. In both episodes the illness appeared to be chemical irritation of the respiratory tract. The weather circumstances were also similar in that a high pressure system with an inversion layer was present. An inversion, which is discussed further in Chapter 19, is a layer of warm air aloft that prevents the pollutants from escaping and diluting vertically. In addition, inversions are usually accompanied by low surface winds leading to reduced horizontal dilution of pollutants.

In both episodes, the pollutants affected a specific segment of the public— those individuals already suffering from diseases of the cardiorespiratory system. Another observation of great importance is that it was not possible to blame the adverse effects on any one pollutant. This observation puzzled the investigators (industrial hygiene experts) who were accustomed to studying industrial problems in which one could usually relate health effects to a specific pollutant. Today, after many years of study, it is thought that the health problems during the episodes were attributable to the combined action of a particulate matter (solid or liquid particles) and sulfur dioxide, a gas. No one pollutant by itself, however, could have been responsible.

Another episode of historical significance is one involving the accidental release of hydrogen sulfide gas from a refinery at Poza Rica, Mexico, in November 1950. Although the Poza Rica episode is not community air pollution in its usual sense, it is an important example of air pollution that resulted from an industrial accident. The accident occurred when the processes in a refinery went astray and a large quantity of hydrogen sulfide gas escaped. H_2S is a heavy, poisonous gas, and when it escaped, it hugged the ground around the refinery and crept into nearby homes. In all, 22 persons died and 320 became ill.*

Another historically important air pollution health effect is "Yokohama asthma." After World War II, an unusual number of American military personnel and their dependents residing in the Yokohama area of Japan required treatment for attacks of an "asthmalike" condition. These attacks were first noted in the winter of 1948. Investigation showed that the attacks were coincident with conditions of low wind velocity and resulting high pollution levels. Interestingly, only people who were heavy smokers were affected, and the only solution to the problem was reassignment of the victims to another location.

In the United States similar problems affecting asthmatics have been reported in New Orleans. Studies of patients admitted for emergency treatment

---

* Hydrogen sulfide, disguised as "sewer gas," has killed many unsuspecting municipal workers who ventured into large sewers.

for asthma attacks were conducted over a period of years at New Orleans' Charity Hospital. These studies indicated that a sharply increased number of asthmatics required emergency treatment when winds of low speed from the south and southwest prevailed. Investigation further showed that the occurrence of these attacks was associated with a burning dump southwest of New Orleans. A measure of the severity of this problem is illustrated by the October 1962 episode, which caused the deaths of nine asthmatics and required emergency treatment for 300.

Recently, several air pollution episodes have taken place in New York City and other metropolitan areas. The first well-documented episode occurred in November 1953, and air pollution alerts occur every year. The morbidity and mortality* data from these episodes have demonstrated a statistical association between public health and the occurrence of temperature inversions and the resulting increased levels of particulate matter and sulfur dioxide.

Except for these episodes, scientists have very little information from which to evaluate the health effects of air pollution. Laboratory studies with animals are of some help, but the step from a rat to a man (anatomically speaking) is quite large.

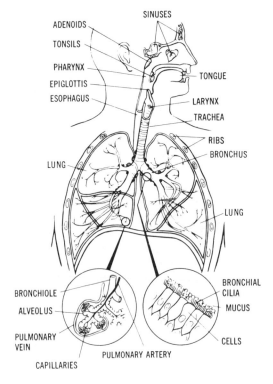

Figure 18–7. The respiratory system. [Courtesy of American Lung Association.]

---

* *Mortality* is of course death (a statistic); *morbidity* is illness not resulting in death.

The Respiratory System

The major target of air pollutants is the respiratory system, pictured in Figure 18–7. Air (and entrained pollutants) enter the body through the throat and nasal cavities and pass to the lungs through the trachea. In the lungs, the air moves through bronchial tubes to the alveoli, small air sacks in which the gas transfer takes place. Pollutants are either absorbed into the bloodstream or moved out of the lungs by tiny hair cells called cilia that are continually sweeping mucus up into the throat. The bronchial cilia can be effectively paralyzed by inhaled smoke, enhancing the synergistic effects between smoking and air pollutants. Pollutant particles small enough to pass the bronchial cilia by are called "respirable particles."

Carbon Monoxide

The effect of carbon monoxide inhalation on human health is directly proportional to the quantity of CO bound to hemoglobin. These effects are summarized in Table 18–4.

Table 18–4. Health Effects of Carbon Monoxide and HbCo

Carbon Monoxide	
Environmental Condition	Effects
9 ppm, 8-hr exposure	Ambient air quality standard
50 ppm, 6-wk exposure	Structural changes in heart and brain of animals
50 ppm, 50-min exposure	Changes in relative brightness threshold and visual acuity
50 ppm, 8- to 12-hr exposure nonsmokers	Impaired performance on psychomotor tests

Carboxyhemoglobin	
HbCO Level (percent)	Effects
<1.0	No apparent effect
1.0–2.0	Some evidence of effect on behavioral performance
2.0–5.0	Central nervous system effects. Impairment of time-interval discrimination, visual acuity, brightness discrimination, and certain other psychomotor functions
>5.0	Cardiac and pulmonary functional changes
10.0–80.0	Headaches, fatigue, drowsiness, coma, respiratory failure, death

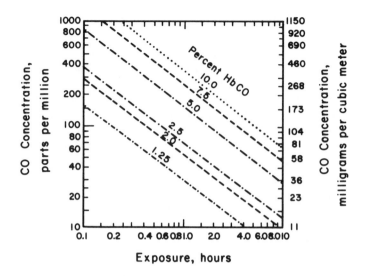

Figure 18–8. HbCO level as a function of exposure. [Source: NAPCA, *Air Quality Criteria for Carbon Monoxide*, AP-62 (Washington, DC: HEW, 1970).]

Oxygen is transported in the blood as oxyhemoglobin (HbO_2), a semistable compound in which O_2 is weakly bound to the Fe^{2+} in hemoglobin in red blood cells. The O_2 is removed for cell respiration, and the regenerated hemoglobin is available for more oxygen transport. CO reduces the oxygen-carrying capacity of the blood by combining with hemoglobin and forming carboxyhemoglobin (HbCO), which is stable. Hemoglobin that is tied up as HbCO cannot be regenerated and is not available for oxygen transport for the life of that particular red blood cell. CO effectively poisons the hemoglobin oxygen transport system.

Hemoglobin has a greater affinity for carbon monoxide than for oxygen. A person who breathes a mixture of CO and oxygen will carry equilibrium concentrations of HbCO and HbO_2 given by:

$$[HbCO]/[HbO_2] = Mp(CO)/p(O_2)$$

where $p(CO)$ = partial pressure of CO
$p(O_2)$ = partial pressure of O_2
M = a constant whose range in human blood is from 200 to 250

Example 18.2

Make an estimate of the saturation value of HbCO in the blood if the CO content of the air breathed is 100 ppm.

Let us assume a value for M of 210. Then since air is 21 percent oxygen,

$$[HbCO]/[HbO_2] = Mp(CO)/p(O_2)$$
$$Mp(CO)/p(O_2) = (210)(100)/210,000$$
$$= 0.1$$

Thus $[HbCO]/[HbO_2] = 0.1$ or 10 percent

HbCO levels as a function of exposure are shown in Figure 18–8.

Particulate Matter

The site and the extent of the deposition of particulates in the respiratory tract is a function of certain physical factors, the most important being particle size.

Alveolar deposition is especially important since that region of the lungs is not provided with cilia to remove particulates. Thus, particles deposited there would remain for a relatively greater length of time.

The cilia (Figure 18–7) that protect the respiratory system by sweeping out particles are also affected by sulfur dioxide. Experiments with rabbits and other animals have shown that the frequency with which cilia beat is decreased in the presence of sulfur dioxide. Thus, sulfur dioxide, in addition to constricting the bronchi, also affects the protection mechanism of the respiratory tract.

Nitrogen dioxide (NO_2) is a pulmonary irritant. Although little is known of the specific toxic mechanism, it is known that nitrogen dioxide at concentrations admittedly greater than those found in community air is an edema producer and also results in pulmonary hemorrhage. Edema is an abnormal accumulation of fluid in body tissues. Pulmonary edema is excessive fluid in the lung tissues.

Nitric oxide (NO), the other common oxide of nitrogen, is not an irritant gas. In vitro nitric oxide combines readily with human hemoglobin to form a highly stable nitric oxide hemoglobin, but interestingly this hemoglobin effect has not been observed in living animals.

Ozone (O_3) is a highly irritating, oxidizing gas. Concentrations of a few parts per million produce pulmonary congestion, edema, and hemorrhage. A 1-hour exposure of human subjects to 2,500 $\mu g/m^3$ can increase the residual lung volume and decrease maximum breathing capacity. The symptoms of ozone exposure are initially a dry throat, followed by headache, disorientation, and altered breathing patterns.*

---

* Ozone in the air we breathe should not be confused with the so-called "ozone layer" in the stratosphere. The stratosphere begins at an altitude of 7 to 10 miles, depending on the latitude and season of the year. Ozone in this thin air absorbs a large part of the sun's ultraviolet radiation. It is believed that vapor from supersonic aircraft flights and fluorocarbon gases from spray cans will cause permanent reduction in stratospheric ozone. This could increase the ultraviolet radiation reaching the earth, raising the incidence of human skin cancer and probably affecting the earth's climate and ecological systems in unpredictable ways. The high ozone layer seems to be maintained by natural processes, mainly the sun's radiation. Lightning discharges are the principal natural source of ozone in the lower atmosphere.

A fascinating observation from animal studies is the development of ozone tolerance. An animal exposed initially to a low concentration of ozone can survive a subsequent exposure to an ordinarily fatal concentration of ozone. To what extent ozone tolerance develops in humans is unknown.

Among the organic gases, formaldehyde (HCHO) is particularly important. Its effects are similar to those of sulfur dioxide, i.e., irritation of mucous membranes and bronchial constriction.

It has also been noted that a linear statistical association exists between aldehyde concentrations and eye irritation. An unsaturated aldehyde, acrolein (CH_2CHCHO), at concentrations of a tenth of a part per million will produce eye irritation comparable to that which occurs in Los Angeles photochemical smog. Acrolein is also irritating to mucous membranes and produces bronchoconstriction.

Some of the chronic diseases believed to be causally related to air pollution are lung cancer and chronic obstructive pulmonary disease (COPD, of which emphysema is one manifestation). Asthma and other respiratory allergies appear to be aggravated by air pollution. The following facts have been cited in support of the hypothesis that lung cancer is related to air pollution: (1) lung cancer mortality is higher in urban than in rural areas, even when cigarette smoking is taken into account; (2) carcinogenic substances for experimental animals (both organic and metallic) are found in polluted atmospheres; (3) carcinogens are relatively stable in the atmosphere; (4) compounds extracted from air samples produce cancers in bioassay animals; (5) atmospheric irritants affect the protective action of cilia and mucus flow; (6) high workplace concentrations of some air pollutants have produced specific types of lung cancer in exposed workers; and (7) both the carcinogens and irritants have physical characteristics compatible with the postulated effects. However, even this staggering statistical and experimental evidence does not conclusively prove that air pollution *does* cause cancer. The actual cause-and-effect relationship is still a medical mystery.

Chronic bronchitis is a disorder characterized by excessive mucus secretion in the bronchial tubes. It is manifested by a chronic or recurrent productive cough. In Great Britain, chronic bronchitis is the second most common cause of death in men aged 40 to 55 and is the leading cause of disability in this age group.* Studies in Britain have shown chronic bronchitis mortality to be associated with the level of air pollution. Chronic bronchitis morbidity in postmen in Great Britain has been shown to be related to the amount of pollution to which they are exposed.

Emphysema is the breakdown and destruction of the alveolar walls in the lung. The destruction of these alveolar sacs leads to great difficulty in breathing.

---

* The high incidence of chronic bronchitis in Great Britain may in part be due to the more frequent diagnosis of chronic bronchitis as a disease. In the United States physicians tend to diagnose emphysema and other respiratory diseases more frequently. The distinction between many of these diseases is rather fuzzy.

Available evidence suggests a relationship between emphysema and polluted air. In the United States, for example, the disease is more common in urban than in nonurban areas and is a leading cause of death, especially for cigarette smokers.

In addition to the chronic diseases, air pollution may have a serious influence on acute diseases such as the common cold and pneumonia. Statistical evidence has shown that both of these illnesses can be aggravated by breathing dirty air.

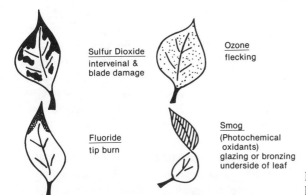

Sulfur Dioxide
interveinal &
blade damage

Ozone
flecking

Fluoride
tip burn

Smog
(Photochemical
  oxidants)
glazing or bronzing
underside of leaf

Figure 18–9.   Typical air pollutant injury to vegetation.

Figure 18–10.   Sulfur dioxide injury to oak.

EFFECTS ON VEGETATION

Vegetation is injured by air pollutants in three ways: (1) necrosis (collapse of the leaf tissue), (2) chlorosis (bleaching or other color changes), and (3) alterations in growth. The types of injury caused by various pollutants differ markedly (Figure 18–9).

Sulfur dioxide produces marginal or interveinal blotches which are white to straw in color on broad-leafed plants (Figure 18–10). Grasses injured by sulfur dioxide show a streaking (light tan to white) on either side of the midvein. Brown necrosis occurs on the tips of conifer needles with adjacent chlorotic areas. Alfalfa, barley, cotton, wheat, and apple are among the plants most sensitive to sulfur dioxide. Sensitive species are injured at concentrations of 780 $\mu g/m^3$ for 8 hours.

In conifers and grasses, fluoride exposure produces an injury known as tip burn. In broad-leafed plants the fluoride injury is a necrosis at the periphery of the leaf. Among the plants most sensitive to fluorides are gladiolus, Chinese apricot, Italian prune, and pine.

At sufficient concentrations ozone produces tissue collapse and markings of the upper surface of the leaf. These markings are known as stipple (pigmented red-brown) and flecking (bleached straw to white). One to two hours of

Figure 18–11. Barren landscape around a copper smelter resulting from acute SO_2 damage.

exposure at air levels of about 300 $\mu g/m^3$ produces injury in sensitive species. Sensitive varieties include tomato, tobacco, bean, spinach, and potato.

Peroxyacyl nitrates (PAN) are present in photochemical smog and produce typical smog injury. The smog-produced injury is a bronzing on the underside of the leaves of vegetables. In grasses the collapsed tissue shows up as bands bleached tan to yellow. Smog exposure has also been shown to produce early maturity or senescence in plants. Even low PAN concentrations will injure sensitive species. Among the most sensitive species are petunia, romain lettuce, pinto bean, and annual bluegrass.

Acute devastation to vegetation has occurred around plants emitting pollutants, such as the SO_2 damage around the copper smelter shown in Figure 18–11.

EFFECTS ON DOMESTIC ANIMALS

As might be expected, air pollutants affect animals other than people. During the Donora episode, 20 percent of the canaries and 15 percent of the dogs were stricken.

At Poza Rica an unknown number of canaries, chickens, cattle, pigs, geese, ducks, and dogs became ill or died as a result of the hydrogen sulfide exposure. All canaries exposed to the H_2S died. (The canary has long been used in British mines as a bioassay method for toxic gases.)

In the 1952 London episode, of 351 cattle on the ground floor at Smithfield Cattle Show, 52 became seriously ill, 5 died, and 9 others were slaughtered later.

Chronic poisoning usually results from ingesting forage contaminated by the pollutant. Pollutants important in this connection are the heavy metals arsenic, lead, and molybdenum. At Anaconda, Montana, in 1902, extensive poisoning of cattle, horses, and sheep occurred in the vicinity of a copper smelter. Of a flock of 3,500, about 620 sheep died after feeding on vegetation some 15 miles from the smelter. Horses remote from the smelter but fed hay from the smelter area also died. The grass and moss ingested by the sheep were found to contain 52 and 405 $\mu g/g$ respectively, of arsenic trioxide. The hay ingested by the horses contained 285 $\mu g/g$ of arsenic trioxide.

Lead poisoning of cattle and horses was reported from Germany as recently as 1955. Cattle and horses gazing within a radius of 5 km of each of two lead and zinc foundries were affected, and some had to be slaughtered. Dust samples collected in the vicinity of the sources showed lead concentrations ranging from 17 to 45 percent and zinc concentrations ranging from 5 to 23 percent.

In 1954 cattle grazing at a distance of 0.1 km from a steel plant in Sweden were poisoned by molybdenum. An analysis of the pasture vegetation showed a molybdenum concentration of 230 $\mu g/g$ of dry matter.

Perhaps the best known of the pollutants affecting livestock is fluoride. The problem of *fluorosis* of livestock is an old one. Farm animals, particularly cattle, sheep, and swine, are susceptible to fluorine poisoning. Horses and poultry, on the other hand, seem to have a high level of resistance. Fluorosis is characterized by mottled teeth and a condition of the joints known as exostosis leading to lameness and ultimately death.

EFFECTS ON MATERIALS

Perhaps the most familiar effect of air pollution on materials is soiling of building surfaces, clothing, and other articles. Soiling results from the deposition of smoke (fine particles of approximately 0.3 μm diameter) on surfaces. Over time this deposition becomes noticeable as soiling, a discoloring or darkening of the surface. Damage to the surface, of course, results from the cleaning operation. In the case of exterior building materials, sandblasting is often required to clean the surface, and part of the surface is removed in the process of cleaning.

Another effect of air pollution is that of accelerating the corrosion of metals. For example, it has been observed that in the presence of sulfur dioxide many materials corrode much faster than they would otherwise.

One of the early noted effects of the Los Angeles smog was rubber cracking. Indeed, the effect of ozone, a principal ingredient of smog, on rubber is so specific that rubber cracking may be used to measure ozone concentrations (Chapter 20). The economic significance of rubber cracking is apparent.

Fabrics are also affected by air pollutants. A few years ago women in Toronto, Canada, and Jacksonville, Florida, noted that the nylon hose they were wearing seemed to disintegrate. This effect was due to a fine sulfuric acid mist present in the air of those cities. Other effects of pollutants on fabrics include bleaching and discoloration, both of which lead to an unacceptable deterioration of products made of fabrics.

The action of hydrogen sulfide on lead-base paints is well known. Hydrogen sulfide, in the presence of moisture, reacts with lead dioxide in paint to form lead sulfide, producing a brown to black discoloration. As might be imagined, this is unattractive on a white house.

EFFECTS ON ATMOSPHERE

The ability of air pollutants, especially particulates, to reduce visibility is well known to passengers on airplanes approaching nearly any large urban area in the United States. The visibility reduction results from light scattering rather than obscuration of light. The particles primarily responsible for this effect are quite small—in the range of 0.3 to 0.6 μm in diameter.

Although carbon dioxide is usually not considered a pollutant, CO_2 concentrations in the atmosphere have been increasing in recent years, attributable to increased combustion of fossil fuels. Since CO_2 absorbs strongly at about

15 μm (heat energy) it retards the radiative cooling of the earth. The CO_2 concentrations will increase from the 1968 level of 0.032 percent to about 0.038 percent by the year 2000. This would theoretically result in an increase of 0.5°C and is often referred to as the "greenhouse effect."

In contrast to the CO_2 effect is the observation that the temperature near the earth's surface has declined in recent years. Some investigators attribute this decline to an increase in the earth's albedo (reflectivity) as a result of an increased atmospheric aerosol from pollution. Measurements of atmospheric turbidity in Washington, D.C., and Davos, Switzerland, indicate increases of particulates of up to 57 and 70 percent, respectively, during the first half of this century. These observations support the suggestion of increased albedo or reflectivity. Is it possible that the aerosol effect is keeping pace with the CO_2 effect?

CONCLUSION

It is important to realize that while the effect of air pollutants on animals, vegetation, and materials is easy to determine, the health effects on humans can only be estimated on the basis of epidemiological evidence. Owing to moral and ethical considerations that preclude deliberate exposure of human subjects to concentrations that might result in disease, this is the only avenue of study open. Epidemiological studies have been criticized because the results are expressed as statistical associations. It should be pointed out that it is probably the best evidence we will ever have. In addition, these studies are with people in the real world, exposed to real pollution, not to a synthetic atmosphere concocted in the laboratory.

One thing is certain—all of the evidence, both experimental and epidemiological, suggests that air pollution is indeed a serious threat to our health and well-being.

PROBLEMS

18.1 If you drive a 1974 car an average of 1,000 miles/month, how much CO and HC would be emitted during the year? (Note: The EPA 1974 standards are 3.4 g/mile for HC and 39 g/mile of CO.) How long would it take to achieve a lethal concentration of CO in a common double-car garage?

18.2 A 2.5 percent level of CO in hemoglobin (HbCO) has been shown to cause impairment in time-interval discrimination. The level of CO on crowded city streets sometimes hits 100 ppm CO. An approximate relationship between CO and HbCO (after prolonged exposure) is

$$\text{HbCO}(\%) = 0.5 + 0.16 \times \text{CO (ppm)}$$

What level of HbCO would a traffic cop be subjected to during a working day directing traffic on a city street?

18.3 Draw a graph showing the concentration of NO, NO_2, HC, and O_3 in the Los Angeles area during a sunny, smoggy day. Superimpose on this graph, in another color, the curves as they appear on a cloudy day.

18.4 Give three examples of synergism in air pollution.

18.5 If SO_2 is so soluble in water, how can it get to the deeper reaches of the lung without first dissolving in the mucus?

18.6 Match the pollutants with the appropriate description.

A. Hydrogen fluoride (HF)   —a secondary pollutant in photochemical smog
B. Hydrogen sulfide (H_2S)   —cause of "weather fleck"
C. Carbon monoxide (CO)   —moved by cilia
D. Carbon dioxide (CO_2)   —one major source is power production
E. Sulfur dioxide (SO_2)   —combines with hemoglobin
F. Particulates   —a pungent, toxic gas often found in sewers
G. Oxides of nitrogen (NOx)   —result of phosphate fertilizer production
H. Peroxyacyl nitrate (PAN)   —produced naturally by decaying matter
I. Ozone (O_3)   —produced only by high-compression combustion.

18.7 The state of Florida has set an annual average ambient SO_2 standard of 0.003 ppm. At ordinary ambient temperatures, what would this be in $\mu g/m^3$?

18.8 The state of Michigan set ambient H_2S standards at 2 ppb. Assuming that the compound is diluted by a factor of 10,000 after emission from the stack and that the stack gas is at a temperature of 100°C, what is the allowed H_2S emission from a kraft pulp mill, expressed in $\mu g/m^3$?

REFERENCES

1. Schrenk, H.H., et al. "Air Pollution in Donora, Pa.," U.S. Public Health Service Bulletin No. 306 (1949).
2. Chanlett, E. "History of Sanitation," in *History of Environmental Sciences and Engineering*, P.A. Vesilind, ed. (University of North Carolina, Department of E.S.E., 1975).
3. Schoefield, C., Cornell University, as reported in the *Washington Post* (7 January 1981).
4. Puget Sound Air Pollution Control Agency. *Annual Report on Air Quality in the Puget Sound Air Quality Region*, 1985.

Chapter 19
Meteorology and Air Quality

One would think that with the earth's atmosphere being about 100 miles deep, we should be able to dilute effectively all the garbage thrown into it. But actually, about 95 percent of the total air mass is within only 12 miles of the surface. This layer, called the *troposphere*, is where we have our weather and air pollution problems.

The science of meteorology has great bearing on air pollution. An air pollution problem involves three parts: the source, the movement of the pollutant, and the recipient (Figure 19–1). Whereas Chapter 18 covers sources and effects, this chapter covers the transport mechanism—how the pollutants travel through the atmosphere.

BASIC METEOROLOGY

Weather changes in the form of fronts, which may be warm or cold. Warm fronts usually are associated with steady rain and drizzle; cold fronts bring heavy local rain.

Another way of picturing weather is in terms of barometric pressure. Low-pressure systems are associated with both hot and cold fronts. The air movement around low-pressure systems is counterclockwise (in the northern hemisphere) and vertical winds are upward, where condensation and precipitation take place. High-pressure systems bring sunny and calm weather, with the wind spiraling clockwise and downward. There are no fronts associated with high-pressure systems, which represent stable atmospheric conditions. The low- and high-pressure systems, commonly called *cyclones* and *anticyclones*, respectively, are illustrated in Figure 19–2. Because of the high stability in anticyclones, these are usually precursors to air pollution episodes.

Figure 19–3 shows a weather map for the eastern United States for 18 January 1968. Note the high-pressure system over the entire area, with clear skies and little or no wind. The air pollutants emitted into the atmosphere on

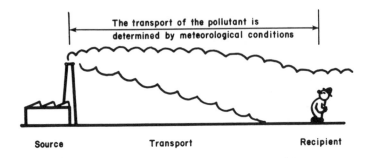

Figure 19–1. Meteorology of air pollutants.

this day would not be dispersed (no wind) and since there is no rain, they would not be removed. Such stable conditions commonly result in a buildup of pollutants. In this case, had the cold front in the middle of the United States not come through, this weather condition might have produced a serious air pollution episode.

A primary objective of air quality management is to attain low air pollutant concentrations in community air. This may be accomplished, of course, by reducing the amount of pollutants emitted. It is also possible to reduce concentrations by attaining adequate dispersal and dilution in the atmosphere. Such dispersal is controlled by atmospheric conditions, and may be vertical as well as horizontal. The principles governing both vertical and horizontal movement of air are therefore important in air quality management.

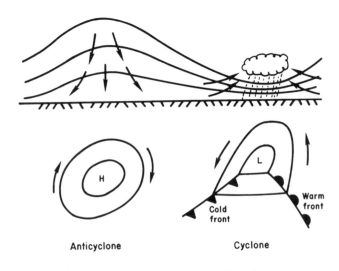

Figure 19–2. Anticyclone and cyclone.

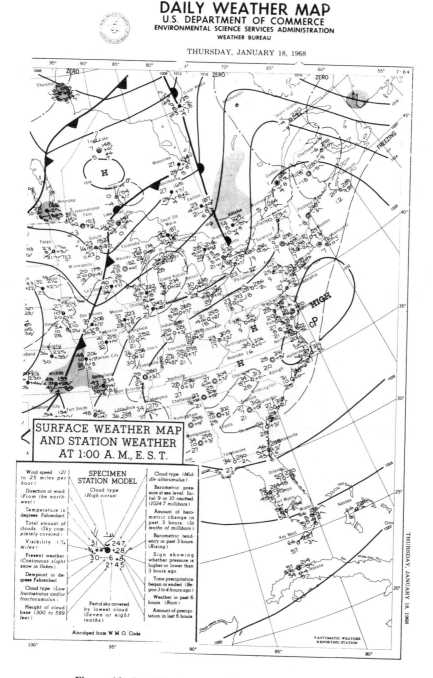

DAILY WEATHER MAP
U.S. DEPARTMENT OF COMMERCE
ENVIRONMENTAL SCIENCE SERVICES ADMINISTRATION
WEATHER BUREAU

THURSDAY, JANUARY 18, 1968

SURFACE WEATHER MAP
AND STATION WEATHER
AT 1:00 A. M., E. S. T.

Wind speed (21 to 25 miles per hour)	SPECIMEN STATION MODEL	Cloud type (Middle altocumulus)
Direction of wind (From the northwest)	Cloud type (High cirrus)	Barometric pressure at sea level Initial 9 or 10 omitted. (1024 7 millibars)
Temperature in degrees Fahrenheit		Amount of barometric change in past 3 hours (In tenths of millibars)
Total amount of clouds (Sky completely covered)		Barometric tendency in past 3 hours (Rising)
Visibility (3/4 miles)		Sign showing whether pressure is higher or lower than 3 hours ago
Present weather (Continuous slight snow in flakes)		Time precipitation began or ended. (Began 3 to 4 hours ago)
Dewpoint in degrees Fahrenheit		Weather in past 6 hours. (Rain)
Cloud type (Low fractostratus and/or fractocumulus)		Amount of precipitation in last 6 hours
Height of cloud base (300 to 599 feet)	Part of sky covered by lowest cloud (Seven or eight tenths)	

Abridged from W M O. Code

Figure 19–3. Weather map for 18 January 1968.

THURSDAY, JANUARY 18, 1968

HORIZONTAL DISPERSION OF POLLUTANTS

The earth acts as a wave converter, accepting the sun's light energy (high-frequency waves) and converting this to heat energy (low-frequency waves), which is then radiated back to space. The heat transfer from the earth to space is by radiation, conduction, and convection. Radiation involves the transfer of heat by energy waves, a minor effect on the atmosphere; conduction is the transfer of heat by physical contact; convection is the process of heating by the movement of air masses.

If the earth did not rotate, the air near the equator would be heated, then it would rise and move toward the poles, where it would sink (Figure 19–4). But the earth does rotate and thus always presents new areas for the sun to shine on and to warm. Accordingly, a pattern of winds is set up around the world, some seasonal (e.g., hurricanes) and some permanent. Local conditions and cloud cover further complicate the picture.

For example, land masses heat and cool faster than water, thus shoreline winds blow out to sea during the night and inland during the day. Valley and slope winds are caused by the cooling of air on mountain slopes. In cities, brick and stone absorb and hold heat, creating a *heat island* about the city during the night (Figure 19–5). This effect sets up a self-contained circulation pattern called a *haze hood* from which the pollutants cannot escape.

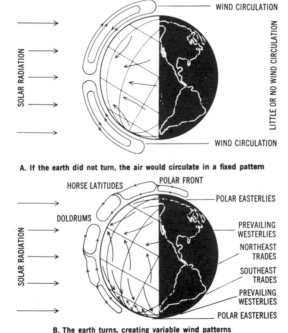

Figure 19–4. Global wind patterns. [Courtesy of American Lung Association.]

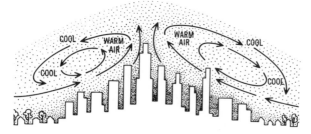

Figure 19-5.   Heat island formed over a city.

The horizontal motion of winds is measured as wind velocity. These wind velocity data are plotted as a *wind rose*, a graphic picture of the direction and velocity *from which the wind came.* The wind rose in Figure 19-6 shows that the prevailing winds were from the southwest.

Air pollution engineers often use a variation of the wind rose, called a *pollution rose*, to determine the source of a pollutant. Instead of plotting all winds on a radial graph, only those days during which the concentration of a pollutant is above a certain minimum are used. Figure 19-7 is an actual plotting of such pollution roses. Only winds carrying SO_2 levels greater than 250 $\mu g/m^3$ were plotted. Note how the fingers of the roses point to plant 3. Pollution roses may be plotted for other pollutants as well and are useful for pinpointing sources of atmospheric contamination.

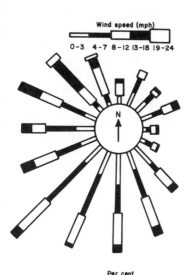

Figure 19-6.   Typical wind rose.

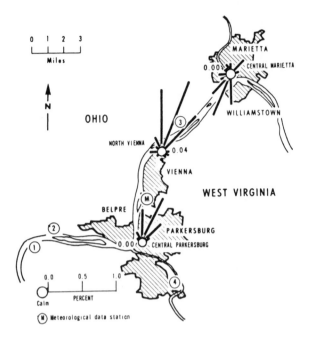

Figure 19–7. Pollution roses, with SO_2 concentrations greater than $250\mu g/m^3$. The major suspected sources are the four chemical plants, but the data indicate that plant 3 is the primary culprit.

VERTICAL DISPERSION OF POLLUTANTS

As a parcel of air rises in the earth's atmosphere, it experiences lower and lower pressure from surrounding air molecules, and thus it expands. This expansion lowers the temperature of the air parcel. The rate at which dry air cools as it rises is called the *dry adiabatic lapse rate* and is independent of the ambient air temperature. The term "adiabatic" means that there is no heat exchange between the parcel of air under consideration and the surrounding air. The dry adiabatic lapse rate may be calculated directly from the First Law of Thermodynamics.

$$(dT/dz)_{dry\ adia} = -9.8°C/km = -5.4°F/1,000\ ft \cong -1°C/100\ m$$

The actual temperature-evaluation measurements are called *prevailing lapse rates* and may be classified as shown in Figure 19–8.

When the prevailing lapse rate is exactly the same as the dry adiabatic lapse rate, it is referred to as a *neutral* lapse rate, or neutral stability. A *superadiabatic* lapse rate, also called a strong lapse rate, occurs when the atmospheric temperature drops more than 1°C/100 m. A *subadiabatic lapse rate*, also called a *weak lapse rate*, is characterized by a drop of less than 1°C/100 m. A special case of the weak lapse rate is the *inversion*, a condition that has warmer air above colder air.

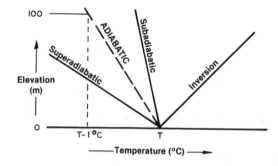

Figure 19–8. Prevailing lapse rates, and the dry adiabatic lapse rate.

During a superadiabatic lapse rate the atmospheric conditions are unstable; a subadiabatic and especially an inversion characterizes a stable atmosphere. This may be demonstrated by depicting a parcel of air at 500 m (see Figure 19–9A). If the temperature at 500 m is 20°C, during a superadiabatic condition the temperature at ground level might be 30°C and at 1,000 m it might be 10°C (a change of more than 1°C/100 m).

If the parcel of air at 500 m is moved upward to 1,000 m, what would be its temperature? Remember that assuming adiabatic condition, the parcel would cool 1°C/100 m. The temperature of the parcel at 1,000 m is thus 5°C less than 20°C or 15°C.

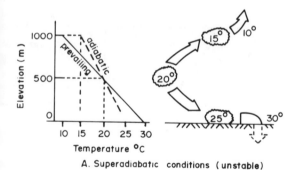

A. Superadiabatic conditions (unstable)

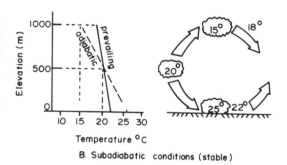

B. Subadiabatic conditions (stable)

Figure 19–9. Stability and vertical air movement.

The prevailing temperature, however, is 10°C, and the air parcel finds itself surrounded by cooler air. Will it rise or fall? Obviously, it will rise, since warm air rises. Once the parcel of air under superadiabatic conditions is displaced upward, it keeps right on going.

Similarly, if a parcel is displaced downward, say to ground level, the air parcel is 20°C + [500 m × (1°C/100 m)] = 25°C. It finds the air around it a warm 30°C, and thus the cooler air parcel would just as soon keep going down if it could. Superadiabatic conditions are thus unstable, characterized by a great deal of vertical air movement and turbulence.

The subadiabatic condition shown in Figure 19–9B is by contrast a very stable system. Consider again a parcel of air at 500 m and at 20°C. A typical subadiabatic system has a ground level temperature of 21°C and 19°C at 1,000 m. If the parcel is displaced to 1,000 m, it will cool by 5 to 15°C. But, finding the air around it a warmer 19°C, it will fall right back to its point of origin. Similarly, if the air parcel were brought to ground level, it would be at 25°C, and finding itself surrounded by 21°C air, it would rise back to 500 m. Thus the subadiabatic system would tend to dampen out vertical movement and is characterized by a very limited vertical mixing.

An inversion is an extreme subadiabatic condition, and thus the vertical air movement within an inversion is almost nil.

Figure 19–10 is an actual temperature sounding for Los Angeles. Note the beginning of the inversion at about 1,000 ft. This puts an effective cap on the

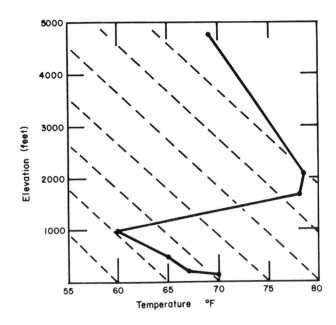

Figure 19–10.   Actual temperature sounding for Los Angeles, October 1962, 4 P.M. The dotted lines show the dry adiabatic lapse rate, 5.4° F/1,000 ft.

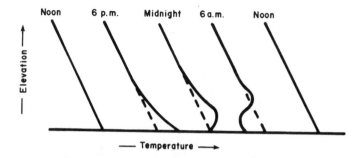

Figure 19–11. Typical prevailing lapse rates during a sunny day and clear night.

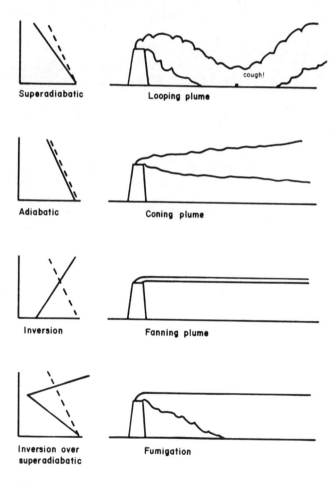

Figure 19–12. Plume shapes and atmospheric stability.

Figure 19–13.   Iron oxide dust looping plume from a steel mill.

city and holds in the air pollution. The Los Angeles inversion is called a *subsidence inversion,* caused by a large warm air mass subsiding over the city.

A more common type of inversion is the *radiation inversion,* caused by the radiation of heat to the atmosphere from the earth. During the night, as the earth cools, the air close to the ground loses heat, thus causing an inversion (Figure 19–11). The pollution emitted during the night is caught under this lid and does not escape until the earth warms sufficiently to break the inversion.

Atmospheric stability may often be recognized by the shapes of plumes emitted from smokestacks (Figures 19–12 and 19–13). One potentially serious condition is called *fumigation,* in which the pollutants are caught under an inversion and are mixed owing to a strong lapse rate. A looping plume also may be dangerous owing to a very high ground level concentration of pollutants as the plume touches ground.

If we assume perfect adiabatic conditions in a plume, we can estimate how far it will rise (or sink) and what type of plume it will be during any given atmospheric temperature condition. This is illustrated in Example 19.1.

Example 19.1
    A stack 100 m tall emits a plume at 20°C. The prevailing lapse rates are shown in Figure 19–14. How high will the plume rise (assuming perfect adiabatic conditions), and what type of plume will it be?

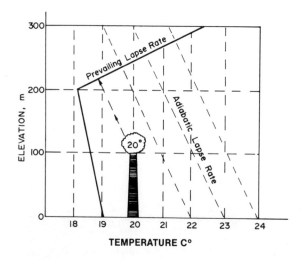

Figure 19–14. Atmospheric conditions for an emission at 20°C, for Example 19.1.

Note that the prevaling lapse rate is subadiabatic to 200 m and that an inversion exists above 200 m. The smoke at 20°C finds itself surrounded by colder (18.5°C) air, and thus rises. As it rises it cools, so that at 200 m it is 19°C. At about 220 m, the surrounding air is at the same temperature as the smoke (about 18.7°C) and the smoke ceases to rise.

Below 220 m, the plume would have been stable and slightly coning. It would not, however, have penetrated 220 m and thus there would have been a cap on the plume.

Effect of Water in the Atmosphere

Thus far we have assumed that there was no water in the atmosphere, hence we speak of the *dry* adiabatic lapse rate. Water will of course condense and evaporate, and in so doing emit and absorb heat, making the calculations of stability quite a bit more complicated.

In general, as a parcel of air rises, the water vapor in that parcel will condense, releasing heat. The wet adiabatic lapse rate will thus be less negative than the dry adiabatic lapse rate. The wet adiabatic lapse rate has been observed to vary from −6.5°C to −3.5°C per 1,000 m.

Water also affects air quality in other ways. Fogs are formed when moist air cools and the moisture condenses. Aerosols provide the condensation nuclei, and fogs thus tend to occur more readily in urban areas.

In addition to inversions, serious air pollution episodes are almost always accompanied by fogs. These tiny droplets of water are detrimental in two ways. In the first place, fog makes it possible to convert SO_3 to H_2SO_4. Secondly, fog sits in valleys and prevents the sun from warming the valley floor and breaking inversions, thus often prolonging air pollution episodes.

ATMOSPHERIC DISPERSION

Dispersion is the process of spreading out the emission over a large area and thus reducing the concentrations of specific pollutants. The plume spreads both horizontally and vertically. If it is a gaseous plume, the motion of the molecules in the plume follows the laws of gaseous diffusion.

Rigorous derivation of the most commonly used model for the dispersion of gaseous air pollutants is beyond the scope of this text. The principles on which this model is based are:

- The predominant force in pollution transport is the wind; pollutants move predominantly downwind.
- The greatest concentration of pollutant molecules is along the plume centerline.
- Molecules diffuse spontaneously from regions of higher concentration to regions of lower concentration.
- The pollutant is emitted continuously, and the emission and dispersion process is steady state, i.e., $dC/dt = 0$.

With reference to Figure 19–15, for a source of pollutant emission the ambient pollutant concentration downwind from the source is given by

$$C(x, y, z) = Q / \pi \bar{u} \sigma_y \sigma_z \qquad \text{if } y = z = 0$$

where $C(x, y, z)$ = ambient concentration at x distance downwind, y distance off the plume centerline, and at elevation z
   $\bar{u}$ = average wind speed
   σ_y, σ_z = standard deviation of the plume in the lateral (y) and vertical (z) direction
   Q = rate of pollutant emission

If Q is measured in grams per second, the wind speed in meters per second, and the standard deviations in meters, the units of the concentration will be grams per cubic meter.

The standard deviations σ_y and σ_z are measures of the plume spread in the lateral and vertical directions. They depend both on atmospheric stability and on distance from the source. Stability is classified in categories A through F, called stability classes. Table 19–1 provides a key to the stability classes and the wind speeds that are consistent with each. In terms of lapse rates, classes A and B are associated with superadiabatic, unstable meteorological conditions; class C, with near neutral stability; class D, with subadiabatic, slightly stable conditions; and classes E and F, with weak and strong inversions, respectively. We should note that urban, populated regions rarely achieve greater stability

Table 19–1. Atmospheric Stability Key for Figure 19–15

| Surface Wind Speed (at 10 m) (m/sec) | Day[a] Incoming Solar Radiation (Sunshine) | | | Night[a] | |
	Strong	Moderate	Slight	Thinly Overcast or 4/8 Low Cloud	3/8 Cloud
<2	A	A–B	B		
2–3	A–B	B	C	E	F
3–5	B	B–C	C	D	E
5–6	C	C–D	D	D	D
>6	C	D	D	D	D

[a]The neutral category, D, should be assumed for overcast conditions during day or night.

than class D, because of the heat island effect. Stability classes E and F are generally found in rural or unpopulated areas.

Values for the horizontal and vertical standard deviations are given in Figure 19–15.

Example 19.2

An oil pipe leak results in emission of 100 gm/hr of H_2S. When the wind speed is 3.0 m/sec and under stability class C, what will be the concentration of H_2S 1.5 km directly downwind from the leak?

From Figure 19–16, at x = 1,500 meters,

$$\sigma_y = 160 \text{ m and } \sigma_z = 80 \text{ m}$$

$$Q = 100 \text{ gm/hr} = 2.8 \text{ gm/sec}$$

$$C(1500, 0, 0) = (2.8 \text{ gm/sec})/\pi(3.0 \text{ m/sec})(160 \text{ m})(80 \text{ m})$$

$$= 2.3 \times 10^{-5} \text{ gm/m}^3 \text{ or } 23 \text{ } \mu g/m^3$$

The plume cross section, in both the lateral and vertical dimensions, may be represented by a Gaussian probability curve. Moreover, the pollution concentration off the plume centerline may be expressed as

$$C(x, y, z) = (Q/\pi \bar{u} \sigma_y \sigma_z) \exp\{-\tfrac{1}{2}[(y/\sigma_y)^2]\} \quad \text{if } z = 0$$

where y = the perpendicular distance from the plume centerline

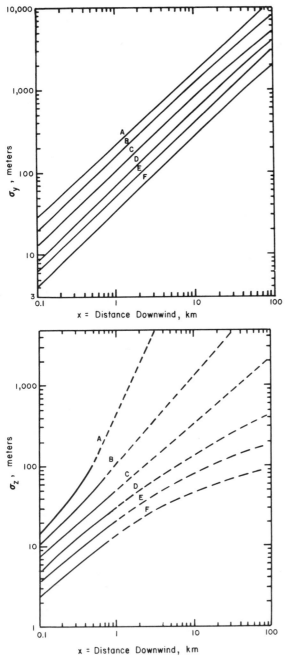

Figure 19–15. Dispersion coefficients.

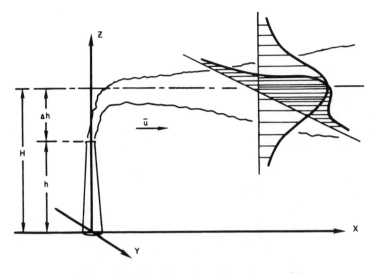

Figure 19–16.   Gaussian dispersion model.

Emission from stacks occurs well above ground level. Consider the geometric arrangement of source, wind, and plume in Figure 19–16. We can construct a Cartesian coordinate system with the bottom of the stack (the emission source) at the origin and the wind direction along the x-axis. Vertical plume dispersion is along the y-axis, with the Gaussian cross sections as shown in Figure 19–16. Since stack gases are generally emitted at temperatures higher than ambient, the plume travels vertically for some distance above the stack before beginning to travel downwind. The sum of this vertical travel distance and the geometric height of the stack is H, the effective stack height. The source of the pollutant plume may thus be placed at

$$z = H$$

and the concentration equation for this elevated source is written

$$C(x, y, z) = (Q/2\pi\bar{u}\sigma_y\sigma_z) \exp\{-\tfrac{1}{2}[(y/\sigma_y)^2]\}$$
$$\times \{\exp(-\tfrac{1}{2}[(z-H)^2/\sigma_z^2]) + \exp(-\tfrac{1}{2}[(z+H)^2/\sigma_z^2])\}$$

We usually measure and model ambient pollutant concentration at ground level (z = 0), so that the concentration equation reduces to

$$C(x, y, z) = (Q/\pi\sigma_y\sigma_z\bar{u}) \exp\{-\tfrac{1}{2}[(y/\sigma_y)^2 + (H/\sigma_z)^2]\} \quad \text{if } z = 0$$

This equation takes into account the reflection of gaseous pollutants from the surface of the ground.

Example 19.3

Given a sunny summer afternoon with average wind, $\bar{u} = 4$ m/sec, emission, $Q = 0.01$ kg/sec, and the effective stack height 20 m, find the ground level concentration at 200 meters from the stack.

Using the above equation, and from Figure 19–16 finding that at 200 meters, $\sigma_y = 36$ and $\sigma_z = 20$ for an unstable superadiabatic strong solar radiation (Table 19–1), the atmospheric conditions are type B, and noting that maximum concentrations occur on the plume centerline, at $y = 0$

$$C(200, 0, 0) = \frac{Q}{2\pi\bar{u}\sigma_y\sigma_z} (\exp\{(-\tfrac{1}{2}[(y/\sigma_y)^2]\})$$

$$\times (\exp(-\tfrac{1}{2}[(z - H)^2/\sigma_z^2]) + \exp(1 - \tfrac{1}{2}[(z + H)^2/\sigma_z^2]))$$

$$= \frac{0.01}{2(3.14)(4)(36)(20)} (\exp(-\tfrac{1}{2}[(0/36)^2]))$$

$$\times (\exp\{(-\tfrac{1}{2}[(0 - 20)^2/20^2])\} + \exp\{-\tfrac{1}{2}[(0 + 20)^2]\})$$

$$= (5.53 \times 10^{-7})(1)(e^{-\frac{1}{2}} + e^{-\frac{1}{2}})$$

$$= (5.53 \times 10^{-7})(0.6 + 0.6) = 6.64 \times 10^{-7} \text{ kg/m}^3$$

$$= 664 \ \mu\text{g/m}^3$$

Among the many assumptions made in this model is the accurate estimation of a uniform, steady, and unidirectional wind. None of these are realistic, of course. Wind direction shifts, and wind speed varies with time (unsteady) as well as with elevation (nonuniform). The latter may be approximated by assuming a parabolic wind velocity profile and calculating the velocity $\bar{u}_1$ at any given elevation h_1 as a function of a measured velocity $\bar{u}_2$ at another elevation, h_2 as

$$\bar{u}_1 = \bar{u}_2 \left(\frac{h_1}{h_2}\right)^n$$

where $n = 0.14$ for unstable and $n = 0.33$ for stable atmospheric conditions.

The effective stack height is the height above ground at which the plume begins to travel downwind: the effective release point of the pollutant and origin of its dispersion. A number of empirical models exist for calculating the plume rise Δh—the height above the stack to which the plume rises. Three equations have been developed by Moses and Carson, for different stability conditions, which give a reasonably accurate estimate of plume rise. These are:

For superadiabatic conditions:

$$\Delta h = 3.47V_s d/u + 5.15\,(Q_h)^{0.5}/u$$

For neutral stability:

$$\Delta h = 0.35V_s d/u + 2.64\,(Q_h)^{0.5}/u$$

For subadiabatic conditions:

$$\Delta h = -1.04V_s d/u + 2.24\,(Q_h)^{0.5}/u$$

where V_s = stack gas exit speed, m/sec
  d = stack diameter, m
  Q_h = heat emission rate from the stack, kJ/sec

As before, length is in meters and time is in seconds, and the heat emission rate is measured in kilojoules per second.

Example 19.4

A power plant has a 2-m-diameter stack and emits gases with a velocity of 15 m/sec and a heat emission rate of 4,800 kJ/sec. The wind speed is 5 m/sec. Stability is neutral. Estimate the plume rise. If the stack has a geometric height of 40 m, what is the effective stack height?

$$\Delta h = 0.35V_s d/u + 2.64(Q_h)^{0.5}/u$$

$$\Delta h = 0.35(15)(2)/5 + 2.64(4,800)^{0.5}/5$$

$$\Delta h = 38.7 \text{ m}$$

$$H = h_g + \Delta h$$

$$H = 40 \text{ m} + 38.7 \text{ m} = 78.7 \text{ m}$$

The accuracy of plume rise and dispersion analysis is not very good. Uncalibrated models predict ambient concentrations to within an order of magnitude at best. To ensure reasonable validity and reliability, a modeler must calibrate the dispersion model with measured ground-level concentrations.

CLEANSING THE ATMOSPHERE

Since we obviously have not yet all suffocated, and yet prodigious amounts of pollutants have been thrown into the air over the millions of years, there must exist a series of processes by which air is cleansed. These include the effect of gravity, contact with the earth's surface, and removal by precipitation.

Gravity

Particulates, if of sufficient size, are removed by simple settling out under the influence of gravity. Unfortunately, the settling velocities of common particulates are very small. For example, a 1-μm particle will have a settling velocity of about 1 cm/sec, in ideal quiescent conditions. Practically, owing to turbulence in the atmosphere, particles smaller than 20 μm will seldom settle out by gravity. Gases are removed by gravity only if they are adsorbed onto particulates. Sulfur dioxide, for example, is readily adsorbed and is thus partially removed by gravity.

Particulate matter is not dispersed in the air in precisely the same way as gaseous pollutants are. The dispersion equation must be modified by consideration of the settling velocity of small particles.

Particles larger than a millimeter in diameter fall out of the air too rapidly to be dispersed over any great distance. The particles of concern as air pollutants, which are small enough to stay in the air for appreciable periods of time but large enough to be dispersed differently from gases, are generally between 1 and 100 μm in diameter. The settling velocity of particles in this size range follows Stokes' Law:

$$V_t = gd^2\rho/18\ \mu$$

where V_t = settling or terminal velocity
   g = acceleration due to gravity
   d = particle diameter
   ρ = particle density
   μ = viscosity of air

The settling velocity thus becomes a modification to the vertical dispersion in the Gaussian dispersion equation. For small particles, the dispersion equation that gives the ground-level concentration is written

$$C(x, y, z) = (Q/\pi\sigma_y\sigma_z\bar{u}) \exp\{-\tfrac{1}{2}[(y/\sigma_y)^2 + (H - V_t x/\bar{u})^2/\sigma_z^2]\}$$

The factor of 1/2 in the first term arises because falling particles are not reflected at the ground surface.

Along the plume centerline, this reduces to

$$C(x, y, z) = (Q/\pi\sigma_y\sigma_z\bar{u}) \exp\{-\tfrac{1}{2}(H - V_t x/\bar{u})^2/\sigma_z^2\}$$

It is sometimes more meaningful to determine the rate at which particulate matter is being deposited on the ground, instead of the concentration of particles in the air at ground level. This deposition rate is related to ambient concentration as follows:

$$\text{deposition rate } w = \text{mass transport rate/area}$$

or

$$w = (\text{volume rate})(\text{concentration})/\text{area}$$

or

$$w = V_t C$$

Along the plume centerline

$$w = (QV_t/\pi\sigma_y\sigma_z)\exp[-[H-(V_tx/u)]^2/2\sigma_z^2]$$

Example 19.5

Using the data of Example 19.3, assume that the emission consists of particles 10 μm in diameter and having a density of 1 gm/cm^3 (10^6 gm/m^3). The viscosity of air is 0.0185 g/m-sec at 25°C. The settling velocity

$$V_t = (9.8 \text{ m/sec}^2)(10^{-10} \text{ m}^2)(10^6 \text{ g/m}^3)/18(0.0185 \text{ g/m-s})$$

$$V_t = 2.9 \times 10^{-3} \text{ m/sec} = 2.9 \text{ cm/sec}$$

From Examples 19.3

$$Q = 0.01 \text{ kg/sec} = 10 \text{ gm/sec}$$

and

$$C(200, 0, 0) = [10/\pi(4)(36)(20)] \times \exp\{-\tfrac{1}{2}(20 - [(0.0029)(200)/4]^2/(36)^2)\}$$

$$C(200, 0, 0) = 408 \ \mu\text{g/m}^3$$

As expected, this result is smaller than the concentration obtained in Example 19.3. The deposition rate in this case is

$$w = (0.0029 \text{ m/sec})(408 \times 10^{-6} \text{ g/m}^3)$$

$$w = 1.18 \ \mu\text{g/m}^2\text{-sec}$$

Surface Sink

Many of the gases are absorbed by the earth's surface, including stone, vegetation, and other materials. Some gases such as SO_2 are readily dissolved in surface waters.

Precipitation

The third major removal mechanism is by precipitation. Two types of removal occur, the first being an "in-cloud" process called *rainout*, in which submicron particles become nuclei for the formation of rain droplets that will grow and eventually fall as precipitation. The second mechanism is called *washout* and is a "below-cloud" process in which the rain falls through the air pollutants, and the pollutants are impinged or dissolved in the droplets and then carried to earth.

The relative importance of these removal mechanisms was illustrated by a study of SO_2 emissions in Great Britain, where the surface sink accounted for 60 percent of the SO_2, 15 percent was removed by precipitation, and 25 percent left Great Britain, heading northwest toward Norway and Sweden.

CONCLUSION

Air pollution episodes are the results of high emissions and a combination of meteorological factors. Some of these factors are:

1. little horizontal wind movement
2. stable atmospheric conditions, resulting in very limited vertical air movement
3. fog, which promotes the formation of secondary pollutants and hinders the sun from warming the ground and breaking inversions
4. high-pressure areas resulting in downward vertical air movement and absence of rain for washing the atmosphere

It would seem reasonable, therefore, that episodes can be predicted on the basis of meteorological data, provided the potential exists. The EPA, in cooperation with the Weather Bureau, has indeed established procedures for evaluating meteorological data to provide early warning for impending episode conditions and has developed emergency plans (including shutting down industries) should the conditions warrant.

PROBLEMS

19.1   Given the following temperature soundings:

Elevation (m)	Temperature (°C)
0	20
50	15
100	10
150	15
200	20
250	15
300	20

what type of plume would you expect if the exit temperature of the plume were 15°C and the smoke stack were

a. 50 m tall?
b. 150 m tall?
c. 250 m tall?

19.2 Consider a prevailing lapse rate that has these temperatures: ground = 21°C, 500 m = 20°C, 600 m = 19°C, 1,000 m = 20°C. If we released a parcel of air at 500 m and at 20°C, would it tend to sink, rise, or remain where it was? If a stack is 500 m tall, what type of plume would you expect to see?

19.3 Draw a map with X and Y coordinates (X horizontal, Y vertical) and place on the map the following:

Industrial plant A at   X = 3, Y = 3
Industrial plant B at   X = 3, Y = 1
Industrial plant C at   X = 8, Y = 8
Air sampling station at X = 5, Y = 5

The data at the air sampling station are:

Day	Wind Direction	Particulates ($\mu g/m^3$)	SO_2 ($\mu g/m^3$)
1	N	80	80
2	NE	120	20
3	NW	30	30
4	N	90	40
5	NE	130	20
6	SW	20	180
7	S	30	100
8	SW	40	200
9	E	100	60
10	W	10	100

Draw pollution roses to show which plant is guilty of the air pollution.

19.4 If your job were to continuously analyze meteorological data and watch for conditions that might lead to the occurrence of an "episode," what specific conditions would you be looking for? (That is, what are the meteorological criteria for the formation of an episode?)

19.5 Take a photograph of a smoke plume and describe: (a) the time of day, (b) the climatological conditions, (c) the atmospheric stability, and (d) the type of plume. You may submit a photograph or a 2 × 2 slide. Put your name on the picture and staple or tape it to a sheet of paper.

19.6   A power plant burns 1,000 tons of coal per day, 2 percent of which is sulfur, and all of this is emitted from the 100-m stack. For a wind speed of 10 m/sec, calculate: (a) the maximum ground-level concentration SO_2 10 km downwind from the plant, (b) the maximum ground-level concentration and the point at which this occurs for stability categories A, C, and F.

19.7   A power plant emits 20 metric tons of particulates per hour out of a 100-m (effective height) stack. At 50 m downwind at a wind velocity of 2 m/sec, at plume centerline, what is the ground-level concentration of particulates if we use stability category E? Assume that the particles have a density of 1.5 gm/cm^5 and diameter of 20 μm.

19.8   Given a wind velocity of 2 m/sec, calculate the maximum expected SO_2 concentration at ground level in a town 10 km downwind from a power plant that burns 1,000 metric tons of coal per day (sulfur content of 1 percent). The stack is 100 m high.

19.9   Consider the following atmospheric temperature soundings:

Elevation (ft)	Temperature (°F)
0	70.0
200	68.0
400	66.0
600	72.0
800	70.0
1,000	68.0

a. Indicate below the type of lapse rate involved at
   0 to   400 ft _____
   400 to   600 ft _____
   600 to 1,000 ft _____
b. Indicate below the plume type if the stack were
   300 ft tall _____
   500 ft tall _____
   700 ft tall _____
c. What would be the ground-level concentration of the pollutant (negligible, moderate, high) if the stack were
   300 ft tall _____
   500 ft tall _____
   700 ft tall _____

19.10   The odor threshold of H_2S is about 0.7 μg/m^3. If an industry emits 0.08 g/sec of H_2S out of a 40-m stack during an overcast night with a wind speed of 3 m/sec, estimate the area (in terms of x and y coordinates) where H_2S would be detected. (Do this on a computer.)

19.11   If the maximum ground-level concentration of SO_2 is to be limited to 50 $\mu g/m^3$ on a clear day with a wind speed of 3 m/sec and an emission rate of 100 g/sec, what is the required effective stack height? (Do this on a computer.)

19.12   A copper smelter processes 1,000 metric tons per day of ore, principally $CuFeS_2$, producing SO_2 as a pollutant. The stack height is 280 m with an exit velocity of 15 m/sec through a 2-m-diameter stack and temperature of 250°C. The wind speed at a height of 5 m is 4 m/sec. During an overcast night (10°C):

    a. Determine the maximum ground-level concentration.

    b. Estimate the percent SO_2 removal necessary to reduce that maximum to below EPA primary ambient air quality standards, 24-hr exposure (Chapter 22).

(Do this on a computer.)

19.13   The concentration of H_2S at a location 200 m downward from an abandoned oil well is 3.2 $\mu g/m^3$. What is the emission rate on a partially overcast afternoon? (Wind speed is 2.5 m/sec.)

19.14   A power plant has a 500-m stack with an outlet diameter of 10 m. The plant burns 15,500 metric tons of coal per day, with a sulfur content of 3.5 percent. The flue gas exit temperature is 200°C, and the weather is overcast on a winter afternoon, 0°C. The plant used 8 kg of air per kg of coal. Wind measurements at 3 m from the ground show an average velocity of 4 m/sec, from due north. Plot a graph of ground-level SO_2 with distance downwind. At the 10-km point, plot the SO_2 at a right angle to the wind, using perspective (see Figure 19–15) in both y and z directions.

19.15   What is the expected SO_2 concentration (Problem 19.14) if the sampling site is SE of the plant at a distance of 8 km?

19.16   A coal-fired power plant emits SO_2 at a rate of 1,500 g/sec. What is the centerline concentration of SO_2 3 km downwind on a clear summer afternoon if the wind speed measured at 20 m is 4 m/sec? The effective stack height is 110 m.

19.17   A power plant burns coal with a 2 percent sulfur content and has a stack with a height of 70 m and a diameter of 1.5 m. The exit velocity and temperature are 15 m/sec and 340°C, respectively. The atmospheric conditions are 28°C and a barometric pressure of 90 kPa, and the weather is a clear summer afternoon with a wind speed of 2.8 m/sec at 10 m. At a point 2.5 km downwind and 0.3 km off plume centerline, the ground-level concentration of SO_2 was found to be 143 $\mu g/m^3$. How much coal does the plant burn?

19.18   The power plant in Problem 19.17 burns coal with a 10 percent ash, and 20 percent of the ash is emitted into the air. The ash has an average density of 1.0 gm/cm^3 and an average diameter of 2 μm. Calculate the deposition rate at the point indicated in Problem 19.17.

LIST OF SYMBOLS

C = concentration of pollutant, kg/m^3

C_p = specific heat, kJ/kg-°K

d = stack exit diameter, m

H = effective height of the stack, m

h = stack elevation, m

m = mass flow rate, kg/sec

Q = emission rate, kg/sec

Q_h = heat emission rate, kJ/sec

T = temperature

$\bar{u}$ = average wind speed, m/sec

v = stack exit velocity, m/sec

σ_y = standard deviation, y direction, m

σ_z = standard deviation, z direction, m

Chapter 20

Measurement of Air Quality

It was previously observed that air quality measurement, like water quality measurement, is complicated by the lack of knowledge as to what is "clean" and by the difficulty in defining "quality."

"Pure air" is ordinarily defined as containing only the naturally occurring gases, but "pure air" doesn't exist in nature. Pollen, dust, fog, etc., are all contaminants, although they occur without any assistance from industrialized man. No attempt is usually made to differentiate between natural and man-made pollutants, and air quality measurements are designed to measure all types of contaminants.

The measurements of air quality generally fall into three classes:

1. *Measurement of emissions.* This is called *stack sampling* when a stationary source is analyzed. A hole is punched into the stack and samples are drawn out for on-the-spot analyses. Sampling moving sources is more difficult.
2. *Meteorological measurements.* The measurement of meteorological factors is necessary if we are to know how and why the pollutants travel from the source to the recipient. Some of these are discussed in the previous chapter.
3. *Ambient air quality.* The quality of ambient air is of course of major concern. Almost all evidence of health effects is based on these measurements. Monitoring air quality may also provide data for recognizing episodes and thus may provide some warning of impending health problems.

The development of air quality instrumentation may be divided into three distinct phases or generations. The original, or first-generation, devices were developed by the air pollution control engineers and scientists and were needed for obtaining quantitative information on air quality. There was little precedent for measuring minute quantities of various gases in the atmosphere, nor was there much money for development of the necessary instrumentation. Accordingly, first-generation devices were simple, inexpensive, and required no external power for operation. They were also inconvenient, very slow, and of questionable accuracy.

Second-generation measurements evolved when more accurate and rapid data were required. These devices use a power source (usually electrical power to drive an air pump) and thus can sample more air in a shorter time. The techniques for gas measurement generally involved wet chemistry, in that the gas was either dissolved into or reacted with a collecting fluid.

Third-generation devices differ from their predecessors in that they provide a continuous readout. It is thus possible to get a continuous graph of levels of various pollutants with the measurement occurring almost instantaneously.

Examples of all three types of instrumentation are discussed below, for both particulates and gaseous pollutants.

MEASUREMENT OF PARTICULATES

First-generation devices for measuring particulates involved the determination of how much dust settles to the earth. Such *dustfall* measurements are by far the simplest means of evaluating air quality. Dust is collected either in open buckets or on sticky tapes wrapped around jars. The sampling period for dustfall buckets is generally 30 days; sticky tapes are usually read in 7 days and give a qualitative indication of the direction of the particulate pollution. The dustfall jars are dried to remove moisture and weighed to determine the amount of dust in the jar, which is reported commonly as tons (2,000 lb) of dust settled per square mile in 30 days.*

The measurement of particulates by dustfall jars is subject to many problems (not the least of which is uncooperative pigeons) and only one data point per month is obtained. The second-generation particulate sampling device, the *high-volume sampler* (or hi-vol) is a substantial improvement since its sampling time is normally only 24 hours.

The high-volume sampler, the workhorse of particulate sampling (Figure 20–1), operates much like a vacuum cleaner by simply forcing over 2,000 m^3 (70,000 ft^3) of air through the filter in 24 hours. The analysis is gravimetric; the filter is weighed before and after, and the difference is the particulates collected.

The air flow is measured by a small flow meter, usually calibrated in cubic feet of air per minute. Because the filter gets dirty during the 24 hours of operation, less air goes through the filter during the latter part of the test than in the beginning, and the air flow must therefore be measured at both the start and end of the test period and the values averaged.

---

* A more realistic method of reporting dustfall data might have been as grams of dust per square centimeter per 30 days. The *tons* is, however, much more impressive, and this consideration no doubt lead to the use of the more awkward tons per square mile per month.

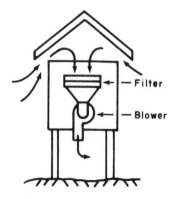

Figure 20–1.   High-volume sampler.

Example 20.1

A clean filter is found to weigh 10.00 g. After 24 hours in a hi-vol, the filter plus dust weighs 10.10 g. The air flow at the start and end of the test was 60 and 40 ft³/min, respectively. What is the particulate concentration?

$$\text{Weight of the particulates (dust)} = (10.10 - 10.00) \text{ g} \times 10^6 \ \mu\text{g/g}$$

$$= 0.1 \times 10^6 \ \mu\text{g}$$

$$\text{Average air flow} = \frac{60 + 40}{2} = 50 \text{ ft}^3/\text{min}$$

Total air through the filter $= 50 \text{ ft}^3/\text{min} \times 60 \text{ min/hr} \times 24 \text{ hr/day} \times 1 \text{ day}$

$$= 72,000 \text{ ft}^3$$

$$= 72,000 \text{ ft}^3 \times 28.3 \times 10^{-3} \text{ m}^3/\text{ft}^3$$

$$= 2,038 \text{ m}^3$$

$$\text{Total suspended particulates} = \frac{0.1 \times 10^6 \ \mu\text{g}}{2,038 \text{ m}^3}$$

$$= 49 \ \mu\text{g}/\text{m}^3$$

The particulate concentration thus measured is often referred to as *total suspended particulates* (TSP) to differentiate it from other measurements of particulates.

Another widely used measure of particulates in the environmental health area is of *respirable particulates*, or those particulates that would be respired into lungs. These are generally defined as being less than 0.3 μm in size, and the measurements are done with stacked filters. The first filter removes only particulates >0.3 μm and the second filter, having smaller spaces, removes the small respirable particulates.

Another device for collecting airborne particulates is the *cascade impactor* (Figure 20–2). The impactor consists of four tubes, each with a progressively smaller opening, thus forcing higher and higher velocities. A particle entering the device may be small enough to follow the streamline of flow without hitting the microscope slide. At the next nozzle, however, the velocity may be suffic-

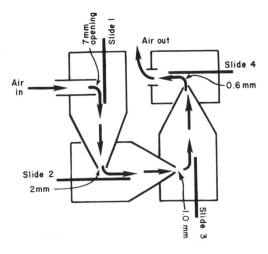

Figure 20–2. Cascade impactor.

iently high to prevent the particle from quite negotiating the turn, and it will impinge on the slide.

Interestingly, no third-generation particulate-measuring devices have yet been developed and accepted. The difficulty, of course, is that the measurement must be gravimetric, and it is difficult to construct a device that continuously weighs minute quantities of dust.

Since 1977, measurements of decrease in visibility have become important in air quality assessment. Fine particles interfere with visibility by scattering light. This scattered light can be measured with a *nephelometer*, an instrument that measures emitted or scattered light at 90° to incident light. The air nephelometer can be calibrated directly in units of percent visibility decrease.

MEASUREMENT OF GASES

As is the case with particulates, the first-generation methods for measuring the concentration of gaseous pollutants are simple yet ingenious gadgets, using no power, but providing a quantitative measure of the atmospheric concentration of specific gases. Two of the most interesting (and important) gases for which first-generation devices were developed are ozone (O_3) and sulfur dioxide (SO_2).

The original ozone-measuring technique takes advantage of the fact that ozone attacks and cracks rubber. In this test, specially prepared and weighed strips of rubber are hung outside, and the cracks formed in the rubber strip are measured and related to ozone levels in the atmosphere.

Sulfur dioxide is measured by impregnating filter papers with chemicals that react with SO_2 and change color. For example, lead peroxide reacts as

$$PbO_2 + SO_2 \rightarrow PbSO_4$$

forming a dark lead sulfate. The extent of this reaction is estimated by the dark areas on the filter paper.

These devices are obviously slow, inaccurate, and subject to many interferences. Second-generation techniques are by contrast much faster (hours instead of days) and more accurate. These techniques almost all involve the use of a *bubbler*, shown in Figure 20–3. The gas is literally bubbled through the liquid, which either reacts chemically with the gas of interest or into which the gas is dissolved. Wet chemical techniques are then used to measure the concentration of the gas.

A simple (but now seldom used) bubbler technique of measuring SO_2 is to bubble air through hydrogen peroxide, so that the following reaction occurs:

$$SO_2 + H_2O \rightarrow H_2SO_4$$

The amount of sulfuric acid formed can be determined by titrating the solution with a base of known strength.

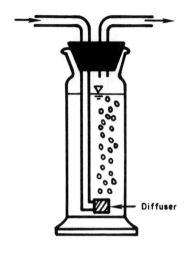

Figure 20–3. A typical bubbler used for measurement of gaseous air pollutants.

One of the better methods of measuring SO_2 is the colorimetric pararosaniline method, in which SO_2 is bubbled into a liquid containing tetrachloromercurate (TCM). The SO_2 and TCM combine to form a stable complex. Pararosaniline is then added to this complex, with which it forms a colored solution. The amount of color is proportional to the SO_2 in the solution, and the color is measured with a spectrophotometer at a wavelength of 560 nm. (See Chapter 4 for ammonia measurement—another example of a colorimetric technique.) Figure 20–4 graphically illustrates the pararosaniline method for SO_2 measurement.

Most bubblers are not 100 percent efficient, for not all of the gas bubbled in will be absorbed by the liquid, and some will escape. Obviously, this creates problems since it is usually necessary to have quantitative determinations. Most

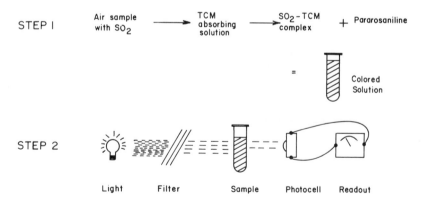

Figure 20–4. Schematic of the pararosaniline method for measuring SO_2.

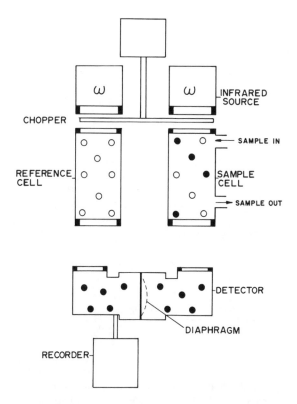

Figure 20–5. Nondispersive infrared spectrometry for CO measurement.

bubblers are thus tested to establish their efficiencies, and these values are used as contrasts by which the measured values are multiplied to obtain the actual concentrations.

Hundreds of different techniques have been used in third-generation gaseous measurement. One widely used device is the *nondispersive infrared analyzer*, used for carbon monoxide measurement. This technique relies on the fact that CO absorbs infrared radiation of a specific frequency. Schematically, as shown in Figure 20–5, the detector consists of two chambers, with the air sample pumped into one and the other containing an inert reference gas (e.g., nitrogen). The two infrared lamps shine through the cells, and the CO absorbs a part of that energy, directly proportional to the amount of CO in the chamber. After passing through the two cells, the radiant energy is absorbed by the gas in two detector cells, each one containing CO. This absorption causes the gas in the detector cells to be heated, but the detector under the reference cell that receives more energy is heated more. The hotter gas expands, moving a diaphragm separating the two detector cells. This movement is detected electrically, and the signal is read out on a recorder, thus giving a continuous trace of CO in the sample.

Table 20–1.   Standard EPA Reference Methods for Air Quality Measurements

Pollutant	Reference Method	Comments
Particulates (TSP)	High-volume sampler	Note that this is a second-generation device, giving 24-hour readings.
Sulfur dioxide	Pararosaniline	
Carbon monoxide	Nondispersive infrared spectrometry	
Nitrogen dioxide	Chemiluminescence	This method was adopted after the original reference method was judged unreliable (and years of NO_2 monitoring data became worthless).
Photochemical oxidants (O_3)	Chemiluminescence	This method measures only O_3.
Hydrocarbons (nonmethane)	Flame ionization	Methane is not measured because it is a nonreactive hydrocarbon, naturally occurring.

REFERENCE METHODS

Because of the great number of methods available for measuring air pollutants, the U.S. Environmental Protection Agency (EPA) has chosen a series of reference methods (Table 20–1). These are not necessarily absolutely accurate, but they are judged the best available, and results from other methods are to be compared with these. The existence of reference methods is especially important when compliance with air quality standards is in question.

GRAB SAMPLES

Often it is necessary to obtain a sample of a gas for future analysis in the laboratory. Whereas in water pollution work the collection of samples does not present a serious problem, in air quality measurements it is a real challenge.

One method is the use of evacuated containers, drawing a vacuum in a glass jar and releasing it when the air is to be sampled. Unfortunately, it is not possible to evacuate the jar 100 percent, and contamination is always a possibility.

Bubbling gas into a container with water in it and displacing the water is an effective method only if the gases of interest are not at all soluble in water.

Plastic and aluminum bags have been used extensively. Usually the gas is pumped in and allowed to escape through a hole. By thus displacing two or three volumes of the bag, contamination problems are avoided.

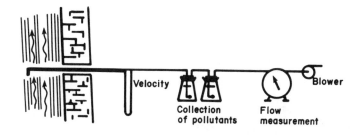

Figure 20–6. A stack sampling train.

STACK SAMPLING

Stack sampling is an art worthy of individual attention. As the name implies, this involves the sampling of gas from a smokestack and is necessary for evaluating compliance with emission standards and determining efficiencies of air pollution control equipment.

The most serious problem with stack sampling is the risk of obtaining an unrepresentative sample. Accordingly, a thorough survey is usually made of the flow, temperature, and pollutant concentration in a stack. It is also usually desirable to obtain samples from a series of locations within a stack to be assured of accurate measurements. A train of instruments (Figure 20–6) is often utilized for stack sampling so that a number of measurements may be determined at each positioning of the intake nozzle.

SMOKE AND OPACITY

Air pollution has historically been associated with smoke—the darker the smoke, the more pollution. We now know that this isn't necessarily true, but many regulations (e.g., for municipal incinerators) are still written on the basis of smoke density. The opacity of a smoke plume is still the only method for enforcing air quality standards that may be used without the emitter's knowledge. Most citations for air pollution violations are written for opacity violations.

The density of black or gray smoke is measured on the Ringelmann scale. One end of the scale, Ringelmann 0, is complete transparency (no visible plume); the other end, Ringelmann 5, is a completely opaque black plume. A typical opacity standard for a stationary source is Ringelmann 1 (20 percent opacity), with allowances for very short periods of Ringelmann 2 (40 percent opacity). Ringelmann 1 is a barely visible plume. The Ringelmann test was, at one time, conducted by comparing the blackness of a card (such as those shown in Figure 20–7) with the blackness of the observed plume. Modern practice involves the training of enforcement agents (in a "smoke school") to recognize

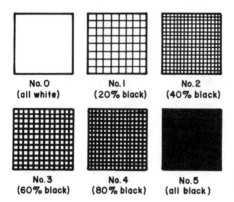

| No. 0 (all white) | No. 1 (20% black) | No. 2 (40% black) |
| No. 3 (60% black) | No. 4 (80% black) | No. 5 (all black) |

Figure 20–7. Ringlemann scale for measuring the opacity of smoke.

Ringelmann opacities by repeated observation of smoke of predetermined opacity. The opacity of white smoke is reported as "percent opacity" rather than by Ringelmann number.

Opacity may also be measured continuously by installing a photometer in the stack breach and calibrating the emitted smoke by the Ringelmann or opacity scale.

CONCLUSION

As with water pollution, the analytical tests of air quality can be only as good as the samples or sampling techniques used. In addition, the prevailing analytical techniques leave a great deal to be desired in both precision and accuracy. It is important to remember, therefore, that most measurements of environmental quality, and especially air quality, are at best reasonable estimates and should not be believed to the fourth decimal point.

PROBLEMS

20.1 An empty 6-in.-diameter dustfall jar weighed 1,560 g. After sitting out a prescribed amount of time, it weighed 1,570 g. Report the dustfall in the usual manner.

20.2 A hi-vol clean filter weighs 20.0 g, and the dirty filter weighs 20.5 g. The initial and final air flows were 70 and 50 ft^3/min, respectively. What volume of air went through the filter in 24 hours? What was the level of particles in the atmosphere? How does this compare with the National Air Quality Standards? (See Chapter 22.)

20.3 A high-volume sampler draws air in at an average rate of 70 ft^3/min. If the particulate reading is 200 μg/m^3, what was the weight of the dust on the filter?

Chapter 21
Air Pollution Control

It is not difficult to see how air has become the ubiquitous wastebasket. For ages people have been dumping wastes into the atmosphere, and these pollutants have "disappeared" with the wind. It is a difficult transition, therefore, suddenly to force people to limit emissions, and even more difficult to get them to pay for it. Nevertheless, our industrialized society must now prevent the emission of what would, a few years ago, have been wasted into the atmosphere.

The control of emissions may be realized in a number of ways. Five separate possibilities for control are pictured in Figure 21–1. Dispersion is covered in Chapter 19, and the four remaining control points are discussed individually in this chapter.

SOURCE CORRECTION

Often the easiest solution to an air pollution problem is to stop or change the guilty process. Once the decision has been made that a product or process is necessary, the engineer must consider the possibility of controlling emissions by changing the process. For example, if automobiles are blamed for high lead levels in urban air, the most reasonable solution is elimination of the lead in gasoline. Similarly, some removal of sulfur from coal and oil is possible before the fuel is burned. In these cases, the source has been corrected and the problem solved.

In addition to a change of raw material, a modification of the process might also be used to achieve a desired result. For example, municipal refuse incinerators have been known to stink. The odors may often be readily controlled if the incinerators are operated at a high enough temperature to oxidize completely the organics that cause the odor. Provided this higher-temperature operation is possible, it is a very reasonable process change to obtain the desired air pollution control.

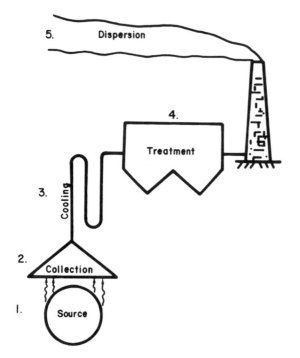

Figure 21–1. Points of possible air pollution control.

Strictly speaking, such measures as process change, raw material conversion, or equipment modification to meet emission standards are known as *controls*. In contrast, *abatement* is the term used for all devices and methods for decreasing the quantity of pollutant reaching the atmosphere, once it has already been emitted from the source. For the sake of simplicity, however, we refer to all of the procedures as controls.

COLLECTION OF POLLUTANTS

Often the most serious problem in air pollution controls is collection of the pollutants to provide treatment. Automobiles are notorious polluters, but only because their emissions cannot be readily collected. If we could channel the exhausts from automobiles to some central facilities, their treatment would be much more reasonable than controlling each individual car.

One success in collecting pollutants has been the recycling of blow-by gases in the internal combustion engine (see page 374). By reigniting these gases and emitting them through the car's exhaust system, the necessity of installing a separate treatment device for the car has been eliminated.

Air pollution control engineers have their toughest problems when the pollutants from an industry are not collected but are emitted from windows, doors, and cracks in the walls. It is not, therefore, always possible to solve a problem by simply installing some piece of control equipment. Often a complete overhaul of the air flow in the entire plant is required.

COOLING

The exhaust gases to be treated are sometimes too hot for the control equipment, and the gases must first be cooled. This can be done in three general ways: dilution, quenching, or heat-exchange coils (Figure 21–2). Dilution is acceptable only if the total amount of hot exhaust is small. Quenching has the added advantage of scrubbing out some of these gases and particulates, but this method may result in a dirty and hot liquid that must be disposed of. The cooling coils are probably the most widely used and are especially appropriate when heat can be conserved.

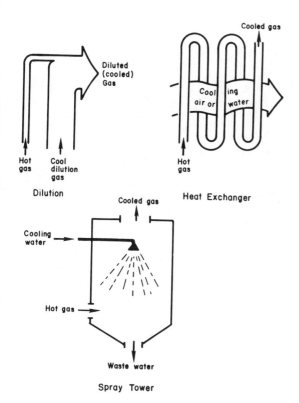

Figure 21–2. Cooling hot waste gases.

TREATMENT

Selection of the correct treatment device requires matching characteristics of the pollutant with features of the control device. It is important to realize that the sizes of air pollutants range many orders of magnitude, and it is therefore not reasonable to expect one device to be effective and efficient for all pollutants. In addition, the types of chemicals in emissions often will dictate the use of some devices. For example, a gas containing a high concentration of SO_3 could be cleaned by water sprays, but the resulting sulfuric acid might present serious corrosion problems.

The various air pollution control devices are conveniently divided into those applicable for controlling particulates and those used for controlling gaseous pollutants. The reason, of course, is the difference in the size. Gas molecules have diameters of about 0.0001 μm; particulates range from 0.1 μm and up. In any case, however, the efficiency, η, of a collection device is evaluated according to the following relationship (see Figure 21–3):

$$\eta = \frac{M_c}{M_a} \times 100$$

where M_c = amount of pollutant collected, kg/sec
M_a = amount of pollutant that was collectible, kg/sec

In other words, M_a is the percentage of the pollutant that could have been collected were the devices perfect. The terms M_c and M_a are in terms of mass per unit time.

If M_b is the pollutant that *escapes* capture, then

$$M_a = M_b + M_c$$

and efficiency could also be expressed as

$$\eta = \frac{M_c}{M_b + M_c} \times 100$$

If the concentration of the pollutants and the rate of air flow are known, the efficiency is of course expressed as

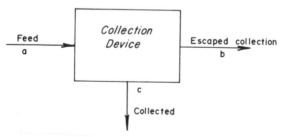

Figure 21–3.   Definition sketch of a collecting device.

$$\eta = \frac{Q_c C_c \times 10}{Q_a C_a}$$

where Q = air flow rate, m^3/sec
C = concentration of pollutant, kg/m^3

and the subscripts refer to inflow and captured pollutant (Figure 21–3). In situations in which the air flow rates are not known and only concentrations in the three streams are measured, the efficiency can be shown to be equal to

$$\eta = \frac{C_c(C_a - C_b) \times 100}{C_a(C_c - C_b)}$$

Settling Chambers

The simplest devices for controlling particulates are settling chambers consisting of nothing more than wide places in the exhaust flue where larger particles can settle out, usually with a baffle to slow the gas stream. Obviously, only very large particulates ($>100\ \mu\text{m}$) can be efficiently removed in settling chambers.

Cyclones

Possibly the most popular, economical and effective means of controlling particulates is the cyclone. Figure 21–4 shows a simple schematic and Figure 21–5 is a more detailed picture of a cyclone. The dirty air is blasted into a conical cylinder, but off the centerline. This creates a violent swirl within the cone, and the heavy solids migrate to the wall of the cylinder where they slow down owing to friction and exit at the bottom of the cone. The clean air is in the middle of the cylinder and exits out the top. Cyclones are widely used as precleaners to remove heavy material before further treatment.

Cyclones are sized on the basis of a *separation factor*, which is developed as follows. Consider a particle "settling" radially in a cyclone (from the centerline to the outside wall). The velocity of this movement, if we assume laminar flow (not a good assumption) and spherical particles (likewise) is

$$v_R = \frac{(\rho_s - \rho) r \omega^2 d^2}{18\ \mu}$$

where v_R = radial velocity, m/sec
ρ_s = density of particle, kg/m^3
ρ = density of air, kg/m^3
r = radial distance, m
ω = rotational velocity, rad/sec
d = diameter of particle, m
μ = viscosity of air, kg/m-sec

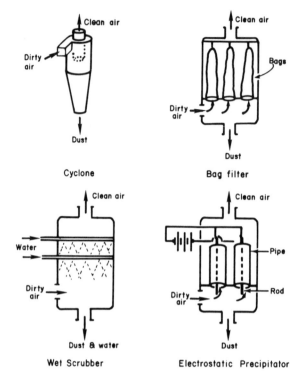

Figure 21–4. Four methods of controlling particulates from stationary sources.

Since $r\omega^2$, the centrifugal acceleration, equals v_{tan}^2/r, the tangential velocity squared divided by the radius,

$$v_R = \left[\frac{(\rho_s - \rho)d^2}{18\,\mu} \right] \cdot \frac{v_{tan}^2}{r}$$

The term in brackets is the gravitational velocity with the gravitational acceleration, g, missing. Hence

$$v_R = \left(\frac{v}{g} \right)\left(\frac{v_{tan}^2}{r} \right)$$

The separation factor, S, is defined as

$$S = \frac{v_R}{v} = \frac{v_{tan}^2}{rg}$$

A large separation factor requires high tangential velocities, thus small diameters and high velocities, leading to high pressure drops. A large-diameter cyclone, on the other hand, would have a low separation factor. The value of S may range from 5 to as high as 2,500. (Note: S is dimensionless.)

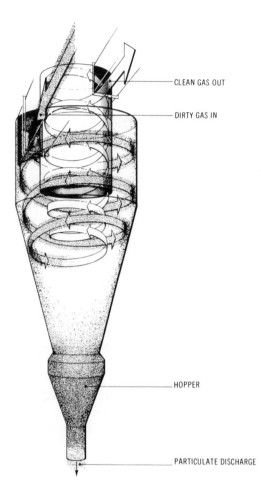

CLEAN GAS OUT

DIRTY GAS IN

HOPPER

PARTICULATE DISCHARGE

Figure 21–5.   Cyclone. [Courtesy of American Lung Association.]

The pressure drop through a cyclone can be calculated by an empirical relationship:

$$\Delta P = \frac{3{,}950 \, KQ^2 P\rho}{T}$$

where ΔP = pressure drop, m of water
   Q = m³/sec of gas
   P = absolute pressure, atm
   ρ = gas density, kg/m³
   T = temperature, °K
   K = factor, function of cyclone diameter. Approximate values are shown in Table 21–1.

Typical pressure drops in cyclones are from 1 to 8 inches of water.

Table 21–1.   Values of K for Calculating Pressure Drop in Cyclones

Cyclone diameter (in.)	K
29	10^{-4}
16	10^{-3}
8	10^{-2}
4	10^{-1}

The efficiency of a cyclone can be estimated by using the concept of a *cut diameter*, or that particle diameter at which 50 percent of the particles are removed by the cyclone. This is defined as

$$d_c = \left[\frac{9 \mu b}{2 \pi N v_i (\rho_s - \rho)} \right]^{1/2}$$

where μ = gas viscosity, kg/m-hr
$\quad\quad$ b = cyclone inlet width, m
$\quad\quad$ N = effective number of outer turns in the cyclone (normally about 4)
$\quad\quad$ v_i = inlet gas velocity, m/sec
$\quad\quad$ ρ_s = particle density, kg/m^3
$\quad\quad$ ρ = gas density, kg/m^3 (usually estimated as zero)

This cut diameter can be used to establish the collection efficiency for any other diameter particle, d, as shown in Figure 21–6. The number of turns (N) can also be approximated as

$$N = \frac{\pi}{H} (2L_1 + L_2)$$

where   H = height of the inlet, m
$\quad\quad$ L_1 = length of cylinder, m
$\quad\quad$ L_2 = length of cone, m

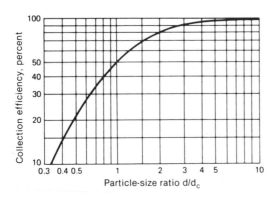

Figure 21–6.   Cyclone efficiency.

Example 21.1
    A cyclone has an inlet width of 10 cm and four effective turns. The gas has a temperature of 350°K and an inlet velocity of 10 m/sec, and the particles have an average size of 8 μm and a density of 1.5 g/cm³. What is the efficiency of collection?

$$d_c = \left[\frac{9 \mu b}{2 \pi N v_i (\rho_s - \rho)} \right]^{1/2}$$

The viscosity of air at 350°K is about 0.0748 kg/m-hr. We can assume $\rho = 0$ and N = 4. All parameters are in m-hr-kg.

$$d_c = \left[\frac{(9)(0.0748)(0.1)}{2(3.14)(4)(36,000)(1.5 \times 10^3)} \right]^{1/2}$$

$$= 7.12 \times 10^{-6} \text{ m} = 7.12 \ \mu\text{m}$$

Then, $d/d_c = 8/7.12 = 1.12$. From Figure 21–6, the expected removal efficiency is about 55 percent.

Bag (or Fabric) Filters

The filters used for controlling particulates (Figures 21–4 and 21–7) operate like the common vacuum cleaner. Fabric bags are used to collect the dust, which

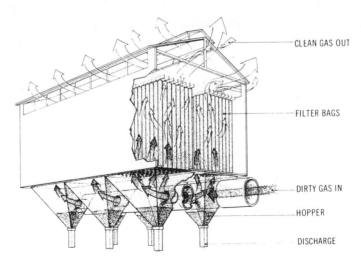

CLEAN GAS OUT

FILTER BAGS

DIRTY GAS IN

HOPPER

DISCHARGE

Figure 21–7. Fabric filters. [Courtesy of American Lung Association.]

must be periodically shaken out of the bags. The fabric will remove nearly all particulates, including submicron sizes. Bag filters are widely used in many industrial applications, but are sensitive to high temperatures and humidity.

The basic mechanism of dust removal in fabric filters is thought to be similar to the action of sand filters in water quality management (see Chapter 6). The dust particles adhere to the fabric owing to entrapment and surface forces. They are brought into contact by impingment or Brownian diffusion. Since fabric filters commonly have an air space-to-fiber ratio of 1:1, the removal mechanism cannot be simple sieving.

As the particles adhere to the fabric, the removal efficiency increases, but so does the pressure drop. The latter is thus the sum of the pressure drop owing to the fabric and the drop owing to the caked or adhered particles. This is expressed as

$$\Delta P = v \mu \left(\frac{x_f}{K_f} + \frac{x_p}{K_p} \right)$$

where ΔP = overall pressure drop, m of water
   v = superficial gas velocity through the fabric, m/sec
   μ = gas viscosity, poise
   x = thickness of filter (f) and particle layer (p), m
   K = filter (f) and particle layer (p) permeability

The values of K must be obtained by experiment for each type of fabric and type of dust to be collected.

Wet Collectors

The simple spray tower, pictured in Figures 21–4 and 21–8, is an effective method for removing large particulates. More efficient scrubbers promote the contact between air and water by violent action in a narrow throat section into which the water is introduced. Generally, the more violent the encounter, and hence the smaller the gas bubbles or water droplets, the more effective the scrubbing.

A commonly used high-energy wet collector is a *venturi scrubber* (see Figure 21–9) in which the gas flow is constructed through a venturi throat section and water is introduced as high-pressure streams perpendicular to the gas flow. The venturi scrubber is essentially 100 percent efficient in removing particles >5 μm. The pressure drop is estimated by an empirical equation as

$$\Delta P = v_g^2 L \times 10^{-6}$$

where ΔP = pressure drop across the venturi, cm of water
   v_g = gas velocity in the throat, cm/sec
   L = water-to-gas volume ratio, L/m^3

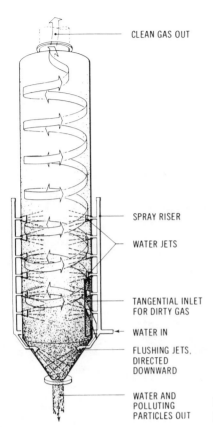

CLEAN GAS OUT

SPRAY RISER

WATER JETS

TANGENTIAL INLET
FOR DIRTY GAS

WATER IN

FLUSHING JETS,
DIRECTED
DOWNWARD

WATER AND
POLLUTING
PARTICLES OUT

Figure 21–8. Scrubber. [Courtesy of American Lung Association.]

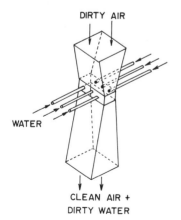

DIRTY AIR

WATER

CLEAN AIR +
DIRTY WATER

Figure 21–9. Venturi scrubber.

Wet scrubbers are efficient devices, but have two major drawbacks:

1. They produce a visible plume, albeit only water vapor. The lay public seldom differentiates, and hence public relations often dictate no visible plume.
2. The waste is now in liquid form and some manner of treatment is necessary.

Electrostatic Precipitators

Today, electrostatic precipitators are widely used in power plants, mainly because the power is expensive and readily available. The particulate matter is removed by first being charged by electrons jumping from one high-voltage electrode to the other, and then migrating to the positively charged collecting electrode (electrons are negatively charged). The type of electrostatic precipitator shown in Figure 21–4 consists of a pipe with a wire hanging down the middle. The particulates collect on the pipe and must be removed by banging the pipes with hammers. Figure 21–10 shows a plate-type precipitator, with flat plates acting as the collection electrodes. Electrostatic precipitators have no moving parts, require only electricity to operate, and are extremely effective in removing submicron particulates. They are also expensive.

The efficiency of an electrostatic precipitator is commonly estimated by an empirical equation

$$\eta = 1 - \exp\left(\frac{-Av_d}{Q}\right)$$

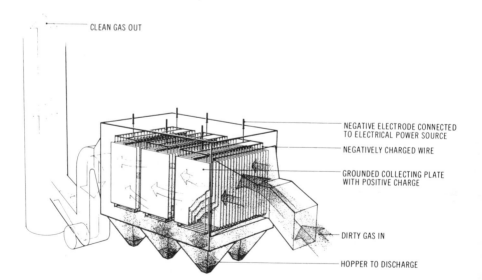

CLEAN GAS OUT

NEGATIVE ELECTRODE CONNECTED TO ELECTRICAL POWER SOURCE

NEGATIVELY CHARGED WIRE

GROUNDED COLLECTING PLATE WITH POSITIVE CHARGE

DIRTY GAS IN

HOPPER TO DISCHARGE

Figure 21–10.   Electrostatic precipitator. [Courtesy of American Lung Association.]

where   A = area of the collection electrodes (in the case of a pipe-type unit, the inside area of the pipes), m^2

Q = flow rate of gas through the pipe, m^3/sec

v_d = the drift velocity, m/sec

The drift velocity is the velocity of the particles toward the collecting electrode. It can be calculated theoretically by equating the electrostatic force on the charge particle in an electrical field and a drag force as the particle moves through the gas. This is perfectly analogous to the terminal settling velocity in a wastewater settling tank, except that in the latter case the force is due to gravity. Practically, the drift velocity can be estimated by simply relating it to particle size as

$$v_d = 0.5 \, d$$

where   d = particle size, μm

v_d = drift velocity, m/sec

Commonly, drift velocities are on the order of 0.03 to 0.2 m/sec.

A major problem with electrostatic precipitators in that as the dust layer builds up inside the pipe, the resistivity increases, thus decreasing the drift velocity. Ironically, one means of reducing the insulating effect of the dust is to inject sulfur (!) as SO_2 or sulfuric acid into the gas.

Figure 21–11 shows the effectiveness of an electrostatic precipitator in controlling emission from a power plant. The large white boxes in the fore-

Figure 21–11. Effectiveness of electrostatic precipitators on a coal-fired power plant. The electrostatic precipitator on the right has been turned off.

ground are the electrostatic precipitators. The one on the right, leading to the two stacks on the right, has been turned off to show the effectiveness by comparison with the almost undetectable emission from the stacks on the left.

Comparison of Particulate Control Devices

The efficiencies and costs of the various control devices obviously vary widely. Figure 21–12 shows approximate collection efficiency curves, as a function of particle size, for the various devices discussed.

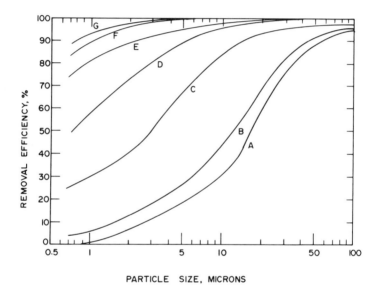

PARTICLE SIZE, MICRONS

Figure 21–12. Comparison of removal efficiencies: (A) baffled settling chamber, (B) sample cyclone, (C) high-efficiency cyclone, (D) electrostatic precipitator, (E) spray tower wet scrubber, (F) venturi scrubber, (G) bag filter.

CONTROL OF GASEOUS POLLUTANTS

The control of gases involves the removal of the pollutant from the gaseous emissions, a chemical change in the pollutant, or a change in the process producing the pollutant.

Wet scrubbers, such as already discussed, can remove gaseous pollutants by simply dissolving them in water. Alternatively, a chemical may be injected into the scrubber water which then reacts with the pollutants. This is the basis for most SO_2 removal techniques, as discussed below. Adsorption is a useful method when it is possible to bring the pollutant into contact with an efficient adsorber like activated carbon. This method is effective for many organic pollutants (Figure 21–13).

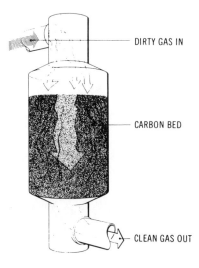

DIRTY GAS IN

CARBON BED

CLEAN GAS OUT

Figure 21–13. Adsorber for control of gaseous pollutants. [Courtesy of American Lung Association.]

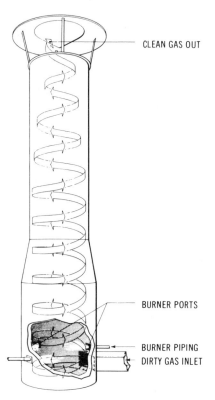

CLEAN GAS OUT

BURNER PORTS

BURNER PIPING
DIRTY GAS INLET

Figure 21–14. Incinerator for controlling gaseous pollutants. [Courtesy of American Lung Association.]

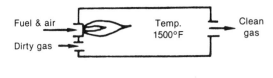

Incineration

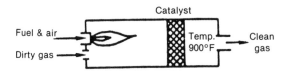

Catalytic Combustion

Figure 21–15. Catalytic combustion compared with simple incineration.

Incineration (Figure 21–14) or flaring is used when an organic pollutant can be oxidized to CO_2 and water. A variation of incineration is *catalytic combustion* in which the temperature of the reaction is lowered by the use of a catalyst that mediates the reaction. Figure 21–15 is a comparison of simple incineration and catalytic combustion. The catalytic "reactor" is used in many cars to reduce HC and CO emissions by oxidizing them to CO_2 and H_2O (see below).

Control of Sulfur Oxides

We noted earlier that sulfur oxides (SO_2 and SO_3) are serious and yet ubiquitous air pollutants. The major source of sulfur oxides (or SO_x as they are often referred to in shorthand) is coal-fired power plants. The increasingly strict standards for SO_x control have prompted the development of a number of options and techniques for reducing emissions of sulfur oxides. Among these options are:

1. *Change to Low-Sulfur Fuel.* Natural gas is extremely low in sulfur. Oil burned for industrial heating and electric power production ranges in sulfur content from 0.5 to 3 percent, and coal varies from 0.3 to 4 percent. Low-sulfur fuels are an expensive and sometimes uncertain option.
2. *Coal Desulfurization.* Sulfur in coal may be both inorganically bound (as iron pyrite, FeS_2) or organically bound. Pyrite can be removed by pulverizing the coal and washing it with a detergent solution. Organically bound sulfur can be removed by washing with very concentrated acid. Preferred methods are coal gasification, which produces pipeline-quality gas, or solvent extraction, which produces low-sulfur liquid fuel.

3. *Tall Stacks.* A short-sighted method, albeit locally economical, of SO_2 control is to build incredibly tall smokestacks and disperse the SO_x. This option has been used in Great Britain and is in part responsible for the acid rain problem plaguing Scandinavia.

4. *Flue-Gas Desulfurization.* The last option is to reduce the SO_x emitted by cleaning the gases coming from the combustion, the so-called flue gases. Many systems have been used, and a great deal of research is under way to make these processes more efficient.

Since the 1940s, industrial SO_2 emitters have used the condensation of SO_2 to sulfuric acid to remove sulfur dioxide from stack gas. This control method has two limitations: the stack gas must be cleaned of particulate matter before entering the acid plant, and acid formation is energetically favorable only for a fairly concentrated gas stream (at least 3 percent—30,000 ppm—SO_2). The reactions for acid formation are

$$SO_2 + 1/2O_2 = SO_3$$

$$SO_3 + H_2O = H_2SO_4$$

A double-contact acid plant can produce good industrial-grade 98 percent sulfuric acid.

The nonferrous smelting and refining industry has made the most use of this method of control. Analogous reactions can be carried out to form $(NH_4)_2SO_4$, which is used as fertilizer, and $CaSO_4$ (gypsum).

Flue gas from the fossil fuel combustion is too dilute to permit trapping the SO_2 as acid or commercial fertilizer or gypsum. Coal combustion off gases are also too dirty. However, the principle of trapping the SO_2 as a sulfate was incorporated in the use of a lime-limestone mixture to absorb the SO_2. The reaction with lime is

$$SO_2 + CaO \rightarrow CaSO_3$$

or if limestone is used,

$$SO_2 + CaCO_3 \rightarrow CaSO_4 + CO_2$$

Both calcium sulfite and sulfate (gypsum) are solids that have low solubilities and thus can be separated in gravity settling tanks. The three commonly used options, illustrated in Figure 21–16, are injecting limestone into the scrubber, calcining the limestone first and injecting lime into the scrubber, and adding limestone to the boiler. In the last two cases the limestone reacts to form quicklime

$$CaCO_3 \overset{\Delta}{\rightarrow} CaO + CO_2$$

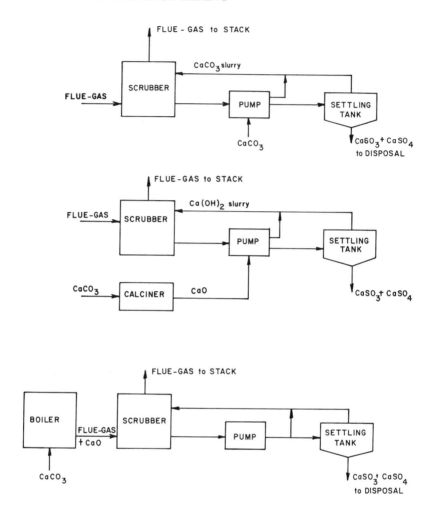

Figure 21–16. Three options for desulfurization of flue gases by using limestone.

In the one case the quicklime so produced is hydrated to $Ca(OH)_2$, which is then injected into the scrubber. In the last system the quicklime is produced in the boiler without the need of a separate calciner, and the lime is carried to the scrubber with the exit gases.

The calcium salts thus formed represent a staggering disposal problem. If, according to U.S. Environmental Protection Agency (EPA) estimates, the total emission of SO_2 from all sources in the United States is about 40×10^6 metric tons per year, and if 90 percent of this is removed, 35×10^6 metric tons of SO_2 or 76×10^6 metric tons of $CaSO_4$ would be produced. This is in slurry form after settling in the tank, at about 20 percent solids, so that the total slurry disposal problem is 380×10^6 metric tons per year, or enough flue-gas desulfurization sludge to cover about 400 square miles one foot thick—every year! In three years it would cover the entire state of Rhode Island.

During the past 10 years, flue-gas desulfurization methods have been developed that trap SO_2 as the sulfite rather than the sulfate. The advantage of these methods is that the absorbing material can be regenerated, so that the waste disposal problem is somewhat mitigated. A typical sulfite control method is single alkali scrubbing, for which the reactions are

$$SO_2 + Na_2SO_3 + H_2O = 2NaHSO_3$$

Sodium sulfite is then regenerated from the bisulfite by heating:

$$2NaHSO_3 = Na_2SO_3(s) + H_2O + SO_2 \text{ (conc)}$$

The concentrated SO_2 that is recovered from this process can be used industrially (e.g., in pulp and paper manufacture or for sulfuric acid manufacture). Figure 21–17 shows a simplified schematic diagram for single alkali scrubbing.

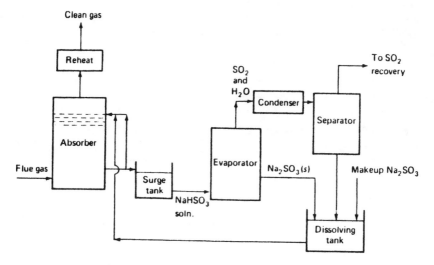

Figure 21–17. Simplified schematic for single alkali scrubbing of flue gas with regeneration. [*Source*: Megonnel, H.G., *J. Air Poll. Control Assoc.* 28, 1978.]

Control of Nitrogen Oxides

Nitrogen oxides from fuel combustion are an important contributor to air pollution. Wet scrubbing, particularly with an alkaline scrubbing solution, will absorb NO_2. However, this method is not a feasible one to install only for NO_2 control, as in oil-burning power plants.

A moderately effective method, when scrubbing is not feasible, is off-stoichiometric combustion. This method controls NO formation by limiting the amount of air (or oxygen) in the combustion process to just a bit more than is needed to burn the hydrocarbon fuel in question. For example, the reaction for burning natural gas is

$$CH_4 + 2O_2 = 2H_2O + CO_2$$

Nitrogen in the air used for combustion will compete with the natural gas for oxygen, in the competitive reaction

$$N_2 + O_2 = 2NO$$

The stoichiometric ratio of oxygen needed in natural gas combustion is

$$32\,g \text{ of } O_2/16\,g \text{ of } CH_4$$

If the air used for combustion provides oxygen in slight excess of this ratio, virtually all of the oxygen will combine with the fuel rather than with the nitrogen in the air. In practice, off-stoichiometric combustion is achieved by adjusting the air flow to the combustion chamber so that any visible plume just disappears.

Control of Volatile Organics and Odors

Volatile organic compounds and odors are controlled by thorough oxidation, either by incineration or by catalytic combustion. Before moving on to the next topic, it might again be useful to reiterate the importance of matching the type of pollutant to be removed with the proper process. Figure 21–18 re-emphasizes

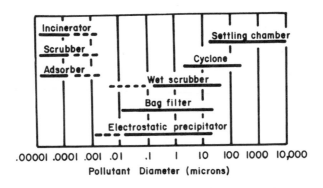

Figure 21–18. Comparison of control options by particle size.

the importance of particle size in the application control equipment. Other properties, however, may be equally important. A scrubber, for example, removes not only particulates, but gases that can be dissolved in the water. Thus SO_2, which is readily soluble, but not NO, which is poorly soluble, would be removed in a scrubber. The selection of the proper control technology is an important component of the environmental engineering profession.

CONTROL OF MOVING SOURCES

Although many of the above control techniques apply to moving sources as well as stationary ones, one very special moving source—the automobile—deserves special attention. As noted in Chapter 19, the automobile has many potential sources of pollution. Realistically, however, there are only a few important points requiring control (Figure 21–19):

A. evaporation of hydrocarbons (HC) from the fuel tank

B. evaporation of HC from the carburetor

C. emissions of unburnt gasoline and partially oxidized HC from the crankcase

D. the NO_x, HC, and CO from the exhaust

The evaporative losses from the gas tank and carburetor have been eliminated by storing any vapors emitted in an activated-carbon canister. This would normally occur when the engine is turned off and the hot gasoline in the carburetor vaporizes. The vapors may later be purged by air and burned in the engine, as shown schematically in Figure 21–20.

The third source of pollution, the crankcase vent, can be eliminated by closing off the vent to the atmosphere and recycling the blowby gases into the intake manifold. The positive crankcase ventilation (PCV) valve is a small check valve that prevents the buildup of pressure in the crankcase.

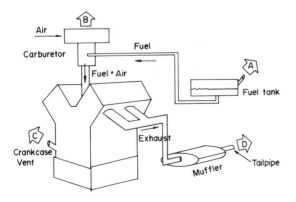

Figure 21–19. Internal combustion engine showing four major emission points.

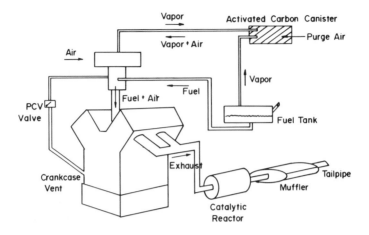

Figure 21–20.   Internal combustion engine showing methods of controlling emissions.

The most difficult control problem is the exhaust, which accounts for about 60 percent of the HC and almost all of the NO_x, CO, and lead. One immediate problem is how to measure these emissions. It is not as simple as sticking a sampler up the tailpipe, since the quantity of pollutants emitted changes with the mode of operation. The effect of operation on emissions is illustrated as Table 21–2. Note that when the car is accelerating, the combustion is efficient (low CO and HC) and the high compression produces a lot of NO_x. On the other hand, decelerating results in low NO_x and very high HC owing to partially burned fuel.

Because of these difficulties the EPA has instituted a standard test for measuring emissions. This test procedure includes a cold start, acceleration and cruising on a dynamometer to simulate a load on the wheels, and a hot start.

Emission control techniques for the internal combustion engine include tune-ups, exhaust gas recirculation, engine modifications, and catalytic reactors. A tune-up may have a significant effect on emissions. For example, a high

Table 21–2.   Effect of Engine Operation on Automotive Exhaust Characteristics, Shown as Fraction of Idling Emissions

	Component		
	CO	HC	NO_x
Idling	1.0	1.0	1.0
Accelerating	0.6	0.4	100
Cruising	0.6	0.3	66
Decelerating	0.6	11.4	1.0

air/fuel ratio (a lean mixture) will reduce both CO and hydrocarbons, but will increase NO emission. A well-tuned car is nonetheless the first line of defense for emission control.

Both CO and hydrocarbons can be reduced by as much as 60 percent by recirculating the exhaust gas through the engine. There are two major drawbacks to this control method: the resulting control does not meet the 1983 U.S. standards, and the recirculated gas must be cooled considerably before recycling, or the piston surfaces will become deformed. Even with cooling, there is unusual wear on the engine when this method is used, and it has largely been superseded.

The catalytic reactor ("catalytic converter") must perform two pollution control functions: *oxidation* of CO and hydrocarbons to CO_2 and water, and *reduction* of NO to N_2. Reduction of NO is accomplished by burning a fuel-rich mixture and thus depleting the oxygen to the catalyst, then introducing air and oxidizing the CO and HC at reduced temperatures, so that production of NO will be suppressed. In modern engines, a platinum-rhodium three-stage catalyst is used, and both the air/fuel mixture and the temperature are carefully controlled.

The drawbacks of catalytic converters are cost and need for maintenance, susceptibility to lead poisoning (catalytic converters require the use of unleaded gas), and the formation of SO_2 from sulfur compounds in gasoline. Nonetheless, they are the most successful pollution control devices to date, and are presently standard on gasoline-driven vehicles sold in the United States.

A wide range of engine modifications is possible. CO and HC can be reduced by making the mixture leaner, and NO can be reduced simultaneously by retarding the spark, but the resulting engine is virtually undriveable. Injection of water can reduce NO emissions, and fuel injection (bypassing the carburetor) can reduce CO and HC emissions. The stratified-charge engine represents the most sophisticated use of this system: this engine operates with a very lean mixture and has good temperature control. The engine cylinder has two compartments: the first receives and ignites the rich mixture, and the second provides a broad flame for an efficient burn. CO reduction of 99 percent has been achieved with this type of engine.

An emission-free internal combustion engine is something of a contradiction in terms. Drastic lowering of emissions to produce a virtually pollution-free engine would require use of an external combustion engine. Laboratory tests indeed indicate that such engines can achieve better then 99 percent control of all three major exhaust pollutants. However, although work began in 1968 on a mobile external combustion engine for modern cars, a working model has yet to be built. The unsolved problem is finding a working fuel (the "steam" in a "steam engine") that will permit ready acceleration but that is not flammable. Water, the working fuel in steam engines, and in the old Stanley Steamer automobile, has a high heat capacity and heat of vaporization and thus does not respond quickly enough on acceleration. Organic fluids, which have the right combination of thermodynamic properties, are usually flammable.

Natural gas may be used as a fuel in cars, but there are so many competing uses for a limited supply that a changeover to natural gas as an automobile fuel would not be feasible. Electric cars are clean, but can store only limited power, thus their range is limited. In addition, the electricity used to power such vehicles must be generated, thus creating more pollution.

The diesel engines used in trucks and buses also are important sources of pollution. There are, however, fewer of these vehicles in operation than gasoline-powered cars, and the emissions may not be as harmful owing to the nature of the diesel engines. The main problems associated with diesel engines are the visible smoke plume and odors, two characteristics that have led to considerable public irritation with diesel-powered vehicles. But from a public health viewpoint, diesel exhaust does not constitute the problem that the exhaust from gasoline-powered cars does. Diesel engines in passenger cars can, in fact, readily meet strict emission control standards.

CONCLUSION

This chapter is devoted mainly to the description of air pollution control alternatives by "bolt-on" devices. It should be re-emphasized that this is usually the most expensive method of control. A general environmental engineering truism is that the least expensive and most effective control point is always the farthest up the process line. It is at the beginning of the process or, better yet, consideration of alternatives to the process, where the most effective control is achieved. This is obvious in the case of flue-gas desulfurization. We should not seek to bury Rhode Island, but rather ask if all the electricity is really necessary. Are there other power sources? Not only is this good engineering technology and economics, but it is also sensitive and enlightened analysis of our lifestyle and its impact on the environment.

PROBLEMS

21.1   Taking into account cost, ease of operation, and ultimate disposal of residuals, what type of control device would you suggest for the following emissions?

    a. A dust with particle range of 5 to 10 μm?
    b. A gas containing 20 percent SO_2 and 80 percent N_2?
    c. A gas containing 90 percent HC and 10 percent O_2?

21.2   A stack emission has the following characteristics: 90 percent SO_2, 10 percent N_2, no particulates. What treatment device would you suggest and why?

21.3   How big is 10,000 μm in inches? How big is 1 μm in inches?

21.4   An industrial emission has the following characteristics: N_2—80 percent, O_2—15 percent, CO_2—5 percent. You are called in as a consultant to advise on the type of air pollution control equipment required. What would be your recommendation?

21.5   A whiskey distillery has hired you as a consultant to design the air pollution equipment for their new plant, to be built in a residential area. What problems would you encounter, and what would be your control strategy?

21.6   A dust has a particle size analysis as follows:

Mean Size (μm)	Percentage of Particles by Weight
0.05	10
0.1	20
0.5	25
1.0	35
5.0	10

The particle specific gravity is 1.2.

A 0.51-m-diameter cyclone with an inlet width of 15 cm and an inlet length of 25 cm is operating at five effective turns and 2.0 m^3/sec air flow.

a. What is the efficiency of removal?

b. What is the pressure drop (assume normal temperature and pressure of the air)?

c. What is the separation factor? Is this a high- or low-efficiency cyclone?

21.7   A plate-type electrostatic precipitator is to remove a particulate with a diameter of 0.5 μm at a flow rate of 2 m^3/sec. The 40 plates are 5 cm apart and 3 m high. How deep must the plates be to achieve removal (ideally)?

21.8   A dust has particles with a drift velocity of about 0.15 m/sec. For a total air flow of 60 m^3/sec, what must be the number of 10-m × 10-m collecting plates to achieve 90 percent removal?

21.9   Design a start/stop/drive test for measuring the emission of pollutants from an automobile. Justify your choice.

21.10   Calculate the removal efficiencies for the dust described in Problem 21.6 for a standard cyclone, an electrostatic precipitator, and a bag filter, using the curves in Figure 21–12.

21.11   Determine the efficiency of a cyclone of diameter 0.5 m, at a flow rate of 0.4 m^3/sec and a gas temperature of 25°C. The inlet width is 0.13 m^2, and the area of the entrance is 0.04 m^2. The particulates have a diameter of 10 μm and a density of 2 g/cm^3.

21.12   An electrostatic precipitator has the following specifications:
height: 7.5 m
length: 5 m
number of passages: 3
plate spacing: 0.3 m

The flow rate is 18 m^3/sec, the particle size is 0.35 μm, and the particles have a drift velocity of 0.162 m/sec. Find the expected efficiency.

21.13   For the problem above, what would be the increase in efficiency by
a. doubling the length?
b. doubling the number of plates and reducing the flow rate by one-half?

LIST OF SYMBOLS

A = area, m^2
b = cyclone inlet width, m
C = concentration, kg/m^3
d = particle diameter, m
d_c = cut diameter, m
H = height of cyclone inlet, m
L = water-to-gas volume ratio, L/m^3
L_1 = length of cylinder in a cyclone, m
L_2 = length of cone in a cyclone, m
M_a = mass of pollutant that could have been collected, kg
M_b = mass of pollutant that escaped capture, kg
M_c = mass of pollutant collected, kg
N = effective number of turns in a cyclone
P = pressure, atm
ΔP = pressure drop, m of water
Q = flow rate, m^3/sec
r = radial distance, m
v_d = drift velocity, m/sec
v_i = inlet gas velocity, m/sec
v_R = radial velocity, m/sec
x = thickness, m
η = collection efficiency, %
μ = gas velocity, poise
ρ = gas density, kg/m^3
ρ_s = particle density, kg/m^3
ω = rotational speed, rad/sec

Chapter 22
Air Pollution Law

As with water pollution, a complex system of laws and regulations governs the use of air pollution abatement technologies. In this chapter, the evolution of air pollution law is described, from its roots in common law through the passage of federal statutory and administrative initiatives. Problems encountered by regulatory agencies and polluters are addressed with particular emphasis on the impacts the system may or may not have on future economic development. Figure 22–1 offers a roadmap to be followed through this maze.

AIR QUALITY AND COMMON LAW

When dealing with common law, an individual or groups of individuals injured by a source of air pollution may cite general principles in two branches of that law which have developed over the years and may apply to their particular damages:

- tort law
- property law

The harmed party, the plaintiff, could enter a courtroom and seek remedies from the defendant for damaged personal well-being or damaged property.

Tort Law

A tort is an injury incurred by one or more individuals. Careless accidents, defamation of character, and exposure to harmful airborne chemicals are the types of wrong included under this branch of common law. A polluter could be held responsible for the damage to human health under three broad categories of tort liability, negligence, and strict liability.

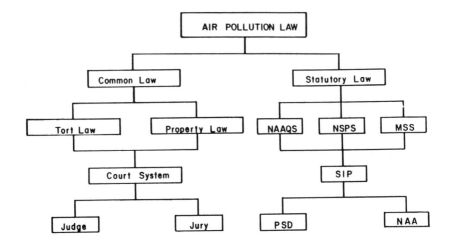

Figure 22–1.   Air pollution law in the United States.

Intentional liability requires proof that somebody did a wrong to another party *on purpose*. This proof is especially complicated in the case of damages from air pollution. The fact that a "wrong" actually occurred must first be established, a process that may rely on direct statistical evidence or strong inference, such as the results of laboratory tests on rats. Additionally, intent to do the "wrong" must be established, which involves producing evidence in the form of written documents or direct testimony from the accused individual or group of individuals. Such evidence is not easily obtained. If intentional liability can be proven to the satisfaction of the courts, actual damages as well as punitive (punishment) damages can be awarded to the injured plaintiff.

Negligence may involve mere inattention by the air polluter who allowed the injury to occur. Proof in the courtroom focuses on the lack of reasonable care taken on the defendant's part. Examples of such neglect in air pollution include failure to inspect the operation and maintenance (O&M) of electrostatic precipitators or the failure to design and size an adequate abatement technology. Again, damages may be awarded to the plaintiff.

Strict liability does not consider the *fault* or state of mind of the defendant. Under certain extreme cases, a court of common law has held that some acts are *abnormally dangerous* and that individuals conducting those acts are strictly liable if injury occurs. The court does tend to balance the danger of an act against the public utility associated with the act.

Again, if personal damage is caused by air pollution from a known source, the damaged party may enter a court of common law and argue for monetary damages to be paid by the defendant or an injunction to stop the polluter from polluting or both. Sufficient proof and precedent is often difficult if not impossible to muster, and in many cases, tort law has been found to be inadequate in controlling air pollution and awarding damages.

Property Law

Property law, on the other hand, focuses on the theories of nuisance and property rights; nuisance is based on the interference with the use or enjoyment of property, and property rights is based on actual invasion of the property. Property law is founded on ancient actions between land owners and involves such considerations as property damage and trespassing. A plaintiff basing a case on property law rolls the dice and hopes the court will rule favorably as it balances social utility against individual property rights.

Nuisance is the most widely used form of common law action concerning the environment. Public nuisance involves unreasonable interference with a right, such as the "right to clean air," common to the general public. A public official must bring the case to the courtroom and represent the public that is harmed by the air pollution. Private nuisance, on the other hand, is based on unreasonable interference with the use and enjoyment of private land. The key to a nuisance action is how the courts define "unreasonable" interference. Based on precedents and the arguments of the parties involved, the common law court balances the equities, hardships, and injuries in the particular case and rules in favor of either the plaintiff or the defendant.

Trespass is closely related to the theory of nuisance. The major difference is that some physical invasion, no matter how minor, is technically a trespass. Recall that nuisance theory demands an "unreasonable" interference with land and the outcome of a particular case depends on how a court defines "unreasonable." Trespass is relatively cut and dried. Examples of trespass include physical walk-ons, vibrations from nearby surface or subsurface strata, and possibly gases and microscopic particles flowing from an individual smoke stack.

In conclusion, common law has generally proven inadequate in dealing with problems of air pollution. The strict burdens of proof required in the courtroom often result in decisions that favor the defendant and lead to smoke stacks that continue to pollute the atmosphere. Additionally, the technicality and complexity of individual cases often limit the ability of a court to act; complicated tests and hard-to-find experts often leave a court and a plaintiff with their hands tied. Furthermore, the absence of standing in a common law courtroom often prevents private individuals from bringing a case before the judge and jury unless the individual actually suffered material or bodily harm from the air pollution.

One key aspect of these common law principles is their degree of variation. Each state has its own body of common law, and individuals relying on the court system are generally confined to using the common laws of the applicable state.

Given these shortcomings inherent in common law, Congress adopted a federal Clean Air Act, with the objective of plugging some of the pollution holes in common law.

STATUTORY LAW

Federal statutory law controlling air pollution began with the 1963 and 1967 Clean Air Acts. Although these laws provided broad clean air goals and research money, they did not apply air pollution controls throughout the entire United States, but only in heavily industrialized, dirty communities. In 1970, however, the Clean Air Act was amended to cover the entire United States, and the Environmental Protection Agency (EPA) was created to promulgate clean air regulations and to enforce the Act. The 1970 Amendments are the basis for the clean air legislation that we have today.

National Ambient Air Quality Standards (NAAQS)

The EPA is empowered to determine allowable ambient concentrations of certain pollutants; these are the NAAQS. The primary NAAQS are intended to protect human health; the secondary NAAQS, to "protect welfare." The latter levels are actually determined as those needed to protect vegetation. These standards are listed in Table 22–1.

The EPA has revised the standards for total suspended particulate matter somewhat. It has been observed that the particles that are most closely correlated with adverse health effects are those having a diameter of 10 μm or less. A NAAQS has been promulgated that includes these features:

- primary standards are set only for particles 10 μm or less in diameter; this is called the PM_{10} standard.
- the primary annual average PM_{10} standard is 50 $\mu g/m^3$.
- the primary 24-hour PM_{10} standard is 150 $\mu g/m^3$.

The secondary 24-hour TSP standard remains the same. A secondary PM_{10} standard is under consideration.

NAAQS are set on the basis of extensive collections of information and data on the effects of these air pollutants and human health, ecosystems, vegetation, and materials. These documents are called "criteria documents" by the EPA, and

Table 22–1.   Selected National Ambient Air Quality Standards (NAAQS)

Pollutant	Primary		Secondary	
	(*ppm*)	(*μg/m³*)	(*ppm*)	(*μg/m³*)
Particulate matter (μg/m³)				
Annual geometric mean		75		60
Max 24-hr concentration		260		150
Sulfur oxides				
Annual arithmetic mean	0.03	80	0.02	60
Max 24-hr concentration	0.14	365	0.1	260
Max 3-hr concentration			0.5	1,300
Carbon monoxide				
Max 8-hr concentration	9	10,000	same	
Max 1-hr concentration	35	40,000	as primary	
Photochemical oxidants				
Max 1-hr concentration	0.12	260	same as primary	
Hydrocarbons				
Max 3-hr concentration	0.24	160	same as primary	
Nitrogen oxides				
Annual arithmetic mean	0.05	100	same as primary	
Lead				
Avg of 3 months		1.0	same as primary	

the pollutants for which NAAQS exist are sometimes referred to as "criteria pollutants." Data indicate that all criteria pollutants have some threshold below which there is no damage. Recent epidemiological information has lowered this threshold for lead; further data might eliminate a threshold for lead entirely.

Under the Clean Air Act, most enforcement power is delegated to the states by the EPA. However, the states must show the EPA that they can clean up the air to the levels of the NAAQS. This showing is made in each state's Air Quality Implementation Plan (AQIP), a document that contains all of that state's regulations governing air pollution control, including local regulations within the state. The AQIP must be approved by the EPA, but once approved, it has the force of federal law.

The 1970 Clean Air Act Amendments envisioned that virtually all of the United States would meet the ambient standards by 1975. When it became evident that the 1975 deadline would not be met, Congress again amended the Act in 1977. Two new deadlines for meeting NAAQS were set: 1987 for mobile source-related criteria pollutants and 1985 for all other criteria pollutants. With the exception of about five communities in the United States, such as Los Angeles and Denver, these deadlines have been met.

Table 22–2.   Some Typical NSPS for SO_2

Facility	NSPS
Coal-burning generating plant	70% reduction and 0.6 lb/10^6 BTU heat input, or 90% reduction and 1.2 lb/10^6 BTU heat input
H_2SO_4 plants	4 lb/ton of acid produced

Regulation of Emissions

Under the Clean Air Act, the EPA may set *emission* standards only for new or modified stationary sources, not for existing sources that are already in operation. Sources of hazardous air pollutants are an exception, as we shall see. Emissions from existing stationary sources are regulated by the state through regulations in the AQIP. The state also has the power to grant variances from emission standards, though usually not for more than one year.

Emission standards for new and modified stationary sources are called the new source performance standards (NSPS). New source performance standards are *not* determined by the need to meet ambient standards, but are set according to the best available emissions control technology, taking into account the cost of such control technology and its energy requirements. In sum, the NAAQS are health based, whereas the NSPS are technology based. As new technology has become available, NSPS for various facilities have been revised. The most notable example of such a change was the revision of NSPS for coal-burning electric generating plants. The 1971 NSPS for SO_2 was 1.2 pounds of SO_2 per

Table 22–3.   Typical NSPS for Particulate Matter

Facility	NSPS
Coal-burning generating plant	0.03 lb/10^6 BTU heat input, and 20% opacity
Portland cement plants	0.30 lb/ton of feed, and 20% opacity
Solid waste incinerators	0.08 gr/dscf corrected to 12% CO_2
Refineries	0.022 gr/dscf
Iron and steel plants	0.022 gr/dscf

dscf = dry standard cubic foot
 gr = grain

million BTU heat input: a standard which could be met by using low-sulfur coal. The 1977 NSPS is 70 percent reduction in SO_2 emissions, which requires flue-gas desulfurization.

The 1977 amendments to the Clean Air Act also prohibit the substitution of tall stacks or curtailment during bad dispersion conditions for actual emission controls.

Prevention of Significant Deterioration

In 1973 the Sierra Club sued the EPA for failing to protect the cleanliness of the air in those parts of the United States where the air was cleaner than the NAAQS, and won. In response, Congress included prevention of significant deterioration (PSD) in the 1977 Clean Air Act amendments.

For PSD purposes, the United States is divided into class I and class II areas, with the possibility of class III designation for some areas. Class I includes the so-called "mandatory class I" areas—all national wilderness areas larger than 5,000 acres and all national parks and monuments larger than 6,000 acres—and any area that a state or Native American tribe wishes to designate class I. The rest of the United States is class II, except that a state or tribe may petition the EPA for redesignation of a class II area to class III.

To date, the only pollutants covered by PSD are sulfur dioxide and particulate matter. PSD limits the allowed increases in these as indicated in Table 22–4. In addition, visability is protected in class I areas.

An industry wishing to build a new facility must show, by dispersion modeling with a year's worth of weather data, that it will not exceed the allowed increment. On making such a showing, the industry receives a PSD permit from the EPA. The PSD permitting system has had considerable impact in siting new facilities.

Where the ambient concentrations are already close to the NAAQS and the PSD limits would allow exceedence of the NAAQS, the PSD limits clearly are moot.

Table 22–4. Maximum Allowed Increases Under PSD (Values are in $\mu g/m^3$)

| Class | Particulate Matter | | Sulfur Dioxide | | |
	Annual mean	24hr max	Annual mean	24-hr max	3-hr max
I	5	10	2	5	25
II	19	37	20	91	512
III	37	75	40	182	700

Hazardous Air Pollutants

The Clean Air Act recognizes that some substances that are emitted into the air are particularly hazardous, and the NAAQS for particulate matter is not restrictive enough. These hazardous air pollutants are listed under Section 112 of the Act, and the EPA is required to set ambient and emission standards at a level that "provides an ample margin of safety to protect the public health." These National Emission Standards for Hazardous Air Pollutants (NESHAPS) are determined for *all* sources—both new and existing—of the particular hazardous pollutant.

A hazardous air pollutant is one that might reasonably be anticipated to cause an increase in deaths or serious, incapacitating illness. To date, the following substances have been designated hazardous air pollutants:

- arsenic
- asbestos
- beryllium
- benzene
- mercury
- radioactive substances
- vinyl chloride monomer

Several of these substances are carcinogens and, as such, are considered to have no threshold of effect. For these pollutants, an "ample margin of safety" cannot be assured, and risk associated with various degrees of control must be assessed.

Mobile Sources

The federal government has pre-empted the setting of emission standards for mobile sources. Emission standards for gasoline-powered vehicles require a reduction in CO, hydrocarbons, and nitrogen oxides beginning with the 1979 model year, and culminating in

- 90 percent reduction in CO and hydrocarbon emission in the 1983 models, and
- 75 percent reduction in NO_x emissions in the 1985 models

This degree of emission control has been achieved by using catalytic converters in back of the exhaust. Since catalysts are poisoned by lead, the Clean Air Act also requires the use of unleaded gasoline in automobiles having catalytic converters. In addition, in 1984, the EPA promulgated a regulation reducing the concentration of lead in *leaded* gasoline by 90 percent, leaving only enough lead to lubricate adequately.

Nonattainment

A region in which the NAAQS for one or more criteria pollutants is exceeded more than once or twice a year is called a "nonattainment area" for that pollutant. The Clean Air Act Amendments of 1977 require the following in a nonattainment area:

- If there is nonattainment of the lead, CO, or ozone standard, a traffic reduction plan and an inspection and maintenance program for exhaust emission control are required. Failure to comply results in the state's loss of federal highway construction funds.
- If there is nonattainment resulting from stationary source emission, an offset program must be initiated. Such a program requires that there be a rollback in emissions from existing stationary sources such that total emissions after the new source operates will be less than before. New sources in nonattainment areas must attain the lowest achievable emission rate (LAER), without necessarily considering the cost of such emission control.

The offset program allows industrial growth in nonattainment areas, but offers a particular challenge to the air pollution control engineer. The offset represents emission reductions that would otherwise not be required.

Types of action that could generate offsets include:

1. tighter controls on existing operations at the same site
2. a binding agreement with another facility to reduce emissions
3. the purchase of another facility to reduce emissions by installation of control equipment or by closing the facility down

PROBLEMS OF IMPLEMENTATION

Both regulatory agencies and the sources of pollution have encountered difficulties in the implementation of air pollution abatement measures. Because most major stationary sources of air pollution now have installed abatement equipment in response to federal pollution control requirements, regulatory agencies must develop programs that promote continuing compliance. However, the track records of the agencies indicate that current programs are not always effective.

Flaws in design and construction of control equipment contribute to significant noncompliance. Problems include the use of improper materials in constructing controls, undersizing of controls, inadequate instrumentation of the control equipment, and inaccessibility of control components for proper operation and maintenance. Numerous design flaws such as these cast doubt on how effectively the air pollution control agencies evaluate the permit applications for these sources. Additionally, many permit reviewers working in government lack the necessary practical experience to fully evaluate proposed controls and tend to rely too heavily on the inadequate technical manuals available to them.

A source permit, as it is typically designed within the regulatory framework, could ensure that an emitter plans to install necessary control equipment. It could also require the emitter to perform recordkeeping functions that facilitate proper operation and maintenance. However, agencies do not always use the permit program to accomplish these objectives. For example, in some states, a majority of the sources fail to keep any operating records at all that would enable an independent assessment of past compliance. In other states, although many sources tend to keep good operating records, they are not required to do so by the control agency.

An agency's ability to detect violations of emission requirements may be critical to the success of its regulatory effort. However, many agencies rely on surveillance by sight and smell as opposed to stack testing or monitoring. For example, air pollution control agencies often issue notices of violation for such problems as odor, dust, and excessive visible emissions, even though it is recognized that this approach neglects other, perhaps more detrimental pollutants.

Many state and local agency programs rely on the voluntary approach to achieving compliance by sources. Although a majority of sources appear to respond to this strategy, in some instances such programs are ineffective. In one state, for example, eight major emitters who agreed to voluntary controls in 1978 were later referred for legal action. This indicates a failure of good faith efforts to comply with their commitments. In addition, cases referred for civil or criminal action rarely result in significant penalties. In the same state six sources received fines in 1977; one was for $2,000 and five ranged from $200 to $500. This compares with a legal maximum of $10,000 per day of violation.

The emitters of air pollution also face problems in complying with regulations. Although most large stationary sources of pollution have installed control equipment, severe emission problems may be caused by design or upsets either with the control equipment or in the production process. Poor design of the control equipment is probably the primary cause of excess emissions in the greatest number of sources, and process upsets and routine component failure appear to be the second major cause of excess emissions. Other causes of excess emissions include improper maintenance, lack of spare parts, improper construction materials, and lack of instrumentation. Additionally, wide variations in the frequency and duration of excess emissions are generally experienced by different industrial categories.

CONCLUSION

Air pollution law is a complex web of common and statutory law. Although common law has offered and continues to offer checks and balances between polluters and economic development, shortcomings do exist. Federal statutory law has attempted to fill the voids and, to a certain extent, has been successful in cleaning the air. Engineers must be aware of the requirements placed on industry by this system of laws. Particular attention must be paid to the siting of new plants in different sections of the nation.

PROBLEMS

22.1 Acid rain is a mounting problem, particularly in the northeastern states. Discuss how this problem can be controlled under the system of common law. Compare this approach with the remedy under federal statutory law.

22.2 Pollution from tobacco smoke can significantly degrade the quality of certain air masses. Develop sample town ordinances to improve the quality of air: (1) inside city buses and (2) over the spectators at both indoor and outdoor sporting events. Rely on both structural and nonstructural alternatives; i.e., solutions that mechanically, chemically, or electrically clean the air, and solutions that prevent all or part of the pollution in the first place.

22.3 Would you favor an international law that permits open burning of hazardous waste on the high seas? Open burning refers to combustion without emission controls. Would you favor such a law if emissions were controlled? Discuss your answers in detail; would you require permits that specify time of day burning, or distance from shore of burning, or banning certain wastes from incineration?

22.4 Emissions from a nuclear power generating facility pose a unique set of problems for local health officials. Such facilities typically have very low emissions, but each has the potential for a catastrophic discharge. What precautions would you take if you were charged with protecting the air quality of nearby residents. How would your set of rules consider: (1) direction of prevailing winds, (2) age of residents, and (3) siting new schools?

22.5 Assume you live in a small town that has two stationary sources of air pollution: a laundry/dry cleaning establishment and a regional hospital. Automobiles and front porches in the town have a habit of turning black literally overnight. What air laws apply in this situation, and given that they presumably are not being enforced, how would you: (1) determine if, in fact, ambient or emissions standards or both were being violated and, if either was, (2) advise the residents to respond to see that the laws are enforced?

22.6 Emission standards for automobiles are set in units of grams per mile. Why not ppm or grams per passenger-mile?

LIST OF SYMBOLS

        EPA = U.S. Environmental Protection Agency
      LAER = lowest achievable emission rate
   NAAQS = National Ambient Air Quality Standards
NESHAPS = National Emission Standards for Hazardous Air Pollutants
      PSD = prevention of significant deterioration
     AQIP = [State] Air Quality Implementation Plan

Chapter 23
Noise Pollution

The ability to make and detect sound provides humans with the facility to communicate with each other as well as to receive useful information from the environment. Sound can provide warning (the first alarm), useful information (a whistling tea kettle), and enjoyment (music).

In addition to such useful and pleasurable sounds there is noise, often defined as unwanted or extraneous sound. What is and is not noise is often in dispute, especially in the area of music.*

We generally think of noise as an unwanted by-product of our civilization and classify such sources as trucks, airplanes, industrial machinery, air conditioners, and similar sound producers as noise.

Urban noise is not, surprisingly, a modern phenomenon. Legend has it, for example, that Julius Caesar forbade the driving of chariots on Rome cobblestone streets during nighttime so he could sleep.[1] Before the widespread use of automobiles, the 1890 noise pollution levels in London were described by an anonymous contributor to the *Scientific American* as[2]:

> The noise surged like a mighty heart-beat in the central districts of London's life. It was a thing beyond all imaginings. The streets of workaday London were uniformly paved in "granite" sets ... and the hammering of a multitude of iron-shod hairy heels, the deafening side-drum tattoo of tyred wheels jarring from the apex of one set (of cobblestones) to the next, like sticks dragging along a fence; the creaking and groaning and chirping and rattling of vehicles, light and heavy, thus maltreated; the jangling of chain harness, augmented by the shrieking and bellowings called for from those of God's creatures who desired to impart information or proffer a request vocally—raised a din that is beyond conception. It was not any such paltry thing as noise.

---

* George Bernard Shaw entered a posh London restaurant, took a seat, and was confronted by the waiter. "While you are eating sir, the orchestra will play anything you like. What would you like them to play?" Shaw's repy? "Dominoes."

Noise may adversely affect humans both physiologically as well as psychologically. It is an insidious pollutant in that damage is usually long range and permanent. And yet it is certainly the pollutant of least public concern, and (except for radioactivity) the least understood.

THE CONCEPT OF SOUND

The man-on-the-street has little if any concept of what sound is or what it can do. This is perhaps best exemplified by a well-meaning industrial plant manager who decided to decrease the noise level in his factory by placing microphones in the plant and channelling the noise through loudspeakers to the outside.[3]

Sound is in effect a transfer of energy. For example, rocks thrown at you would certainly get your attention, but this would require the transfer of mass (rocks). Alternatively, your attention may be gained by poking you with a stick, in which case the stick is not lost, but energy is transferred from the poker to the pokee.[3] In the same way, sound travels through a medium such as air without a transfer of mass. Just as the stick had to move back and forth, so must air molecules oscillate in waves to transfer energy.

The small displacement of air molecules that creates pressure waves in the atmosphere is illustrated in Figure 23–1. As the piston is forced to the right in the tube, the air molecules next to it are reluctant to move and instead pile up on the face of the piston (Newton's First Law). These compressed molecules now act as a spring, and release the pressure by jumping forward, creating a wave of compressed air molecules that moves through the tube. The potential energy has been converted to kinetic energy.

These pressure waves move down the tube at a velocity of 344 m/sec (at 20°C). If the piston oscillates at a frequency of, say, 10 cycles/sec, there would be a series of pressure waves in the tube each 34.4 m apart. This relationship is expressed as

$$\lambda = \frac{c}{f}$$

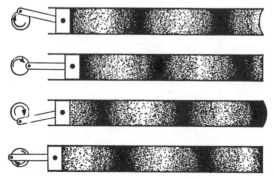

Figure 23–1. A piston creates pressure waves that are transmitted through air.

where λ = wavelength, m
  c = velocity of the sound in a given medium, m/sec
  f = frequency, cycles/sec

Sound travels at different speeds in different materials, depending on the material's elasticity.

Example 23.1
  In cast iron, sound waves travel at about 3,440 m/sec. What would be the wavelength of a sound from a train if it rumbles at 50 cycles/sec and one listens to it placing an ear on the track?

$$\lambda = \frac{c}{f} = \frac{3,440}{50} \cong 69 \text{ m}$$

In acoustics, the frequency as cycles per second is denoted by the name hertz, and written Hz.* The common audible range for humans is between 20 and 20,000 Hz. The middle A on the piano, for example, is 440 Hz. The frequency is one of the two basic parameters that describe sound. Amplitude is the other.

If the amplitude of a pressure wave of a pure sound (a sound with only one frequency) is plotted against time, the wave is seen to produce a sinusoidal trace (Figure 23–2). All other nonrandom (dirty) sounds are made up of a number of suitable sinusoidal waves, as demonstrated originally by Fourier.

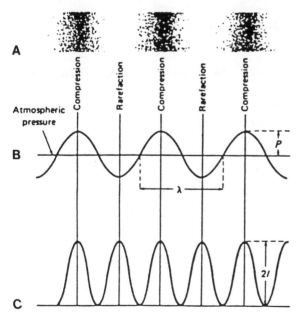

Figure 23–2. Sound waves. All parts of the figure show the spatial variation along the wave at a particular instant of time. (A) Regions of compression and rarefaction in the air. (B) Pressure wave; P = pressure amplitude ($P_{rms} = P/\sqrt{2}$) and λ = wavelength. (C) Intensity wave; the average intensity is denoted I.

---

* In honor of German physicist Heinrich Hertz (1857–1894).

Although the human ear is a remarkable instrument, able to detect sound pressures over 7 orders of magnitude, it is not a perfect receptor of acoustic energy. In the measurement and control of noise, it is therefore important to know not only what a sound pressure is, but also to have some notion of how loud a sound *seems* to be. Before we address that topic, however, we must review some basics of sound.

SOUND PRESSURE LEVEL, FREQUENCY, AND PROPAGATION

Figure 23–2 represents a wave of pure sound: a single frequency. A sound wave is a compression wave, and the amplitude is a pressure amplitude, measured in pressure units like N/m^2. As is the case with other wave phenomena, intensity is the square of the amplitude, or

$$I = P^2$$

The intensity of a sound wave is measured in watts.

When a person hears sounds of different intensities, the total intensity that is heard is not the sum of the intensities of the different sounds. Rather, the human ear tends to become overloaded or saturated with too much sound. Another statement of this phenomenon is that human hearing sums up sound intensities logarithmically rather than linearly. A unit called the *bel* was invented to measure sound intensity. Sound intensity level (IL) in bels is defined as

$$IL_b = \log_{10}(I/I_0)$$

where I = sound intensity in watts

I_0 = intensity of the least audible sound, usually given as $I_0 = 10^{-12}$ watts

The bel is an inconveniently large unit. The more convenient unit, which is now in common usage, is the *decibel (dB)*. Sound intensity level in dB is defined as

$$IL = 10 \log_{10}(I/I_0)$$

Since intensity is the square of pressure, an analogous equation may be written for sound pressure level (SPL) in dB

$$SPL(as\ dB) = 20 \log_{10} \frac{P}{P_{ref}}$$

where SPL (dB) = sound pressure level in dB

P = pressure of sound wave

P_{ref} = some reference pressure, generally chosen as the threshold of hearing, 0.00002 N/m^2

These relationships are also derivable from a slightly different point of view. In 1825, E.H. Weber found that people can perceive differences in small weights, but if a person is already holding a substantial weight, that same increment is not detectable. The same idea is true with sound. For example, a $2\,N/m^2$ increase from an initial sound pressure of $2\,N/m^2$ is readily perceived, whereas the same $2\,N/m^2$ difference is not noticed if it is added to a background of $200\,N/m^2$ sound pressure. Mathematically, this may be expressed as

$$ds = K\,\frac{dW}{W}$$

where   ds = is the minimum perceptible increase in a sensation (e.g., hearing)
      W = the load (e.g., background sound pressure level)
    dW = the change in the load
     K = a constant

The integrated form of this concept is known as the Weber–Fechner Law

$$s = K \log_{10} W$$

This idea is used in the definition of a *decibel*:

$$dB = 10 \log_{10} \frac{W}{W_{ref}}$$

where the power level W is divided by a constant that is a reference value, both measured in watts.

Since the pressure waves in air are half positive pressure and half negative pressure, adding these would result in zero. Accordingly, sound pressures are measured as root-mean-square (rms) values, which are related to the energy in the wave. It may also be shown that the power associated with a sound wave is proportional to the square of the rms pressure. We can therefore write

$$dB = 10 \log_{to} \frac{W}{W_{ref}} = 10 \log_{10} \frac{P^2}{P_{ref}^2}$$

$$= 10 \log_{10}\left(\frac{P}{P_{ref}}\right)^2 = 20 \log_{10} \frac{P}{P_{ref}}$$

where P and P_{ref} are the pressure and reference pressure. If we define $P_{ref} = 0.00002\,N/m^2$, the threshold of hearing, we can define the *sound pressure level* (SPL) as

$$SPL\ (dB) = 20 \log_{10} \frac{P}{P_{ref}}$$

Although it is common to see the sound pressure measured over a full range of frequencies, it is sometimes necessary to describe a noise by the amount of sound pressure present at a specific range of frequencies. Such a *frequency analysis*, shown in Figure 23–3, may be used for solving industrial problems or evaluating the danger of a certain sound to the human ear.

In addition to amplitude and frequency, sound has two more characteristics of importance. Both may be visualized by imagining the ripples created by dropping a pebble into a large, still pond. The ripples, analogous to sound pressure waves, are reflected outward from the source, and the magnitude of the ripples is dissipated as they get farther away from the source. Similarly, sound levels decrease as the distance between the receptor and source is increased.

In summary, then, the four important characteristics of sound waves are:

1. Sound pressure is the magnitude or amplitude of sound.
2. The pitch is determined by the frequency of the pressure fluctuations.
3. Sound waves propagate away from the source.
4. Sound pressure decreases with increasing distance from the source.

But these characteristics ignore the human ear. We know that the ear is an amazingly sensitive receptor, but is it equally sensitive at all frequencies? Can we hear low and high sounds equally well? The answers to these questions lead us to the concept of *sound level*, which is discussed in the next chapter.

The mathematics of adding decibels is a bit complicated. The procedure may be simplified a great deal by using the graph shown as Figure 23–4. As a rule of thumb, adding two equal sounds increases the SPL by 3 dB, and if one

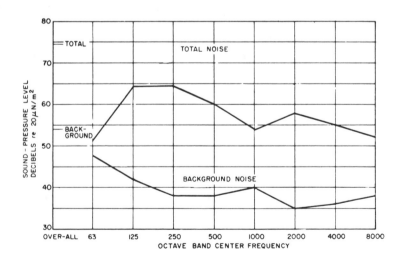

Figure 23–3. Typical frequency analysis of a machine noise with background noise.

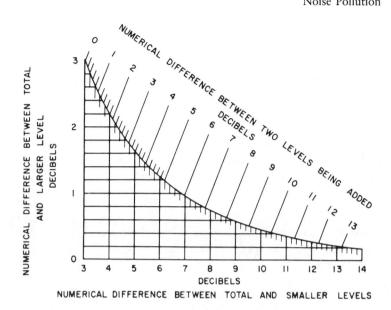

NUMERICAL DIFFERENCE BETWEEN TOTAL AND LARGER LEVEL DECIBELS

NUMERICAL DIFFERENCE BETWEEN TOTAL AND SMALLER LEVELS

DECIBELS

Figure 23–4. Chart for combining different sound pressure levels. For example: Combine 80 and 75 dB. The difference is 5 dB. The 5-dB line intersects the curved line at 1.2 dB, thus the total value is 81.2 dB. [Courtesy of General Radio.]

sound is more than 10 dB louder than a second sound, the contribution of the latter is negligible.

Background noise (or ambient noise) must also be subtracted from any measured noise. Using the above rule of thumb, if the sound level is more than 10 dB greater than the ambient level, the contribution may be ignored.

The covering of a sound with a louder one is known as masking. Speech can be masked by industrial noise, for example, as shown in Table 23–1. These data show that an 80-dB SPL in a factory will effectively prevent conversation. Telephone conversations are similarly affected, with a 65-dB background making communication difficult and 80-dB making it impossible. In some cases, it has been found advisable to use *white noise*, a broad frequency hum, to mask other more annoying noises.

Table 23–1. Sound Levels for Speech Masking

Distance (ft)	Speech Interference Level (dB)	
	Normal	Shouting
3	60	78
6	54	72
12	48	66

Example 23.1

A jet engine has a sound intensity level of 80 dB, as heard from a distance of 50 feet. A ground crew member is standing 50 feet from a four-engine jet. What IL does she hear when the first engine is turned on? the second, so that two engines are running? the third? then all four?

When the first engine is turned on, the ground crew hears 80 dB (provided there is no other comparable noise in the vicinity). To determine, from the chart of Figure 23–3, what they hear when the second engine is turned on, we note that the difference between the two engine intensity levels is

$$80 - 80 = 0$$

From the chart, a numerical difference of 0 between the two levels being added gives a numerical difference of 3 between the total and the larger of the two. The total IL is thus

$$80 + 3 = 83 \, dB$$

When the third engine is turned on, the difference between the two levels being added is

$$83 - 80 = 3$$

yielding a difference from the total of 1.8, for a total IL of

$$83 + 1.8 = 84.8 \, dB$$

When all four engines are turned on, the difference between the sounds being added is

$$84.8 - 80 = 4.8$$

yielding a difference from the total of 1.2, for a total IL of 86 dB.

THE ACOUSTIC ENVIRONMENT

The types of sound around us vary from a Rachmaninoff concerto to the roar of a jet plane. Noise, the subject of this chapter, is generally considered to be unwanted sound, or the sound incidental to our civilization that we would just as soon not have to put up with. The intensities of some typical environmental noises are shown in Figure 23–5.

Laws against noise abound. Most local jurisdictions have ordinances against "loud and unnecessary noise." The problem is that precious few of these

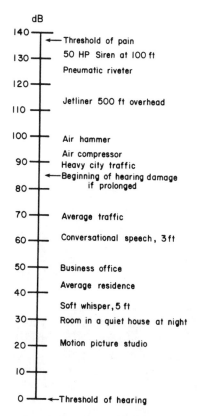

dB

140	←—Threshold of pain
130	50 HP Siren at 100 ft
	Pneumatic riveter
120	
	Jetliner 500 ft overhead
110	
100	Air hammer
	Air compressor
90	Heavy city traffic
	←—Beginning of hearing damage
80	if prolonged
70	Average traffic
60	Conversational speech, 3 ft
50	Business office
	Average residence
40	
	Soft whisper, 5 ft
30	Room in a quiet house at night
20	Motion picture studio
10	
0	←—Threshold of hearing

Figure 23–5. Environmental noise.

ordinances are enforced. Only recently have some cities begun to enforce the laws already on the books. Memphis, for example, gave 10,000 traffic tickets one year for hornblowing. Recently, the churchbells of a Chicago church were muffled because they were beamed directly into nearby apartments. Traffic cops in Illinois are handing our citations to truck drivers for overly loud trucks.

"But officer! I wasn't going too fast."

"No, but you were going too loud."

In the industrial environment, noise is regulated by federal legislation called the Occupational Safety and Health Act (OSHA), which sets limits for noise in the working place. Table 23–2 lists these limits.

There is some disagreement as to the level of noise that should be allowed for an 8-hour working day. Some researchers and health agencies insist that 85 dB(A)* should be the limit. This is not, as might seem on the surface, a minor quibble, since the jump from 85 to 90 dB is actually an increase of about four times the sound pressure!

---

* The designation of dB(A) is explained in the next chapter.

Table 23–2.   OSHA Maximum Permissible Industrial Noise Levels

Sound Level dB(A)	Maximum Duration During Any Working Day (hr)
90	8
92	6
95	4
100	2
105	1
110	$\frac{1}{2}$
115	$\frac{1}{4}$

HEALTH EFFECTS OF NOISE

In the Bronx borough of New York City, one spring evening, four boys were at play, shouting and racing in and out of an apartment building. Suddenly, from a second-floor window, came the crack of a pistol. One of the boys sprawled dead on the pavement. The victim happened to be thirteen years old, son of a prominent public leader, but there was no political implication in the tragedy. The killer confessed to police that he was a nightworker who had lost control of himself because the noise from the boys prevented him from sleeping.

We have only recently become aware of the devastating psychological effects of noise. The effect of excessive noise on our ability to hear, on the other hand, has been known for a long time.

The human ear is an incredible instrument. Imagine having to design and construct a scale for weighing just as accurately a flea or an elephant. Yet this is the range of performance to which we are accustomed from our ears.

A schematic of the human auditory system is shown in Figure 23–6.

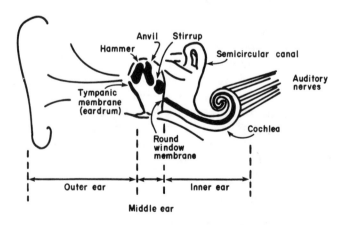

Figure 23–6.   Cut-away drawing of the human ear.

Sound pressure waves caused by vibrations set the eardrum (*tympanic membrane*) in motion. This activates the three bones in the middle ear. The *hammer*, *anvil*, and *stirrup* physically amplify the motion received from the eardrum and transmit it to the inner ear. This fluid-filled cavity contains the *cochlea*, a snail-like structure in which the physical motion is transmitted to tiny hair cells. These hair cells deflect, much like seaweed swaying in the current, and certain cells are responsive only to certain frequencies. The mechanical motion of these hair cells is transformed to bioelectrical signals and transmitted to the brain by the auditory nerves. Acute damage may occur to the eardrum, but this occurs only with very loud sudden noises. More serious is the chronic damage to the tiny hair cells in the inner ear. Prolonged exposure to noise of a certain frequency pattern may cause either temporary hearing loss, which disappears in a few hours or days, or permanent loss. The former is called *temporary threshold shift*, and the latter is known as *permanent threshold shift*. Literally, your threshold of hearing changes, so you are not able to hear some sounds.

Temporary threshold shift is generally not damaging to your ear unless it is prolonged. People who work in noisy environments commonly find that they hear less well at the end of the day. Performers in rock bands are subjected to very loud noises (substantially above the allowable OSHA levels) and commonly are victims of temporary threshold shift. In one study, the results of which are shown in Figure 23–7, the players suffered as much as 15 dB temporary threshold shift after a concert.

Repeated noise over a long time leads to permanent threshold shift. This is especially true in industrial applications in which people are subjected to noises of a certain frequency. Figure 23–8 shows data from a study performed on workers at a textile mill. Note that the people who worked in the spinning and weaving parts of the mill, where noise levels are highest, suffered the most severe loss in hearing, especially at around 4,000 Hz, the frequency of noise emitted by the machines.

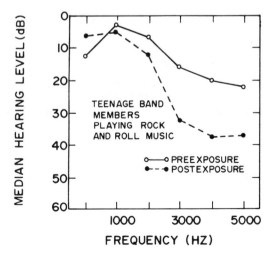

Figure 23–7. Temporary threshold shift for rock band performers. [Source: Data by the U.S. Public Health Service. See reference 4.]

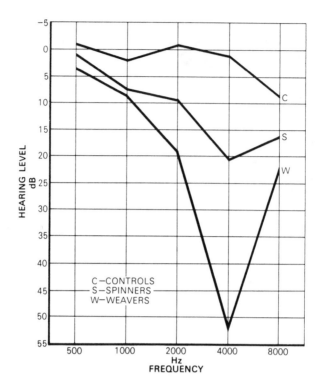

Figure 23–8. Permanent threshold shift for textile workers. [Source: Burns, W. et al. "An Exploratory Study of Hearing and Noise Exposure in Textile Workers," *Ann. Occup. Hyg.* 7:323 (1958).

As people get older, hearing becomes less acute simply as one of the effects of aging. This loss of hearing, called *presbycusis*, is illustrated in Figure 23–9. Note that the greatest loss occurs at the higher frequencies. Speech frequency is about 1,000 to 2,000 Hz, and thus older people commonly accuse others of "whispering."

In addition to presbycusis, however, there is a serious loss of hearing owing to environmental noise. In one study,[4] 11 percent of ninth graders, 13 percent of twelfth graders, and 35 percent of college freshmen had a greater than 15 dB loss of hearing at 2,000 Hz. The study concluded that this severe loss resulted from exposure to loud noises such as motorcycles and rock music and that as a result "the hearing of many of these students had already deteriorated to a level of the average 65-year-old person."

Noise also affects other bodily functions such as the cardiovascular system. Noise alters the rhythm of the heartbeat, makes the blood thicker, dilates blood vessels, and makes focusing difficult. It is no wonder that excessive noise has been blamed for headaches and irritability. Noise is especially annoying to people who do close work, like watchmakers.

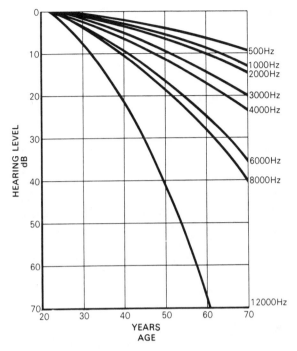

Figure 23–9. Hearing loss with age. [Source: Hinchcliffe, R. "The Pattern of the Threshold of Perception of Hearing and Other Special Senses as a Function of Age," *Gerontologica* 2:311(1958).]

All of the above reactions are those that our ancestral caveman also experienced. Noise to him meant danger, and his senses and nerves were "up," ready to repel the danger. In the modern noise-filled world, we are always "up," and it is unknown how much of our physical ills are due to this.

We also know that we cannot adapt to noise in the sense that our body functions no longer react a certain way to excessive noise. People do not, therefore, "get used to" noise in the physiological sense.

In addition to the noise problem, it might be appropriate to mention the potential problems of very high or very low frequency sound, out of our usual 20 to 20,000 Hz hearing range. The health effects of these, if any, remain to be studied.

THE DOLLAR COST OF NOISE

Numerous case histories comparing patients in noisy and quiet hospitals point to increased convalescence time when the hospital was noisy (either from within or owing to external noise). This may be translated directly to a dollar figure.

Recent court cases have been won by workers seeking damages for hearing loss suffered on the job. The Veterans Administration spends many, many millions of dollars every year for care of patients with hearing disorders.

Other costs, such as sleeping pills, lost time in industry, and apartment soundproofing are difficult to quantify. The John F. Kennedy Cultural Center in Washington spent $5 million for soundproofing, necessitated by the jets using the nearby National Airport.

DEGRADATION OF ENVIRONMENTAL QUALITY

It is even more difficult to measure the effect noise has had on the quality of life. How much is noise to blame for irate husbands and grumpy wives, for grouchy taxi drivers and surly clerks?

Children reared in noisy neighborhoods must be taught to listen. They cannot focus their auditory senses on one sound, such as the voice of a teacher.

Within the 80- to 90-dB circle around the Kennedy airport in New York are 22 schools. Every time a plane passes the teacher must stop talking and try to re-establish attention.

The effect of these insults cannot be measured. Yet we intuitively feel that the effect *must* be negative.

CONCLUSION

Noise is a real and dangerous form of environmental pollution. Since people cannot adapt to it physiologically, we are perhaps adapting psychologically instead. Noise may keep our senses "on edge" and prevent us from relaxing. Our mental powers must therefore control this insult to our bodies. Since noise, in the context of human evolution, is a very recent development, we have not yet adapted to it, and must thus be living on our buffer capacity. One wonders how plentiful this is.

PROBLEMS

23.1 If an office has a noise level of 70 dB and a new machine emitting 68 dB is added to the din, what is the combined sound level?

23.2 The OSHA standard for 8-hour exposure to noise is 90 dB(A). The Environmental Protection Agency (EPA) suggests that this should be 85 dB(A). Show that the OSHA level is almost 400 percent louder in terms of sound level than the EPA suggestion.

23.3 A machine with an overall noise level of 90 dB is placed in a room with another machine putting out 95 dB.
   a. What will be the sound level in the room?
   b. Based on OSHA criteria, how long should workers be in the room during one working day?

23.4 The advertisement in Figure 23–10 shows a white noise generator.
   a. What is "white noise"?
   b. Discuss the effects of such white noise on health.

(E) Low-Cost White Sound Conditioner
An ideal "sound screen", this model makes "white sounds" only. Produces a steady, pleasant "breeze-like" sound many find helpful for study, reading & sleep. Uses 115 V AC. 3½" high x 6" wide.
No. 71,980   **$33.95**

Figure 23–10. Advertisement for a white noise generator. [Courtesy of Edmund Scientific.]

23.5 What sound pressure level results from the combination of three sources: 68 dB, 78 dB, and 72 dB?

23.6 If an occupational noise standard were set at 80 dB for an 8-hour day, 5-day-per-week working exposure, what standard would be appropriate for a 4-hour day, 5-day-per-week working exposure?

23.7 What reduction in sound intensity would have been necessary to reduce the takeoff noise of the American SST from 120 dB to 105 dB?

LIST OF SYMBOLS

C = velocity of sound wave, m/sec
dB = decibels
f = frequency of sound wave, cycle/sec
P = pressure, N/m^2
P_{ref} = reference pressure, generally stated as 0.00002 N/m^2
SPL = sound pressure level
λ = wavelength, m

REFERENCES

1. Baron, R.A. *The Tyranny of Noise* (New York: St. Martins Press, 1970).
2. Still, H. *In Quest of Quiet* (Harrisburg, PA: Stackpole Books, 1970).
3. Taylor, R. *Noise* (New York: Penguin Books, 1970).
4. Lipscomb, D.M., reported in White, F.A., *Our Acoustic Environment* (New York: John Wiley & Sons, Inc., 1975).

Chapter 24

Noise Measurement and Control

Sound and noise are unique among environmental pollutants in that they may be measured easily, accurately, and precisely. A number of techniques exist for measuring loudness that, in one way or another, duplicate the logarithmic response of the human ear. These techniques may be used both to establish and to predict profiles of noise sources, such as airports.

SOUND LEVEL

Suppose you were put into a very quiet room and subjected to a pure tone at 1,000 Hz at 40-dB sound pressure level (SPL). If, in turn, this sound was turned off and a pure sound at 100 Hz was piped in and adjusted in loudness until you judged it to be "equally loud" to the 40-dB, 1,000-Hz tone you had just heard a moment ago, you would, surprisingly enough, judge the 100-Hz tone to be equally loud when it was at about 55 dB. In other words, more energy is needed at the lower frequency to hear a tone at the same loudness, indicating that the human ear is rather inefficient for low tones.

It is possible to conduct such experiments for many sounds and with many people and to draw average equal loudness contours (Figure 24–1). These contours are in terms of *phons*, which correspond to the sound pressure level in decibels of the 1,000-Hz reference tone.

Using Figure 24–1, a person subjected to a 65-dB SPL tone at 50 Hz would judge this to be equally as loud as a 40-dB, 1,000-Hz tone. Hence, the 50-Hz, 65-dB sound has a *loudness level* of 40 phons.

Such measurements are commonly called *sound levels* and are not based on basic physical phenomena only, but have a "fudge factor" thrown in that corresponds to the inefficiency of the human ear.

479

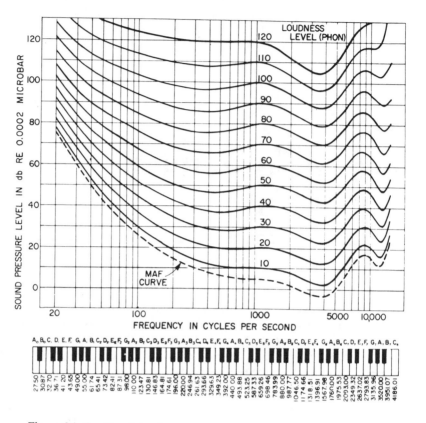

Figure 24–1. Equal loudness contours. [Courtesy of General Radio.]

Sound level (SL) is measured with a *sound level meter* consisting of a microphone, amplifier, a frequency-weighing circuit (filters), and an output scale. The sound level meter is shown schematically in Figure 24–2, and Figure 24–3 shows a typical hand-held sound level meter. The weighing network filters out specific frequencies to make the response more characteristic of human hearing. Through use, three scales have become internationally standardized (Figure 24–4).

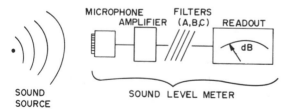

Figure 24–2. Schematic representation of a sound level meter.

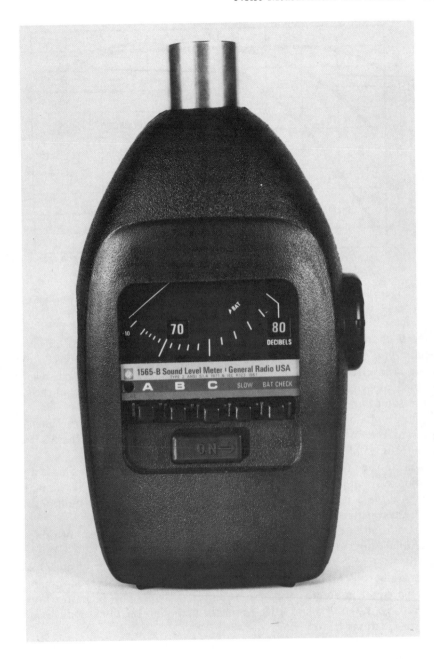

Figure 24–3. A typical sound level meter. [Courtesy of General Radio.]

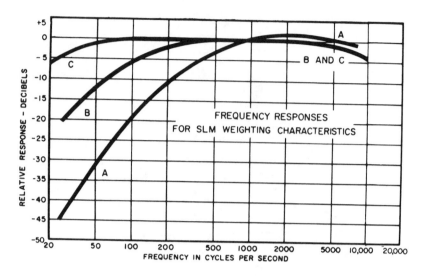

Figure 24–4. The A, B, and C filtering curves for a sound level meter.

Note that the A scale in Figure 24–4 corresponds closely to an inverted 40-phon contour in Figure 24–1. Similarly, the B scale in response approximates an inverted 70-phon contour. The C scale is an essentially flat response, giving equal weight to all frequencies. It approximates the response of the ear to intense sound pressure levels.

The results of noise measurement with the standard sound level meter are expressed in terms of decibels but with the scale designated. If on the A scale the meter reads 45 dB, the measurement is reported as 45 dB(A).

Most noise ordinances and regulations are in terms of dB(A). This is a good approximation of human response for not very loud sounds. For very loud noise the C scale is a better approximation, but because the use of multiple scales complicates matters, scales other than A are seldom used.

In addition to the A, B, and C scales, a new D scale has been introduced to approximate human response to aircraft noise (exclusively).

This is another complication involved in the measurement of noise. We do not generally respond the same way to different types of noise, even though they might be equal in sound level. For example, a symphony orchestra hitting a concert C at 120 dB(A) and a jet engine at an equal 120 dB(A) will draw very different reactions from people.

In response to this problem, a plethora of parameters have been devised, all supposedly "the best" means of quantitatively measuring human response to noise. Among these are:

- Traffic Noise Index (TNI)
- Sones
- Perceived Noise Level (PNdB)
- Noise and Number Index (NNI)

- Effective Perceived Noise Level (EPNdB)
- Speech Interference Level (SIL)

More parameters are suggested all the time.

The reason for this proliferation is that a physical phenomenon (pressure waves) is related to physiological human response (hearing) and then to psychological response (pleasure or irritation). Along the way, the science becomes progressively more subjective.

There is a further complication in measuring some noises, particularly those commonly called "community noise," such as traffic, loud parties, etc. Although we have thus far treated noise as if it were constant in intensity and frequency with time, this obviously is not true for transient noises such as trucks moving past a sound level meter or loud parties.

MEASURING TRANSIENT NOISE

Transient noise is still measured with a sound level meter, but the results must be reported in statistical terms. The common parameter is the *percent of time a sound level is exceeded*, denoted by the letter L with a subscript. For example, $L_{10} = 70$ dB(A) means that 10 percent of the time the noise was louder than 70 dB as measured on the A scale.

Transient noise data are gathered by reading the SL at regular intervals. These numbers are then ranked and plotted, and the L values are read off the graph.

Example 24.1

Suppose the traffic noise data in Table 24–1 were gathered at 10-second intervals. These numbers are then ranked as indicated in the table and plotted as in Figure 24–5. Note that since 10 readings were taken, the lowest reading (Rank No. 1) corresponds to a SL that is equalled or exceeded 90 percent of the time. Hence, 70 dB(A) is plotted versus 90 percent in Figure 24–5.* Similarly, 71 dB(A) is exceeded 80 percent of the time.

One widely used parameter for gauging the perceived level of noise from transient sources is the noise pollution level (NPL), which takes into account the irritability of impulse (sudden) noises. The NPL is defined as

$$\text{NPL in dB(A)} = L_{50} + (L_{10} - L_{90}) + \frac{(L_{10} - L_{90})^2}{60}$$

---

* You could also argue that since 70 dB(A) is the lowest value, it is the SL exceeded 100 percent of the time and, therefore, you could plot 70 dB(A) versus 100 percent. The second value, 71 dB(A), is exceeded 90 percent of the time, etc. Either method is correct, and the error diminishes as the number of data points increases.

Table 24–1. Sample Traffic Noise Data and Calculations for Example 24.1

Data		Rank No.	% of Time Equal to or Exceeded	dB(A)
Time (sec)	dB(A)			
10	71	1	90	70
20	75	2	80	71
30	70	3	70	74
40	78	4	60	74
50	80	5	50	75
60	84	6	40	75
70	76	7	30	76
80	74	8	20	78
90	75	9	10	80
100	74	10	0	84

As defined earlier, the symbol L refers to the percentage of time the noise is equal to or greater than some value. The percentage is indicated by the subscript. L_{10} is thus the dB(A) level exceeded 10 percent of the time.

With reference to Figure 24–5, L_{10}, L_{50}, and L_{90} are thus 80, 75, and 70 dB(A), respectively. These may then be substituted into the equation, and the NPL may be calculated.

It is always advisable to take as much data as possible. The 10 readings illustrated above are seldom sufficient for a thorough analysis. The percentages

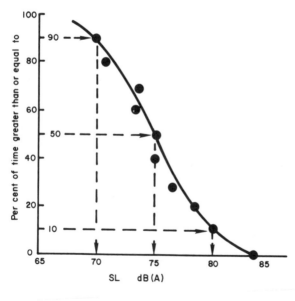

Figure 24–5. Results from a survey of transient noise (data from Table 24–1). The data are plotted as percentage of time the sound level (SL) is exceeded, versus the SL in dB(A).

must be calculated on the basis of the total number of readings taken. If, for example, 20 readings are recorded, the lowest (Rank No. 1) corresponds to 95 percent, the second lowest to 90 percent, etc.

NOISE CONTROL

The control of noise is possible at three levels:

1. reducing the sound produced
2. interrupting the path of the sound
3. protecting the recipient

When we consider noise control in industry, in the community, or in our home, we should keep in mind that all problems have these three possible solutions.

Industrial Noise Control

Industrial noise control generally involves the replacement of noise-producing machinery or equipment with quiet alternatives. For example, the noise from an air fan may be reduced by increasing the number of blades or the pitch of the blades and decreasing the rotational speed (thus obtaining the same air flow).

The second method of decreasing industrial noise is to interrupt the path, for example, by covering a noisy motor with insulating material.

The third method of noise control, often used in industry, is to protect the recipient by distributing earmuffs to the employees. But the problem is thus not really solved, and often the workers refuse to wear earmuffs, considering them sissy. There seems to be something masculine about being able to "take it." Such misdirected masculinity will only end up in deafness.

The Occupational Safety and Health Act applies industrial noise limits to all jobs involving significant federal funds. These limits are tabulated in Chapter 23 as Table 23–2.

Community Noise Control

The three major sources of community noise are aircraft, highway traffic, and construction.

Construction noise must be controlled by local ordinances (unless federal funds are involved). Control usually involves the muffling of air compressors, jack hammers, hand compactors, etc. Since mufflers cost money, contractors will not take it upon themselves to control noise, and outside pressures must be exerted.

Aircraft noise in the United States is the province of the Federal Aviation Administration, which has instituted a two-pronged attack on this problem. First, it has set limits on aircraft engine noise and will not allow aircraft exceeding these limits to use the airports. This has forced manufacturers to design for quiet as well as for thrust.

The second effort has been to divert flight paths away from populated areas and whenever necessary to have pilots use less than maximum power when the takeoff carries them over a noise-sensitive area. Often this approach is not enough to prevent significant noise-induced damage or annoyance, and aircraft noise remains a real problem in urban areas.

Supersonic aircraft present a special problem. Not only are their engines noisy, but the sonic boom may create property damage and mental anguish. The magnitude of this problem will become known only when (and if) supersonic airlines begin regular service over land.

The third major source of community noise is from highways. The car or truck creates noise by a number of means:

1. exhaust noise
2. tire noise
3. engine intake noise
4. gears and transmission
5. aerodynamic (wind) noise

A modern passenger car is so well muffled that its most important contribution, at moderate and high speeds, is tire noise. Sports cars and motorcycles, on the other hand, contribute exhaust, intake, and gear noise.

The worst offender on the highways is the heavy truck. Truck noise is generally from all of the above sources. In most cases, the total noise generated by vehicles may be correlated directly to the truck volume. Figure 24–6 is a

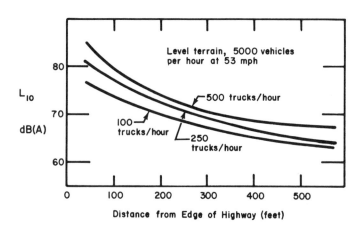

Figure 24–6. The effect of truck density and distance from a highway.

Table 24–2. Design Noise Levels Set by the Federal Highway Administration

Land Category	Design Noise Level, L_{10}	Description of Land Use
A	60 dB(A) exterior	Activities requiring special qualities of serenity and quiet, such as amphitheatres
B	70 dB(A) exterior	Residences, motels, hospitals, schools, parks, libraries
	55 dB(A) interior	Residences, motels, hospitals, schools, parks, libraries
C	75 dB(A) exterior	Developed land not covered in categories A and B
D	No limit	Undeveloped land

typical plot showing sound level as a function of traffic volume (measured in number of trucks per hour). Clearly, truck volume is of great importance. Incidentally, this graph is plotted as "sound level exceeded 10 percent of the time." Peak sound levels could be a great deal higher.

A number of alternatives are available for reducing highway noise. First, the source could be controlled by making quieter vehicles; second, highways could be routed away from populated areas; and third, noise could be baffled with walls or other types of barriers. Other methods include lowering speed limits, designing for nonstop operation, and reducing all highways to less than 8 percent grade.

Vegetation, surprisingly, makes a very poor noise screen. The most effective buffers have been to raise or lower the highway, or to build physical barriers beside the road and thus screen the noise. All of these have limitations. For example, noise will bounce off the walls and create little or no noise shadow. In addition, walls hinder highway ventilation, thus contributing to the buildup of dangerous air pollutants.

The Department of Transportation has established design noise levels for various land uses, as shown in Table 24–2.

Noise in the Home

Private dwellings are getting noisier because of internally produced noise as well as external community noise. The list of gadgets in a modern American home reads like a list of Halloween noisemakers. Some examples of domestic noise are listed in Table 24–3.

Otherwise similar products of different brands often will vary significantly in noise levels. When shopping for an appliance, it is just as important therefore to ask the clerk "How noisy is it?" as it is to ask him "How much does it cost?"

Table 24–3.   Some Domestic Noisemakers

Item	Sound Level dB(A)
Vacuum cleaner (10 ft)	75
Inside quiet car (50 mph)	65
Inside sports car (50 mph)	80
Flushing toilet	85
Garbage disposal (3 ft)	80
Window air conditioner (10 ft)	55
Ringing alarm clock (2 ft)	80
Lawn mower (operator's position)	105
Snowmobile (driver's position)	120
Rock band (10 ft)	115

And if he looks at you as if you had two heads, explain to him that he should know the dB(A) at the operator's position for all of his wares. He may actually bother to find out.

CONCLUSION

As long as noise was considered just another annoyance in a polluted world, not much attention was given to it. We now have enough data to show that noise is a definite health hazard and should be numbered among our more serious pollutants. It is possible, using available technology, to lessen this form of pollution. However, the solution costs money, and private enterprise cannot afford to give noise a great deal of consideration until forced to by either the government or the consuming public.

PROBLEMS

24.1 Given the following noise data, calculate the L_{10} and L_{50}.

Time (sec)	dB(A)	Time (sec)	dB(A)
10	70	60	65
20	50	70	60
30	65	80	55
40	60	90	70
50	55	100	50

24.2   In addition to the data listed in Example 24.1 (Table 24–1), the following SL measurements were taken:

Time (sec)	dB(A)	Time (sec)	dB(A)
110	80	160	95
120	82	170	98
130	78	180	82
140	87	190	88
150	92	200	75

Calculate the L_{50}, L_{10}, and NPL by using all 20 data points.

24.3   Suppose your dormitory is 200 yards from a highway. What truck traffic volume would be "allowable" to stay within the Federal Highway Administration guidelines?

24.4   If the SL were 80 dB(C) and 60 dB(A), would you suspect that most of the noise was of high, medium, or low frequency? Why?

24.5   If the pressure wave was $0.3 \, N/m^2$, what is the SPL in decibels (re $20 \times 10^{-6} \, N/m^2$)? If this is all the information about the sound you have, what can you say about the SL in dB(A)? What data would you need to make a more accurate estimate of the SL in dB(A)?

24.6   Many animals hear better than humans. Dogs, for example, can hear sounds at pressures close to $2 \times 10^{-6} \, N/m^2$. What is this in decibels?

24.7   If you sing at a level 10,000 times greater than the power of the faintest audible sound, at which decibel level are you singing?

24.8   How many times more powerful is a 120-dB sound than a 0-dB sound?

24.9   On a graph of decibels versus hertz (10 to 50,000), show a possible frequency analysis for: (a) a passing freight train, (b) a dog whistle, (c) "white noise."

24.10   Carry a sound level meter with you for one entire day. Measure and record the sound levels as dB(A) in classes, in your room, during sports events, in the dining halls, or wherever you go during the day.

24.11   Seek out and measure the three most obnoxious noises you can think of. Compare these with the noises in Table 24–3.

24.12   In your room measure and plot the sound level in dB(A) of an alarm clock versus distance. At what distance will it still wake you if it requires 70 dB(A) to get you up? Draw the same curve outside. What is the effect of your room on the sound level?

24.13   Construct a sound level frequency curve for a basketball game. Calculate the noise pollution level.

24.14   A noise is found to give the following responses on a sound level meter: 82 dB(A), 83 dB(B), and 84 dB(C). Is the noise of a high or low frequency?

24.15   A machine produces 80 dB(A) at 100 Hz (almost pure sound).

a. Would a person who has suffered a noise-induced threshold shift of 40 dB at that frequency be able to hear this sound? Explain.

b. What would this noise measure on the C scale of a sound level meter?

LIST OF SYMBOLS

dB = decibel

Hz = hertz, cycles/sec

L_x = x percent of the time the stated sound level (L) was exceeded, percentage

NPL = noise pollution level

P = pressure, N/m^2

P_{ref} = reference pressure, N/m^2

s = sensation (hearing, touch, etc.)

SL = sound level

SPL = sound pressure level

W = power level, watts

Chapter 25

Environmental Impact and Economic Assessment

Ideally, engineers are expected to respond to a given problem in a rational manner. Decision-making in the public sector should follow a definite sequence: (1) problem definition, (2) generation of alternative solutions, (3) evaluation of alternatives, (4) implementation of a selected solution, and (5) review and appropriate revision of the implemented solution. This process is even mandated by law. At the federal level the National Environmental Policy Act (NEPA) requires that alternative solutions to a problem be considered by federal agencies. This is also true at the state level where, for example, a similar law in Wisconsin (WEPA) requires state agencies to make a conscious effort to develop alternatives before a solution to a problem is implemented. In some states agencies also operate under the planning, programming, and budgeting system (PPBS), which again stresses the need to consider alternative solutions to problems they address.

In reality, this rational planning sequence is seldom followed. Charles Lindbloom argues that *disjointed incrementalism* ("muddling through") is the more accurate description of the decision-making process, be it at the federal, state, or local level.[1] Nevertheless, more recent national and state policies, such as NEPA, WEPA, and PPBS, are law, and their intent has been to integrate the muddling through with some degree of more rational planning. The mandate to generate and consider all feasible solutions is an example.

The result, for better or worse, has been a hybrid strategy of "muddling through" in which limited groups of alternative solutions are proposed and studied successively, with a choice being made de facto when all relevant parties cease to disagree. In decision-making theory, this particular process has been called *mixed scanning*.[2] Frequently used in administrative agencies by design or default, it has proven practical, with the majority of day-by-day middle-level decisions going unchallenged and the credibility of most agencies remaining stable. However, mixed scanning falls short of achieving the ideal of rational planning, in which a broad range of feasible alternative solutions are reviewed simultaneously rather than successively. Recent trends in court action have resulted in delaying or stopping the implementation of many solutions because alternatives were not adequately developed.

It is granted that administrative constraints of time, limited information, expertise, expense, and often conflicting preordinated priorities make it very hard to realize the ideal of rational planning. What ought not be allowed, however, is the lack of knowledge of how to perform the steps of rational planning. It is one thing to require that alternatives be reviewed simultaneously; it is another to know how to generate and evaluate truly alternative solutions for any given problem. This chapter discusses how alternative solutions to environmental engineering are analyzed in terms of their projected environmental and economic impacts.

ENVIRONMENTAL IMPACT

On January 1, 1970, President Nixon signed into law the National Environmental Policy Act, which declared a national policy to encourage productive and enjoyable harmony between people and their environment. This law established the Council on Environmental Quality (CEQ), which monitors the environmental effects of federal activities and assists the President in evaluating environmental problems and determining the best solutions to these problems. But few people realized the NEPA contained a real sleeper article: Section 102(2)(C), which requires federal agencies to evaluate the consequences of any proposed action on the environment:

> The Congress authorizes and directs that, to the fullest extent possible: (1) the policies, regulations, and public laws of the United States shall be interpreted and administered in accordance with the policies set forth in this chapter, and (2) all agencies of the Federal Government shall include in every recommendation or report on proposals for legislation and other major Federal actions significantly affecting the quality of the human environment, a detailed statement by the responsible official on—
> (i) the environmental impact of the proposed action,
> (ii) any adverse environmental effects which cannot be avoided should the proposal be implemented,
> (iii) alternatives to the proposed action,
> (iv) the relationship between local short-term uses of man's environment and the maintenance and enhancement of long-term productivity, and
> (v) any irreversible and irretrievable commitments of resources which would be involved in the proposed action should it be implemented.

In other words, each project funded by the federal government or requiring a federal permit must be accompanied by an environmental impact assessment

(EIS). Such a published statement must assess in detail the potential environmental impacts of a proposed action and alternative actions. All federal agencies are required to prepare statements for projects and programs (a programmatic EIS) under their jurisdiction. Additionally, the agencies must generally follow a detailed and often lengthy public review of each EIS before proceeding with the project or permit. In some instances, legislation allows substitution of a slightly less rigidly prescribed environmental assessment for an EIS.

The original idea of the EIS is to introduce environmental factors into the decision-making machinery. The purpose of the EIS is not to provide justification for a construction project, but rather to introduce environmental concerns and have them discussed in public before the decision on a project is made. However, this objective is difficult to apply in practice. Historically, interest groups in and out of government articulated plans to their liking, the sum of which provided the engineer with a set of alternatives to be evaluated. In many other instances, the engineer is left to create his own alternative or to participate in a group decision-making process like the Delphi technique. In either case, there are normally one or two plans that, from the outset, seem eminently more feasible and reasonable, and these may be legitimatized by juggling time scales or standards of enforcement patterns just slightly and calling them alternatives (as they are in a limited sense). As a result, nondecisions are made,[3] i.e., wholly different ways of perceiving the problems and conceiving the solutions have been overlooked, and the primary objective of an EIS has been circumvented. Over the past few years, court decisions and guidelines by various agencies have, in fact, helped to mold this procedure for the development of environmental impact statements.

Ideally, an EIS must be thorough, interdisciplinary, and as quantitative as possible. The writing of an EIS involves three distinct phases: *inventory*, *assessment*, and *evaluation*. The first is a cataloging of environmentally susceptible areas, the second is the process of estimating the impact of the alternatives, and the last is the interpretation of these findings.

Environmental Inventories

The first step in evaluating the environmental impact of a project or project's alternatives is to inventory factors that may be affected by the proposed action. In this step no effort is made to assess the importance of a variable. Any number and many kinds of variables may be included, such as:

1. the "ologies": hydrology, geology, climatology, and archeology
2. environmental quality: land, surface and subsurface water, air, and sound
3. plant and animal life

This step involves counting, measuring, and describing existing conditions.

Environmental Assessment

The process of calculating projected effects that a proposed action or construction project will have on environmental quality is called environmental assessment. It is necessary to develop a methodical, reproducible, and reasonable method of evaluating both the effect of the proposed project and the effects of alternatives that may achieve the same ends but that may have different environmental impacts. A number of semiquantitative approaches have been used, among them the checklist, the interaction matrix, and the checklist with weighted rankings.

Checklists are listings of potential environmental impacts, both primary and secondary. Primary effects occur as a direct result of the proposed project, such as the effect of a dam on aquatic life. Secondary effects occur as an indirect result of the action. For example, an interchange for a highway will not directly affect a land area, but indirectly it will draw such establishments as service stations and quick food stores, thus changing land use patterns.

The checklist for a highway project could be divided into three phases: planning, construction, and operation. During planning, consideration is given to environmental effects of the highway route and the acquisition and condemnation of property. The construction phase checklist will include displacement of people, noise, soil erosion, water pollution, and energy use. Finally, the operation phase will list direct impacts owing to noise, air pollution, water pollution resulting from runoff, energy use, etc., and indirect impacts owing to regional development, housing, lifestyle, and economic development.

The checklist technique thus simply lists all of the pertinent factors; then the magnitude and importance of the impacts are estimated. The estimation of impact is quantified by establishing an arbitrary scale, such as:

0 = no impact
1 = minimal impact
2 = small impact
3 = moderate impact
4 = significant impact
5 = severe impact

This scale may be used to estimate both the magnitude and the importance of a given item on the checklist. The numbers may then be combined, and a quantitative measurement of the severity of environmental impact for any given alternative may be estimated.

Example 25.1
    A landfill is to be placed in a floodplain of a river. Estimate the impact by using the checklist technique.

First the items impacted are listed, then a judgment concerning both importance and magnitude of the impact is made. In this example, the items are only a sample of the impacts one would normally consider. The numbers in this example are then multiplied and the sum obtained. Thus:

Potential Impact	*Importance* × *Magnitude*
Groundwater contamination	$5 \times 5 = 25$
Surface water contamination	$4 \times 3 = 12$
Odor	$1 \times 1 = 1$
Noise	$1 \times 2 = 2$
Total	40

This total of 40 may then be compared with totals calculated for alternative courses of action.

In the checklist technique most variables must be subjectively judged. Further, it is difficult to predict further conditions such as land-use pattern changes or changes in lifestyle. Even with these drawbacks, however, this method is often used by engineers in governmental agencies and consulting firms, mainly because of its simplicity.

The *interaction matrix* technique is a two-dimensional listing of existing characteristics and conditions of the environment and detailed proposed actions that may impact the environment. This technique is illustrated in Example 25.2. For example, the characteristics of water might be defined as:

- surface
- ocean
- underground
- quantity
- temperature
- groundwater recharge
- snow, ice, and permafrost

Similar characteristics must also be defined for air, land, and other important considerations.

Opposite these listings in the matrix are lists of possible actions. In our example, one such action is labeled *resource extraction*, which could include the following actions:

- blasting and drilling
- surface extraction
- subsurface extraction
- well drilling
- dredging
- timbering
- commercial fishing and hunting

The interactions, as in the checklist technique, are measured in terms of magnitude and the importance. The magnitudes are represented by the extent of the interaction between the environmental characteristics and the proposed actions and typically may be measured. The importance of the interaction, on the other hand, is often a judgment call on the part of the engineer.

If an interaction is present, for example, between underground water and well drilling, a diagonal line is placed in the block. Numbers may then be assigned to the interaction, with 1 being a small and 5 being a large magnitude or importance, and these are placed in the blocks with magnitude above and importance below. Appropriate blocks are filled in, using a great deal of judgment and personal bias, and then are summed over a line, thus giving a numerical grade for either the proposed action or environmental characteristics.

Example 25.2

Lignite (brown) coal is to be surface mined in the Appalachian Mountains. Construct an interaction matrix for the water resources (environmental characteristics) versus resource extraction (proposed actions). We see that the proposed action would have a significant effect on surface water quality and that the surface excavation phase will have a large impact. The value of the technique is seen when the matrix is applied to alternative solutions. The individual elements in the matrix, as well as row and column totals, can then be compared.

Environmental Characteristics	Blasting + drilling	Surface excavation	Subsurface excavation	Well drilling	Dredging	Timbering	Commercial fishing	Total
Surface water	3/2	5/5						8/7
Ocean water								
Underground water		3/3						3/3
Quantity								
Temperature		1/2						1/2
Recharge								
Snow, ice								
Total	3/2	9/10						

Proposed Action (column header group)

This trivial example cannot fully illustrate the advantage of the interaction technique. With large projects having many phases and diverse impacts, it is relatively easy to pick out especially damaging aspects of the project as well as the environmental characteristics that will be most severely affected.

The search for a comprehensive, systematic, interdisciplinary, and quantitative method for evaluating environmental impact has led to the *checklist-with-weighted-rankings* technique. The intent here is to use a checklist as before to ensure that all aspects of the environment are covered, as well as to give these items a numerical rating in common units.

The first step is to construct a list of items that could be impacted by the proposed alternative, grouping them into logical sets. One grouping might be:

- Ecology
    - Species and Populations
    - Habitats and Communities
    - Ecosystems
- Aesthetics
    - Land
    - Air
    - Water
    - Biota
    - Human-Made Objects
- Environmental Pollution
    - Water
    - Air
    - Land
    - Noise
- Human Interest
    - Educational and Scientific
    - Cultural
    - Mood/Atmosphere
    - Life Patterns

Each title might have several specific topics under it; for example, under Aesthetics, "Air" may list: (1) odor, (2) sound, and (3) visual as items in the checklist.

We must now assign ratings to these items, in common units. One procedure is to first estimate the ideal or natural levels of environmental quality (without man-made pollution) and take a ratio of the expected condition to the ideal. For example, if the ideal dissolved oxygen in the stream is 9 mg/L, and the effect of the proposed action is to lower the dissolved oxygen to 3 mg/L, the ratio would be 0.33. This is sometimes called the environmental quality index (EQI). Another option to this would be to make the relationship nonlinear, as shown in Figure 25–1. Lowering the dissolved oxygen by a few milligrams per liter will not affect the EQI nearly as much as lowering it, for example, below 4 mg/L, since a dissolved oxygen below 4 mg/L definitely has a severe adverse effect on the fish population.

EQIs are calculated for all checklist items, and the values are tabulated. Next, the weights are attached to the items, usually by distributing 1,000 parameter importance units (PIU) among the items. The product of EQI and

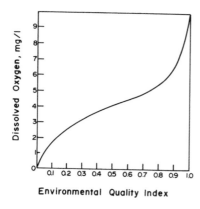

Figure 25–1. Projected environmental quality index curve for dissolved oxygen.

PIU, called the environmental impact unit (EIU), is thus the magnitude of the impact multiplied by the importance.

This method has several advantages. We may calculate the sum of EIUs and evaluate the "worth" of many alternatives, including the do-nothing alternative. We may also detect points of severe impact, for which the EIU after the project may be much lower than before, indicating severe degradation in environmental quality. Its major advantage, however, is that it makes it possible to input data and evaluate the impact on a much less qualitative and a much more objective basis.

Example 25.3

Evaluate the effect of a proposed lignite strip mine on a local stream. Use 10 PIU and linear functions for EQI.

The first step is to list the areas of potential environmental impact. These may be:

- appearance of water
- suspended solids
- odor and floating materials
- aquatic life
- dissolved oxygen

Many other factors could be listed, but these will suffice for this example.

Next, we need to assign EQIs to the factors. If we assume a linear relationship, we can calculate them as follows:

Item	Condition Before Project	Condition After Project	EQI
Appearance of water	10	3	0.30
Suspended solids	20 mg/L	1,000 mg/L	0.02
Odor	10	5	0.50
Aquatic life	10	2	0.20
Dissolved oxygen	9 mg/L	8 mg/L	0.88

Note that we had to put in subjective quantities for three of the items—"appearance of water," "odor," and "aquatic life"—based on an arbitrary scale of decreasing quality from 10 to 1. The actual magnitude is not important since a ratio is calculated. Also note that the sediment ratio had to be inverted to make the EQI indicate improvement, i.e., EQI < 1.

Finally, the EQI indices are weighted by the 10 available PIU, and the environmental impact units (EIU) are calculated.

Item	Project PIU	After Project EQI × PIU = EIU
Appearance of water	1	0.3 × 1 = 0.3
Suspended solids	2	0.02 × 2 = 0.04
Odor	1	0.5 × 1 = 0.5
Aquatic life	5	0.2 × 5 = 1.0
Dissolved oxygen	1	0.88 × 1 = 0.88
Total	10	2.72

The EIU total of 2.72 for this alternative is then compared with the total EIU for other alternatives.

Evaluation

The final part of the environmental impact assessment is evaluation of the results of the preceding studies. Typically, the evaluation phase is out of the hands of the engineers and scientists responsible for the inventory and assessment phases. Decisions made within the responsible governmental agency ultimately use the EIS to justify past decisions or support new alternatives.

SOCIOECONOMIC IMPACT ASSESSMENT

Historically, the President's Council on Environmental Quality has been responsible for overseeing the preparation of EISs, and CEQ regulations have listed what should be included in all EISs developed by federal agencies. For the proposed projects discussed earlier in this chapter two basic issues arose: public health dangers and environmental degradation. Under NEPA and CEQ regulations, both issues must be addressed whenever alternatives are developed and compared.

In many cases, however, consideration of public health and environmental protection alone are not sufficient grounds on which to evaluate a range of alternative programs. Frequently, public acceptability is also a necessary input to an evaluation process. Although an alternative may protect public health and minimize environmental degradation, it may not be publicly acceptable and thus could prove to be an essentially worthless alternative. Factors that influence public acceptability of a given alternative are generally discussed in terms of economics and broad social concerns. Economics includes the nationwide costs of an alternative, including the state, regional, local, and private components, the resulting impacts on user charges and prices, and the ability to finance capital expenditures. Social concerns include public preferences in siting (e.g., no local landfills in wealthy neighborhoods) and public rejection of a particular disposal method (e.g., food-chain landspreading of municipal sludge rejected on "general principle"). Consequently, each alternative that is developed to address the issues of public health and environmental protection must also be analyzed in the context of rigid economic analyses and broad social concerns. Two important considerations that enter an economic impact assessment are discussed below: financing capital requirements and impacts on user charges and prices. Broad social concerns are addressed in Chapter 26.

Financing of Capital Expenditures

A municipality's or industry's inability to finance large capital expenditures will necessarily impact choice among alternatives and possibly affect their ability to comply with environmental regulations. Traditional economic impact assessments examine the amortized capital and operation and maintenance (O&M) costs of a project and the community's ability to pay, but they typically overlook the problems involved in raising the initial capital funds required for implementation. Financing problems are faced by municipalities and industries of all sizes, but may be particularly troublesome for small communities and small firms that face greater institutional barriers to finance debt. Although the following discussion examines the issue of financing capability only with regard to an ability to comply with water quality regulations, parallel issues may also arise for other types of public and private projects.

For relatively small capital needs, communities may make use of bank borrowing or capital improvement funds financed through operating revenues. However, local shares of wastewater treatment facility capital costs are generally raised by long-term borrowing in the municipal bond market. In the absence of other sources of funds, both the availability of funds through the bond market and the willingness of the community to assume the costs of borrowing may impact the availability of high-cost programs. The availability and cost of funds are, in turn, affected by how the financing is arranged.

Bonds issued by a municipality for the purpose of raising capital for a wastewater treatment plant are generally general obligation (GO) bonds or revenue bonds. Both have fixed maturities and fixed rates of interest, but they differ in the security pledge by the issuing authority to meet the debt service requirements, i.e., payments for principal plus interest. General obligation bonds are backed by the basic taxing authority of the issuer; revenue bonds are backed solely by the revenue for the service provided by the specific project.

The GO bond is generally preferred since the overhead costs of financing GO bonds are lower and their greater security allows them to be offered at a lower rate of interest. Some states are, however, constitutionally prevented from issuing GO bonds or are limited in the quantity they can issue. In such cases, cities may have to resort to revenue bonds. Nevertheless, most bond issues for wastewater treatment projects are GOs.

Most municipal bonds carry a credit rating from at least one of the private rating agencies, Standard & Poor's or Moody's Investor Service. Both firms attempt to measure the creditworthiness of borrowers, focusing on the potential for decrease on bond quality by subsequent debt and on the risk of default. Although it is not the sole determinant, the issuer's rating helps to determine the interest costs of borrowing, since individual bond purchasers have little else to guide them, and commercial banks wishing to purchase bonds are constrained by federal regulations to favor investments in the highest rating categories.

Of the rating categories used, only the top ones are considered to be of investment quality. Even among these grades, the difference in interest rates may impose a significantly higher borrowing cost on communities with low ratings. During the 1970s, for example, the interest rate differential between the highest grade (Moody's Aaa) and the lowest investment grade (Moody's Baa) bonds averaged 1.37 percent. Such a differential implies a substantial variation in financing costs for facilities facing extensive borrowing.

To highlight the importance of financing costs, consider the example of a city planning a $2 million expenditure on an incinerator to serve a publicly owned (wastewater) treatment works (POTW) with a capacity of 15 million gallons per day (mgd). Under one set of assumptions, such a facility would support a population of roughly 75,000, which would bear anywhere from 12.5 to 100 percent of the total cost, depending on what portion of the capital expense state and federal agencies agreed to pay. Under the more conservative assumption, this would amount to $250,000. Assuming that the capital is raised through one Aaa bond issue amortized over 25 years at an interest rate of 5.18 percent, the interest payment over the entire borrowing period would total $212,500. If the Baa rate were 6.34 percent, the interest payments would amount to $262,000. As a rule of thumb, total interest payments are roughly equal to the principal and are sensitive to the interest rate.

The rating may also determine the acceptability of bonds on the market. A city with a low or nonexistent rating might find that credit is simply not available. The fiscal problems of small communities are particularly sensitive to the rating process, which places lower ratings on small towns on the basis of an ill-defined "higher risk of default," despite the lack of supporting evidence. Hence, small towns must often endure higher interest payments. The problem is compounded by the higher average cost of small bond issues owing to certain fixed underwriting fees, fees associated with buying and selling the bonds.

Despite the availability of GO financing and good credit, municipalities may face difficulties in raising capital owing to the state of the market, while some borrowers may be precluded from borrowing altogether. Two trends are of particular significance. Periods of high rates of inflation are expected to lead to higher interest rates for all municipal borrowers. The impact is significant for borrowers with marginal credit ratings, who will be squeezed out as the difference in interest rates between high- and low-rated bond issues widens. Second, the combination of expanding bond supply owing to increased municipal borrowing and shrinking demand in periods of slow economic growth is likely to drive interest rates higher, increasing costs for all municipal borrowers.

Clearly, the availability of funding to finance capital expenditures is far from certain. Credit restrictions are not limited to large cities which often suffer from severe budget problems. The financial health of a local government may be analyzed. A two-step procedure is useful as engineers consider the impacts of proposed projects:

1. Define capital requirements of each alternative.
2. Apply financial classification criteria.

Although rating agency analysts use a variety of factors in assessing a city's fiscal strength, current outstanding debt is the principal determinant of the bond rating. Debt is generally expressed as a proportion of assets or on a per capita basis to make comparisons among cities. Three of these municipal debt ratios are of particular significance in determining market acceptance of bond offerings. The ratios and estimates of market-acceptable thresholds are:

• debt/capita ≥$300
• debt/capita as percentage of per capita income ≥7 percent
• debt/full property value ≥4.5 percent.

Empirical studies have shown that bond offerings by cities that exceed all three debt ratios are not likely to be successful.[4]

Increases in User Charges

A second component of an economic impact assessment is the analyses of projected increases in user charges. For example, if a city constructs a large wastewater treatment facility, what is the likely change in household sewer bills in that city? Based on the number of residential users supporting a given facility, the proportion of wastewater generated by them, and the data on alternatives, a projected cost per household may be computed. This cost may then be compared with the existing charge per household to determine the percent increase attributable to the alternative.

The results of these types of user charge analyses often indicate that engineers may find it difficult to make a distinction among alternatives based solely on increases in user charges. For example, an average difference of only 36¢ per year per household is projected between alternatives 1 and 2 in Figure 25–2. Respectively, these two alternatives represent a less stringent and a more

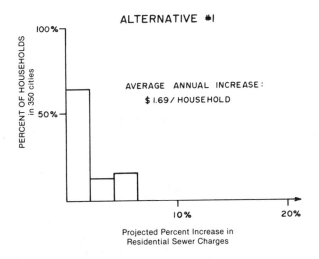

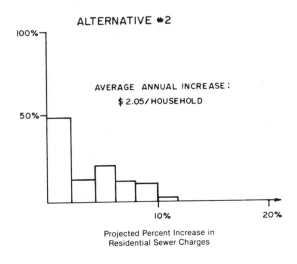

NOTE:
Average annual charge before construction is $50.00 per
year in this sample of 350 cities

Figure 25–2. Projected increases in sewer use charges paid by a household in a sample community if a new wastewater treatment facility is constructed.

stringent regulation of effluent water quality. Figure 25–2 summarizes projected impacts on the user charges in 350 cities across the nation. By almost any standard, 36¢ per year is insignificant. The findings suggest that the selection process, in this instance between alternative 1 (little required reduction in pollution) and alternative 2 (large requirements in pollution reduction), must be based on such other criteria as environmental impact and public health concerns, not on changes in user charges.

Sociological Impacts

Large changes in the population of a community, such as influxes of temporary construction workers or establishment of a military base with immigration of the associated personnel and their families, may have a number of impacts, both positive and adverse. New service jobs may well be created, particularly in small communities, but there may also be increases in the crime rate, need for police and fire protection, etc. Study of such "boom town" phenomena has led to inclusion of these assessments in any EIS.

CONCLUSIONS

Engineers are required to develop, analyze and compare a range of solutions to any given environmental pollution problem. This range of alternatives must be viewed in terms of their respective environmental impacts and economic assessments. A nagging question exists throughout any such viewing: can individuals really measure, in the strict "scientific" sense, degradation of the environment? For example, can we place a value on an unspoiled wilderness area? Unfortunately, qualitative judgments are required to assess many impacts of any project. This balancing of values is discussed in the final chapter.

PROBLEMS

25.1 Develop and apply an interaction matrix for the following proposed actions designed to clean municipal wastewater in your home town: (1) construct a large-scale activated sludge facility, (2) require septic tanks for households and small-scale package treatment plants for industries, (3) construct decentralized, small-scale treatment facilities across town, (4) adopt land application technology, (5) continue direct discharge into the river. Draw conclusions from the matrix.

25.2 Discuss the advantages and disadvantages of a benefit-cost ratio in deciding whether a town should build a wastewater treatment facility. Focus on the valuation problems associated with analyzing the impacts of such a project.

25.3 What conclusions may be drawn about the abilities of the three towns to finance a proposed water treatment facility:

City	Debt per Capita Current	Debt per Capita Projected	Debt per Capita as % per Capita Income Current	Debt per Capita as % per Capita Income Projected	Debt as % Full Valuation Current	Debt as % Full Valuation Projected
A	$946	$950	22.3	22.3	15.0	15.1
B	$335	$337	9.1	9.1	7.8	7.9
C	$ 6	$411	0.1	10.7	0.1	8.6

Discuss the current financial picture of each town as well as the projected incremental impact of this new proposed construction.

25.4 Determine the average *residential* sewer charge for the three cities listed below. What is the incremental impact on these charges if the cost of sludge disposal increases as noted:

City	Wastewater Flow (mgd)	% Flow Residential	Annual Budget for Wastewater Treatment	Projected Increase in Sludge Disposal
A	20.0	60	$5,100,000	$1,500,000
B	5.0	90	$3,600,000	$1,000,000
C	0.1	100	$ 50,000	$ 20,000

Discuss your assumptions.

25.5 Compare the environmental impacts of a coal-fired electricity-generating plant with those of a nuclear power plant. In your presentation, look at the flow of fuel from its natural state to the facility. Finalize your comparison with waste disposal considerations.

LIST OF SYMBOLS

CEQ = Council on Environmental Quality
EIA = economic impact assessment
EIS = environmental impact statement
EIU = environmental impact unit
EPA = Environmental Protection Agency
EQI = environmental quality index
NEPA = National Environmental Policy Act
PIU = parameter importance units
POTW = publicly owned (wastewater) treatment works
PPBS = planning, programming and budgeting system
WEPA = Wisconsin Environmental Policy Act

REFERENCES

1. Lindbloom, C.E. "The Science of Muddling Through," *Public Administration* (Spring 1959).
2. Etzioni, A. "Mixed Scanning—A Third Approach to Decision Making," *Public Administration Review* (December 1967).
3. Bachrach, J., and M.S. Baratz. "The Two Faces of Power," *American Political Science Review* (1962), p. 947–52.
4. Forbes, R. Personal communication. Graduate School of Business Administration, State University of Albany, Albany, NY.

Chapter 26

The Environmental Ethic

It probably won't hurt mankind a whole hell of a lot in the long run if a whooping crane doesn't quite make it.[1]

This statement, made by a director of air and water resources of a large papermill, represents an all-too-common approach to our environment.

There is, however, an opposing view that holds that it would be a major disaster if a bird like the whooping crane became extinct. This concern for nature, no doubt fueled in part by the uncertain survival of humans on this planet, is slowly developing into a new system of values that in aggregate might be called the *environmental ethic*. More and more citizens are considering questions not only from the standpoint of economics, but also from that of environmental quality. Lawyers, economists, and politicians are finding it more difficult to answer such questions as "Is growth really necessary?" or "Should the gross national product always increase?" or "Should we consider the benefits and costs of public projects solely on the basis of dollars?" A few years ago such questions would have been brushed aside as foolish.* They weren't foolish, and they deserve straight answers.

In fact, ecology and economics are on a collision course. Humanity cannot continue to grow in numbers and output (and use of resources) without eventual self-destruction. Present economic systems do not have a feedback loop that slows down the processes detrimental to survival.

The conflict between ecologists and economists has been boiled down to this ditty:

---

* Alfred North Whitehead observed that "The 'silly' question is the first intimation of some totally new development."

> Ecology's uneconomic,
> But with another kind of logic,
> Economy's unecologic.[2]

Technical economists notwithstanding, the environmental ethic has become a permanent part of our lives. It is in its infancy, a new and bold concept that must be given time to mature and develop.

The word "ethic" derives from the Greek "ethos," meaning the character of a person as defined by his actions. This character has been developed during the evolutionary process and has been influenced by the need for adapting to the environment. The "ethic," in short, governs our way of doing things, and this is a direct result of our environment.

When people talk of a "crisis in the environment," they really mean that our way of thinking about and doing things (our ethic) is not adapted to the environment. Humans at one time lived in harmony with the environment, but somehow the species, either as an accident or as a colossal practical joke by some unknown power, changed its lifestyle to the point where it no longer adapted. There is thus no "environmental crisis," but rather a crisis in the recognition that somehow over the years of development we are no longer adapted to our environment.

In the ecological context, such maladaptation results in two options:

1. The organism dies out.
2. The organism evolves to a form and character where it once again is compatible with the environment.

Assuming that humans choose the latter course, how must this change in character (ethic) occur? It obviously cannot take place by the Darwinian natural selection principle, since many, many generations must pass before individuals with the incompatible characteristics are eliminated. Further, our present social system works in opposition to this selection, in that society rewards the despoilers of the environment more than the conservationists.

This selection must, therefore, occur within only a few generations and the

CONCHY   **James Childress**

change must be on two levels—the individual and the system. Individuals must change their character or ethic, and the social system must change to become compatible with the global ecology.[3]

CAUSES OF THE PROBLEM OF OUR INCOMPATIBILITY WITH NATURE

Why is it that the human being—the most intelligent creature that ever evolved (or was created, depending on your beliefs)—is so incompatible with his environment?

There are three basic lines of argument in attempting to answer that question. One view is that humans created religions whose dogmas held the basic seed of incompatibility. The second argument is that the social structure created by humans makes them inherently unable to attain equilibrium. The third view is that the growth of science and technology is responsible for environmental degradation.

It should be obvious that all three—religion, society, and technology—are intertwined and cannot be conveniently separated and individually scrutinized. Nevertheless, for the purposes of our discussion, let us assume that we can pick on each, one at a time.

Religion as the Cause

Although the environment has been severely impacted by human cultures and civilizations, probably the greatest overall environmental damage has been caused in modern times by Western civilization. The major Western religious traditions are rooted in Judaism and Christianity, and these Judeo-Christian traditions have been blamed as the root cause of our environmental problems. And there is clearly some justification in this argument. In the first chapter of Genesis, for example, man is commanded by God to subdue nature, to procreate, and to have dominion over all living things. Such an anthropocentric view of nature runs all through the Judeo-Christian doctrine, a point forcefully made by Lynn White in his 1967 essay "The Historical Roots of Our Ecologic Crisis."[4]

White argues that the people embracing the Judeo-Christian religions are taught to treat nature as an enemy, that the religious dogmas prescribe that nature and natural resources are to be used only to meet the goals of survival and propagation. Western religion, according to White, is responsible for environmental degradation.*

---

* Or, as Mark Twain put it, "Sometimes it seems a shame that Noah and his party did not miss the boat."

The Christian world reacted predictably to White's essay, with books and papers claiming that Christianity is pro-environment and that the environmental ethic can in fact be found in the Bible.[5] This is stretching it a bit, however.

The strongest argument presented to counter Lynn White's indictment of the Judeo-Christian doctrines as a root cause of our environmental problems centers on the notion of *stewardship*. Stewardship represents a view that people were put on the planet as caretakers, to see to the well-being of the earth. This "garden mentality" still relies on the concept of anthropocentrism, however, a world of hierarchical relationships in which man is the noble, managing his property, accountable only to the sovereign. Such a doctrine also requires a substantial measure of faith in a transcendent God and a belief in a reward structure in the afterlife.

Perhaps the defensiveness of the Church toward Lynn White's ideas was unnecessary. White's assertion that religion molds morals is oversimplified, and blaming one religion for our problems seems to be unfair. In fact, during the time that the Christian religion was becoming popular, the people had many religious sects as alternatives, and the Christian ideas and ethics derived from ancient Judaic traditions seemed to fit most comfortably with the needs and existing value systems (whether active or latent). In short, the ethics existed first, and it was the people who developed these ethics into a religion.

Further, the ethics that fit the Judeo-Christian religions so well also spawned a vigorous pursuit of science, a tradition of democracy, and the capitalistic system. Although Christianity was compatible with the idea of individual worth and achievement, it was not the *reason* that these ideas fluorished. It can therefore be argued that the Christian church is not directly responsible for traditions and ethics that promoted the destruction of nature.

According to White, the model that Western civilization has been following is

Ethics and traditions of Judeo-Christian peoples	→	Science, capitalism, technology, democracy	→	Urbanization, money, population, individual ownership	→	Environmental degradation

An equally strong argument, however, can be made for the following model:

Human nature and the search for a comfortable and compatible religion	→	Acceptance of the Judeo-Christian dogma	→	Science, capitalism technology, democracy	→	Urbanization, money, population, individual resource ownership	→	Environmental degradation

If, therefore, the latter model is equally appealing, there is little reason to blame the Judeo-Christian or any other religions for our incompatibility with nature.

Social Structure as the Cause

A second view of the roots of our ecological crisis is that our social structures are responsible. Probably the most damning piece yet written that takes this view is Garrett Hardin's "Tragedy of the Commons."[6]

Hardin illustrates his point by a story of a village that has a common green for the grazing of cattle, surrounded by individual farmhouses. In the beginning, each farmer has one cow, and the green is able to support the herd. It becomes apparent to each farmer, however, that if he gets another cow, the *cost* is negligible to him personally (it's shared by everyone) but the *profits* are his alone. So he gets more cows, reaping greater and greater profits, until the commons are no longer able to support the herd and the system collapses.

Hardin used this parable to illustrate the problem of overpopulation, but it applies equally well to other environmental problems. The social structure in the parable is of course capitalism, the individual ownership of wealth, and the use of that wealth for furthering one's own interest.

A lot has been written about capitalism as the major causative agent for our environmental ills, often with the implication that some form of socialism, Marxist or otherwise, is a superior system.

Unfortunately, the advantages of a centrally controlled (and hence totalitarian) system have not provided the answer. In fact, the environmental devastation in the USSR is substantially more serious than in the West. When *production* is the primary goal of society, the environment and human life take a poor second place.

The only types of socio-political systems that seem to have developed a quasi-steady-state condition are non-industrial societies, such as the Native Americans, the Finno-Ugric people of Northern Europe (Finland and Estonia), the South Sea islanders, and the Amish farmers of Pennsylvania.

To all of these people nature holds within it spirits that are both powerful and friendly (if at times capricious). The spirits in nature do not take human forms (as in Greek and Roman religions). Yet it is possible in some societies to converse freely with the spirits. The old Estonians and Finns, for example, always explained to the spirit of the tree why it was necessary to cut it down.[7] Such a reverence for the closeness with nature is unknown in most modern societies. Imagine the difficulty in the clearing of a forest for a man-made lake if every tree required a special explanation and apology!

There was, in these societies, a camaraderie with all life forms. For example, old Estonians would begin the wheat harvest by cutting a shaft of wheat and placing this aside for the field mice. There did not seem to be any religious significance to this mouse-shaft (hiirevihk), and the only explanation handed down through the generations is that the mice deserve their share of the harvest.[8]

And yet these societies were not all environmentally stable. The Maori first came to New Zealand in the 1300s and proceeded to exterminate the moa, a large ostrichlike bird that was their sole source of meat (the islands had no native mammals, except a bat).

In the face of such failures, it is difficult to argue that we should re-establish such non-industrialized societies. In addition, a society is after all the reflection of the needs and aspirations of the people. People establish societies, and thus it cannot be argued that social systems which the people choose to establish are the root cause of our environmental ills.

Science and Technology as the Cause

It has become somewhat fashionable in some circles to blame our increased knowledge of nature and our related ability to put that knowledge to work for our environmental ills. A decade ago, the "back-to-nature" movement tried to reject technology altogether. The past few years have seen a more reasoned approach, evidenced by the growth of interdisciplinary college programs such as "Science, Technology, and Human Values." The purpose has been, in part, to answer the question: Is technology to blame for our environmental mess?*

If technology is rightfully to blame, we must show first that other, less technologically advanced societies avoided major environmental problems, and second, that modern technology is not value-free.

The first premise is not true. Many less technologically advanced societies were equally or even more destructive than our own. The extermination of the moas by the Maori, overgrazing in Africa, and the destruction of forests by the early Greek civilization are excellent examples. In fact, one can mount a strong argument for technology as a means of being able to *control* the environmentally destructive forces and trends.

Is technology value-free? Is knowledge itself, without the application of it by humans, right or wrong, ethical or unethical? Before we jump to what might seem an obvious answer to most engineers, consider this example. You have it within your power to conduct an experiment and thus prove the feasibility of constructing a simple and inexpensive nuclear bomb. The knowledge may well be used by terrorists in untold acts of violence. Without your experiment, the construction of such a device would not be possible. Should you conduct the experiment, since you know that pure science and technology are value-free and that *you* will never put this knowledge to evil use? But does this knowledge, by its very *existence*, constitute an evil?

Strong arguments have, as you can imagine, been mounted in defense of both views. But even if science is determined to be value-laden, the use of

---

* The distrust of technological advancement is not new. During the industrial revolution in England, the Luddites were people who violently resisted the change from cottage industry to centralized factories. Because the large machines threatened their way of life, they smashed a few factories to make their point, and were hanged for their trouble.

science for curing environmental ills has an equally strong case. We cannot, in short, take the quick and dirty way out and blame science alone for environmental degradation just as we couldn't lay the blame on religion or social structure alone. And indeed, is it not better to seek a solution than to cast blame?

RESOLUTION OF ENVIRONMENTAL CONFLICTS

The development of various systems of ethics stems from the human need to create a unified and universally applicable method of "getting along." The major concern with many ethical theories is first to determine what is *good* and second, what is *right*. In addition, it becomes necessary, if the ethical theory is to have any utility, to strive toward what has been determined to be good and right; once we decide that a system is correct, then we should live by the conclusions we derive from the system.

Many so-called systems of ethics have been proposed over the years.[9]

One of the oldest systems is *hedonism*, espoused by the sophists with whom Socrates debated. Hedonism seeks to maximize pleasure. Whatever gives you pleasure is therefore right. In isolation, this may be a workable philosophy, but not in a society that depends on the general acceptance of the golden rule.

Plato's arguments against hedonism included the concept of a person being governed by reason, and that "good" decisions can only be made in the face of well-considered arguments. He extrapolated this to the idea that since not all people have the opportunity to study philosophy, decisions for the society should be made by a "philosopher-king." In Plato's system, therefore, the decisions are taken out of the hands of the masses and the power is given to a benign dictator who, Plato argued, would make the "correct" decisions.

Immanuel Kant proposed a third alternative: establish a set of rules that everyone agrees to and then follow these no matter what. The *categorical imperative* commits each person to a set of consistent and universally shared required actions (e.g., do not lie under any circumstances; or "dioxin" is bad, so don't allow *any* dioxin out of a municipal waste incinerator). In this system, the focus is on the act itself, not its consequences.

An opposing view holds that consequences are the major concern. *Utilitarianism*, developed in eighteenth century England by Jeremy Bentham, John Stuart Mill, and others, holds that one's actions should strive for the greatest good, defined as individual happiness. Under this system, one could construct a happiness-benefit/cost ratio for every action, and the options with the highest ratio would clearly be the ones of choice. Extending this idea to the social level, we have the concept that governmental actions should follow the dictum of "the greatest good for the greatest number." This concept blends directly into *socialism* developed to a large extent by Karl Marx and Friedrich Engels, in which the *society* is the prime instrument and the individual is no longer very important.* Under socialism, decisions are made on the basis of maximizing production, of increasing the average wealth of the society.

There are other systems of ethics worth noting, none of which have attained the historical stature of the ones discussed above. For example:

- *legalism*—do only what the law tells you to do, and no more. If you break the law, don't get caught. (Is football played under this system?)
- *populism*—let the masses decide.
- *situation ethics*—make up rules as the game goes on, but base these on the golden rule idea.
- *altruism*—always think of the other guy first.
- *egoism*—the opposite of altruism, in that one's behavior is always governed by selfish interest.
- *existentialism*—there is no grand scheme to the world, and one's actions and one's fate are unimportant. This is a doctrine of complete resignation and hence freedom.

A major problem with the application of any of the ethical theories is that they were all developed as means of resolving conflicts among *people*. Our concerns, however, are problems between people and *animals*, people and *plants*, and people and *places*. This severely complicates the issue, since now the conflict exists between two parties, each having a different *value*. As long as the actors are people, the value of a human life is a given constant. Now we have an inequality, and we must seek to place a value on nature.

The concept that nature has a value is fairly modern. With a few noteworthy exceptions,** nature has not been considered through the development of human civilization as an entity possessing value and has thus not entered into the development of ethical theories.

In the past, only dominance and perpetual progress have shaped our relationship with the natural environment. In fact, this has been for years a substitute for an environmental ethic, and these exploitative notions have guided our behavior toward a nature that by itself did not possess value.

But simply stating that nature is valuable is not adequate. *Why* and *how* is it valuable?

This question has emerged as a major point of discussion in philosophical literature. The two primary views that seem to be emerging in this debate may be classified as *instrumental* and *intrinsic*.

---

* Witness the status of the medical arts in Soviet Russia. A physician is paid about the same as a taxi driver or doorman and receives only 3 years of training. In that society, it's not efficient to spend money and resources on medical care for the few workers who are ill. Productivity is enhanced by placing the emphasis on the healthy workers and ignoring the infirm.

** Such as St. Francis of Assisi. As Lynn White points out, the miracle of St. Francis was that he was not burned at the stake as a heretic. St. Francis tried to dethrone man as the center of the universe and preached the importance of *all* of God's creatures.

The instrumental view considers nature for its real value to humans, in terms of life support, material wealth, recreational and aesthetic beauty, etc.[10] This view was first popularized by Ralph Waldo Emerson, who witnessed the wholesale destruction of nature during the nineteenth century "rape of the land" and proposed that in the use of nature's bounty, one should be grateful for the blessings and material wealth nature provides. This is, of course, a utilitarian and anthropocentric view of nature, but it nevertheless represented a first recognition that nature has *value*.

The second view holds that nature has intrinsic value above and beyond what we might be able to measure in dollars.[11] This, however, is a difficult argument to tie down. In a way the belief in the intrinsic value of nature represents a faith—one has to *feel* that it is right. This feeling has been even translated to legalistic terms, most notably in a work entitled *Should Trees Have Standing?—Toward Legal Rights for Natural Objects.*[12] The question we face is whether nature is here for the welfare of humankind, or does nature have its independent claims to exist.

The problem is further complicated by what we mean by nature. It's not difficult to get people upset over the senseless slaughter of whales, or of baby seals. Why? Simply because whales are too much like *humans*; they have many of our own most admirable characteristics and lead a placid, peaceful life, and baby seals are so *cute*!

Can you imagine an equal outcry raised against the destruction of the polio virus, or the rattlesnake? Do these two species also have standing, a right to be left alone and not destroyed?

The issue is of course complex. But not to face it is unacceptable. There is in the intrinsic approach to the value of nature a germ of a positive and intuitively desirable notion. By not striving to develop and foster this idea, might we be freezing our environmental ethic at too low a level? Many widely accepted principles of the past, such as slavery and the divine right of monarchs, have fallen. Albert Schweitzer noted that Europeans long considered people with darker skins to be subhuman. It was at one time stupid to think of them as equals and to treat them humanely. And now this stupidity is a truth. Schweitzer continues:

> Today it is thought as exaggeration to state that a reasonable ethic demands constant consideration for all living things down to the lowest manifestations of life. The time is coming, however, when people will be amazed that it took so long for mankind to recognize that thoughtless injury to life was incompatible with ethics.[13]

René Dubos takes the idea further. Why, Dubos argues, do we limit our concerns with animate beings? He proposes in a thoughtful article entitled "The Theology of the Earth" that everything has its place and reason for being. He suggests that there is a genius of the place, a oneness between an individual and the uniqueness of each locality, be it a city or a grove of trees.[14]

The intrinsic value of nature obviously is on a higher ethical plane than the instrumental view of nature, but it is a much more difficult idea for humanity to accept.

THE FUTURE OF THE ENVIRONMENTAL ETHIC

The birth of the environmental ethic as a force is partly a result of our concern for our own long-term survival, as well as our realization that humans are but one form of life and that we should share our earth with our fellow travelers. It is not a religion, since the ethic is based not only on faith, but also on hard facts and thorough analyses.

One of the first to recognize the degradation of the environment and to voice a concern for nature was Thoreau. His solution to this eloquently stated concern was withdrawal, which was perhaps morally admirable but realistically ineffective. What we have experienced in the last two decades has been the coupling of Thoreau's concern with activism. It is one thing to be concerned, but it's much more effective to take action to promote your concern.

One powerful and original attempt at defining an environmental ethic was made by Aldo Leopold, a naturalist and writer. He proposed that:

> A decision is right when it tends to preserve the integrity, stability, and beauty of the biotic community. It is wrong when it tends otherwise.[15]

Leopold argued convincingly toward the adoption of environmental values other than economic, and his writings have had a major impact on the growth of environmental awareness.

The environmental ethic is very new, and none of the doctrine is cast in immutable decrees and dogma. Some critics have mistaken this maturation of the ethic for transience. The problem is so immense that to think of it as a fad is ludicrous (and fatal!). But herein is the dilemma: Unless there is public awareness as to the true nature of the problems and some realistic solutions, public concern may rapidly fade. For the impetus to survive there must be public confidence, and the environmental scientist or engineer must engender this trust by analyzing and interpreting environmental problems correctly and proposing and developing constructed facilities that are compatible with our ecosystem. Sensationalism and the bandwagon tactics of some "environmentalists," politicians, and other well-meaning citizens may easily destroy the public sentiment so necessary for a successful assault on our common problems.

Education of the public to environmental problems and the solutions (and nonsolutions) is of prime importance. It is necessary for people to be critical and to be literate in the scientific, technological, economic, and legal aspects of controlling environmental pollution. And equally important is the recognition by engineers and scientists that technology is not value-free and that they may have to make important ethical decisions concerning the environment. The introduction of professional ethics, including environmental ethics, into engineering education is mandatory.[16]

And together, regardless of our education or background, we must also live the environmental ethic—recognize the power of nature and feel humble in the realization that we are just one small cog in a wonderful and still mysterious system.

PROBLEMS

26.1 Read Chapter 2 of Genesis. Then study Figure 26–1. Express your thoughts relative to the "stewardship" concept of environmental ethics and the value of nature.

26.2 The three symbols in Figure 26–2 were taken from: (A) a fast food container, (B) an egg carton, and (C) a paper bag.

   a. What purposes did the respective container manufacturers have in mind in placing the symbol on their packages?

   b. Write a critical statement of the basic honesty in the use of the recycling symbol in these three cases.

26.3 The picture in Figure 26–3 was accompanied by a story in the Air New Zealand in-flight magazine:

As an "alarm clock," the Boeing has a "competitor." A topdressing plane sometimes makes a habit of droning across the northern part of the city early in the morning. During the debate on Boeing noise before its introduction in 1975, Invercargill's Mayor Mr. F.R. Miller, made the comment that surely nothing could be as bad as that slow-moving disturber of the peace. Further it "rings" too early for most people, is far less reliable and is one of those "alarm clocks" that jars the nerves, sends the blood pressure skywards and leaves you with the feeling that as you've just gone to bed, why are you being disturbed?

In contrast, the Boeing "alarm clock" contains the right volume of noise to arouse but not to anger and only during bad (unseasonal) weather does it forget to let you know it's time to get up. It is unreachable so it cannot be smashed to silence but perhaps most importantly, like all good alarm clocks, it can be ignored – at your own risk.

What purpose did NAC (now Air New Zealand) have in printing this article? What do you suppose the controversy was that precipitated it? What do you think of the basic integrity of the presentation?

26.4 The following paragraph appeared as part of a full-page ad in a professional journal.

Today's laws.
A reason to act now.

The Resource Conservation and Recovery Act [RCRA] provides for corporate fines up to $1,000,000 and jail sentences up to five years for officers and managers, for

Figure 26–1. Drenched and shivering after nearly drowning, a calf stranded on an island gets a tender toweling from Don Frickie of the Arctic National Wildlife Range. Deposited on the stream's bank, this 2-day-old was quickly reunited with its frantic mother. [Picture and legend courtesy of *National Geographic Magazine*, and the photographer, J. Rearden. The picture appeared in Rearden, J. "Caribou: Hardy Nomads of the North," *National Geographic* 146(6)(1974).]

the improper handling of hazardous wastes. That's why identifying waste problems, and developing economical solutions is an absolute necessity. O'Brien & Gere can help you meet today's strict regulations, and today's economic realities, with practical, cost-effective solutions.

Based on a consideration of environmental ethics, why *else* might "identifying waste problems, and developing economical solutions" be an absolute necessity?

A.   B.   C.

Figure 26–2. The recycling symbol, for Problem 26.2.

Figure 26–3. Invercargill's alarm clock, for Problem 26.3. [Courtesy of Air New Zealand.]

26.5   A major purveyor of auto parts sells a book entitled *How to Bypass Emission Controls—For Better Mileage and Performance.* The ad states that the emission controls rob the car of power and waste gas, and that this book contains easy-to-follow directions for the amateur and professional on how to eliminate the emission controls. (The company, incidentally, would not allow us to reprint the ad or to quote directly from it.) At the bottom of the ad, a statement notes that the book is not available in California and that the buyer should check with the state motor vehicle department before removing the emission controls.

Suppose you are the president of the auto supply company selling this book. How would you ethically justify your decision to sell it? What systems of ethics would you need to adopt to make such a sale morally acceptable?

26.6   You are working as a recruiter for your consulting engineering firm, and an applicant asks you the following question:

"If I work with your firm and I find that I object, on ethical grounds, to a project the firm has been hired to do, can I request not to work on that project, without penalizing my future career with the firm?

Give three responses you might make to such a question, and then respond to these statements according to what you believe the student/interviewee would think. Would he/she believe you?

26.7   The Environmental Advisory Council of Canada published a booklet entitled "An Environmental Ethic—Its Formulation and Implications," (Report No. 2, January 1975, Norma H. Morse, Ottawa). They suggested the following as a concise statement of an environmental ethic.

Every person shall strive to protect and enhance the beautiful everywhere his or her impact is felt, and to maintain or increase the functional diversity of the environment in general.

In a short essay, discuss the validity of this statement as a useful environmental ethic.

26.8   The Dickey and Lincoln dams on the St. John River in Maine were planned with careful consideration to the Furbish lousewort (*Pedicularis furbishiae*), an endangered plant species. An engineer writing in *World Oil* (Jan. 1977, p. 5) termed this concern for the "lousy lousewort" to be "total stupidity." Based on only this information, construct an ethical profile of this engineer.

26.9   The selection of what science studies is the subject of a masterful essay by Leo Tolstoy, "The Superstitions of Science," *The Arena*, 20 (1898), reprinted in *The New Technology and Human Values*, J.G. Burke, Ed. (Belmont, CA: Wadsworth Publishing Company, 1966). He recounts how

... a simple and sensible working man holds in the old-fashioned and sensible way that if people who study during their whole lives, and, in return for the food and support he gives them, think for him, then these thinkers are probably occupied with what is necessary for people, and he expects from science a solution of those questions on which his well-being and the well-being of all people depends. He expects that science will teach him how to live, how to act towards members of his family, towards his neighbors, towards foreigners; how to battle with his passions, in what he should or should not believe, and much more. And what does our science tell him concerning all these questions?

It majestically informs him how many million miles the sun is from the earth, how many millions of ethereal vibrations in a second constitute light, how many vibrations in the air make sound; it tells him of the chemical constitution of the Milky Way, of the new element helium, of microorganisms and their waste tissue, of the points in the hand in which electricity is concentrated, of X-rays, and the like. But, protests the working man, I need to know today, in this generation, the answers to how to live.

Stupid and uneducated fellow, science replies, he does not understand that science serves not utility, but science. Science studies what presents itself for study, and cannot select subjects for study. Science studies everything. This is the character of science.

And men of science are really convinced that this quality of occupying itself with trifles, and neglecting what is more real and important, is a quality not of themselves, but of science; but the simple, sensible person begins to suspect that this quality belongs not to science, but to people who are inclined to occupy themselves with trifles, and to attribute to these trifles a high importance.

Figure 26–4. A massive fish kill in Illinois, 1967, for Problem 26.10. [Courtesy of EPA-Documerica.]

Figure 26–5. Beach at the Anza-Barego State Park, CA, for Problem 26.11. [Courtesy of EPA-Documerica.]

Figure 26–6. New York traffic, near 42nd Street, for Problem 26.12. [Courtesy of EPA-Documerica.]

In the formulation and planning of your career as an engineer, how would you answer Tolstoy's criticism?

26.10   Re: Figure 26–4. Do fish have "standing"?

26.11   Study Figure 26–5. Comment relative to Dubos' idea of the "genius of the place."

26.12   Re: Figure 26–6. Are humans adapting to their environment?

REFERENCES

1. Quoted in Fallows, J.M. *The Water Lords* (New York: Bantam Books, 1971).
2. By Kenneth Boulding, quoted in Potter, V.R., *Bioethics* (Englewood Cliffs, NJ: Prentice-Hall, 1971).
3. Kozlovsky, D. *An Ecological and Evolutionary Ethic* (Englewood Cliffs, NJ: Prentice-Hall, 1975).
4. The article first appeared in *Science* 155:1202 (10 March 1967). It has been reproduced in many books, including Barbour, I.G., ed., *Western Man and Environmental Ethics* (Reading, MA: Addison-Wesley, 1973).
5. See, for example, Scoby, D.R., ed., *Environmental Ethics* (Minneapolis, MN: Burgess, 1971) and Schaeffer, F.A., *Pollution and the Death of Man—A Christian View of Ecology* (Wheaton, IL: Tyndale House Publishers, 1980).
6. Hardin, G. "Tragedy of the Commons," *Science* 162:1243(1968). Reprinted in numerous books on environmental ethics.
7. Paulson, I. *The Old Estonian Folk Religion* (Bloomington, IN: Indiana University Press, 1971).
8. Oinas, F. Personal communication. (Indiana University, 1981).
9. See, for example, MacIntyre, A., *A Short History of Ethics* (New York: Macmillan, 1966), or Pearsall, G.W., "Decision Making in Hazardous Waste Disposal" in Pierce, J.J. and Vesilind, P.A., eds., *Hazardous Wastes* (Ann Arbor, MI: Ann Arbor Science Publishers, 1981).
10. See, for example, Rolston, H., "Values in Nature," *Environmental Ethics* 3(2)(1981) or Kaufman, P.I., "The Instrumental Value of Nature," *Environmental Review* 4(1)(1981).
11. See, for example, Worster, D., "The Intrinsic Value of Nature," *Environmental Review* 4(1)(1981).
12. Stone, C.D., *Should Trees Have Standing?* (Los Angeles: William Kaufman, 1972).
13. Joy, C., "The Animal World of Albert Schweitzer," quoted in Worster, D., "The Intrinsic Value of Nature," *Environmental Review* 4(1)(1981).
14. Dubos, R. "A Theology of the Earth," a lecture reprinted in Barbour, I.G., ed., *Western Man and Environmental Ethics* (Reading, MA: Addison-Wesley, 1973).
15. Leopold, A. *A Sand County Almanac* (New York: Oxford University Press, 1949).
16. Gunn, A.S. and P.A. Vesilind, *Environmental Ethics for Engineers* (Chelsea, MI: Lewis Publishers, 1986).

Appendix
Conversion Factors

Multiply	By	To Obtain
acre	0.404	ha
acre ft	1,233	m^3
atmospheres	14.7	lb/in.^{2}
British thermal units	252	cal
Btu	1.054×10^3	J
Btu/ft^3	8,905	cal/m^3
Btu/lb	2.32	J/g
Btu/lb	0.555	cal/g
Btu/sec	1.05	kW
Btu/ton	278	cal/tonne
calories	4.18	joules
calories	3.9×10^{-3}	Btu
cal/g	1.80	Btu/lb
cal/m^3	1.12×10^{-4}	Btu/ft^3
cal/tonne	3.60×10^{-3}	Btu/ton
centimeters	0.393	in.
feet	0.305	m
ft/min	0.00508	m/sec
ft/sec	0.305	m/sec
ft^2	0.0929	m^2
ft^3	0.0283	m^3
ft^3	28.3	liters
ft^3/sec	0.0283	m^3/sec
ft^3/sec	449	gal/min
ft lb (force)	1.357	joules
ft lb (force)	1.357	newton meters

Multiply	By	To Obtain
gallons	3.78×10^{-3}	m^3
gallons	3.78	liters
gal/day/ft^2	0.0407	m^3/day/m^2
gal/min	2.23×10^{-3}	ft^3/sec
gal/min	0.0631	liter/sec
gal/min	0.227	m^3/hr
gal/min	6.31×10^{-5}	m^3/sec
gal/min/ft^2	2.42	m^3/hr/m^2
million gal/day	43.8	liters/sec
million gal/day	3,785	m^3/day
million gal/day	0.0438	m^3/sec
grams	2.2×10^{-3}	lb
hectares	2.47	acre
horsepower	0.745	kW
inches	2.54	cm
inches of mercury	0.49	lb/in.^{2}
inches of mercury	3.38×10^3	newton/m^2
inches of water	249	newton/m^2
joule	0.239	calorie
joule	9.48×10^{-4}	Btu
joule	0.738	ft lb
joule	2.78×10^{-7}	kWh
joule	1	newton meter
J/g	0.430	Btu/lb
J/sec	1	watt
kilograms	2.2	lb (mass)
kg	1.1×10^{-3}	tons
kg/ha	0.893	lb/acre
kg/hr	2.2	lb/hr
kg/m^3	0.0624	lb/ft^3
kg/m^3	1.68	lb/yd^3
kilometers	0.622	mi
km/hr	0.622	mph
kilowatts	1.341	horsepower
kWh	3,600	kilojoule
liters	0.0353	ft^3
liters	0.264	gal
liters/sec	15.8	gal/min
liters/sec	0.0228	mgd

Multiply	By	To Obtain
meters	3.28	ft
meters	1.094	yd
m/sec	3.28	ft/sec
m/sec	196.8	ft/min
m^2	10.74	ft^2
m^2	1.196	yd^2
m^3	35.3	ft^3
m^3	264	gal
m^3	1.31	yd^3
m^3/day	264	gal/day
m^3/hr	4.4	gpm
m^3/hr	6.38×10^{-3}	gpm
m^3/sec	35.31	ft^3/sec
m^3/sec	15,850	gpm
m^3/sec	22.8	mgd
miles	1.61	km
mi^2	2.59	km^2
mph	0.447	m/sec
milligrams/liter	0.001	kg/m^3
million gallons	3,785	m^3
mgd	43.8	liter/sec
mgd	157	m^3/hr
mgd	0.0438	m^3/sec
newton	0.225	lb (force)
newton/m^2	2.94×10^{-4}	inches of mercury
newton/m^2	1.4×10^{-4}	$lb/in.^2$
newton meters	1	joule
newton sec/m^2	10	poise
pounds (force)	4.45	newton
pounds (force)/in.^{2}	6,895	N/m^2
pounds (mass)	454	g
pounds (mass)	0.454	kg
pounds (mass)/ft^2/yr	4.89	kg/m^2/yr
pounds (mass)/yr/ft^3	16.0	$kg/yr/m^3$
pounds/acre	1.12	kg/ha
pounds/ft^3	16.04	kg/m^3
pounds/in.^{2}	0.068	atmospheres
pounds/in.^{2}	2.04	inches of mercury
pounds/in.^{2}	7,140	newton/m^2
tons (2,000 lb)	0.907	tonne (1,000 kg)
tons	907	kg
ton/acre	2.24	tonnes/ha

Multiply	By	To Obtain
tonne (1,000 kg)	1.10	ton (2,000 lb)
tonne/ha	0.446	tons/acre
yd	0.914	m
yd^3	0.765	m^3
watt	1	J/sec

Index

529